T0207153

Nᴇᴜʀᴏᴍᴇᴛʜᴏᴅs

Series Editor
Wolfgang Walz
University of Saskatchewan
Saskatoon, SK, Canada

For further volumes:
http://www.springer.com/series/7657

Neuromethods publishes cutting-edge methods and protocols in all areas of neuroscience as well as translational neurological and mental research. Each volume in the series offers tested laboratory protocols, step-by-step methods for reproducible lab experiments and addresses methodological controversies and pitfalls in order to aid neuroscientists in experimentation. *Neuromethods* focuses on traditional and emerging topics with wide-ranging implications to brain function, such as electrophysiology, neuroimaging, behavioral analysis, genomics, neurodegeneration, translational research and clinical trials. *Neuromethods* provides investigators and trainees with highly useful compendiums of key strategies and approaches for successful research in animal and human brain function including translational "bench to bedside" approaches to mental and neurological diseases.

Measuring Cerebellar Function

Edited by

Roy V. Sillitoe

Houston, TX, USA

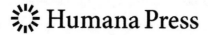 Humana Press

Editor
Roy V. Sillitoe
Houston, TX, USA

ISSN 0893-2336 ISSN 1940-6045 (electronic)
Neuromethods
ISBN 978-1-0716-2028-1 ISBN 978-1-0716-2026-7 (eBook)
https://doi.org/10.1007/978-1-0716-2026-7

This Humana imprint is published by the registered company Springer Science+Business Media, LLC, part of Springer Nature.
The registered company address is: 1 New York Plaza, New York, NY 10004, U.S.A.

Preface

The cerebellum is an exemplary structure for studies of the nervous system. With more than a century of modern work dedicated to uncovering its biology, there is no area of neuroscience that has not benefited from the power of what the cerebellum has revealed and what mysteries it has yet to share with us. The cerebellum has a firm place in embryology, genetics, neurophysiology, neurology, and neurosurgery. The confluence of these different areas has allowed researchers and clinicians to test brain function and behavior and develop therapeutic strategies for patients.

There are now many methods available for studying the cerebellum. Among these are various approaches using stem cell-based techniques, conditional genetics approaches in model systems, neuronal recordings conducted *in vitro* and *in vivo*, and an ever-growing repertoire of behavioral paradigms. The use of these techniques and the richness of the data have allowed for the exciting expansion of cerebellar function into behavioral domains that were traditionally reserved for higher-order cerebral cortical function. Therefore, measures of cerebellar function now include strategies to assess motor functions such as coordination, balance, and learning, but also non-motor functions such as emotion, working memory, and reward. It should not be surprising then that neurologists are on the lookout for cerebellar contributions to autism spectrum disorders, schizophrenia, and attention deficit hyperactivity disorder, among many other neuropsychiatric diseases. The collection of articles in this volume highlights a number of such measures.

The chapters in this volume of Neuromethods have thus brought together experts in the cerebellum field to discuss the current techniques that are used to measure cerebellar function. The chapters provide a reasonably detailed outline of the methods and best practices, including the key steps for successful use of techniques, advantages of the methods and what the measures teach you, and the shortcomings and possible resolutions to keep in mind when using them. The chapters also provide example applications where the approaches have been recently used.

The contributing author list was chosen to represent the breadth and depth of the field, while at the same time providing the reader with tools that range across model systems including *in vitro* techniques such as iPSCs and animal models including mouse, rat, zebrafish, and nonhuman primate. We have also included a chapter describing the neurologist's perspective of measuring cerebellar dys(function) in the clinic. The chapters touch on both motor and cognitive functions.

This volume will be an invaluable resource not only for the cerebellar enthusiast but also for any scientist interested in learning how this beautiful structure has come to be the "go to" system for driving technological advances that unlock brain function and animal behavior. At the heart of the success of this collection are the dedicated researchers and clinicians that continue to devote their efforts to learning about the cerebellum. On that note, I would like to extend my deepest appreciation to each and every author that contributed to what I feel is an elegant collection of chapters from a diverse group of colleagues. I truly admire all the hard work that you put into your chapters and I know that our readers will benefit from your scholarship for years to come.

Houston, TX, USA   *Roy V. Sillitoe*

Preface to the Series

Experimental life sciences have two basic foundations: concepts and tools. The *Neuro-methods* series focuses on the tools and techniques unique to the investigation of the nervous system and excitable cells. It will not, however, shortchange the concept side of things as care has been taken to integrate these tools within the context of the concepts and questions under investigation. In this way, the series is unique in that it not only collects protocols but also includes theoretical background information and critiques which led to the methods and their development. Thus it gives the reader a better understanding of the origin of the techniques and their potential future development. The *Neuromethods* publishing program strikes a balance between recent and exciting developments like those concerning new animal models of disease, imaging, *in vivo* methods, and more established techniques, including, for example, immunocytochemistry and electrophysiological technologies. New trainees in neurosciences still need a sound footing in these older methods in order to apply a critical approach to their results.

Under the guidance of its founders, Alan Boulton and Glen Baker, the *Neuromethods* series has been a success since its first volume published through Humana Press in 1985. The series continues to flourish through many changes over the years. It is now published under the umbrella of Springer Protocols. While methods involving brain research have changed a lot since the series started, the publishing environment and technology have changed even more radically. Neuromethods has the distinct layout and style of the Springer Protocols program, designed specifically for readability and ease of reference in a laboratory setting.

The careful application of methods is potentially the most important step in the process of scientific inquiry. In the past, new methodologies led the way in developing new disciplines in the biological and medical sciences. For example, Physiology emerged out of Anatomy in the nineteenth century by harnessing new methods based on the newly discovered phenomenon of electricity. Nowadays, the relationships between disciplines and methods are more complex. Methods are now widely shared between disciplines and research areas. New developments in electronic publishing make it possible for scientists that encounter new methods to quickly find sources of information electronically. The design of individual volumes and chapters in this series takes this new access technology into account. Springer Protocols makes it possible to download single protocols separately. In addition, Springer makes its print-on-demand technology available globally. A print copy can therefore be acquired quickly and for a competitive price anywhere in the world.

Saskatoon, SK, Canada   *Wolfgang Walz*

Contents

Contributors

RICHARD APPS • *School of Physiology, Pharmacology, and Neuroscience, University of Bristol, Bristol, UK*

ELIZABETH J. APSLEY • *Nuffield Department of Clinical Neurosciences, University of Oxford, Oxford, UK*

ERIC AVILA • *Center for Neural Science, New York University, New York, NY, USA*

ALEKSANDRA BADURA • *Netherlands Institute for Neuroscience, Amsterdam, The Netherlands; Department of Neuroscience, Erasmus MC, Rotterdam, The Netherlands*

ESTHER B. E. BECKER • *Nuffield Department of Clinical Neurosciences, University of Oxford, Oxford, UK*

GERARD JOEY BROUSSARD • *Princeton Neuroscience Institute, Princeton University, Princeton, NJ, USA*

AMANDA M. BROWN • *Department of Pathology & Immunology, Baylor College of Medicine, Houston, TX, USA; Jan and Dan Duncan Neurological Research Institute at Texas Children's Hospital, Houston, TX, USA; Department of Neuroscience, Baylor College of Medicine, Houston, TX, USA*

NADIA L. CERMINARA • *School of Physiology, Pharmacology, and Neuroscience, University of Bristol, Bristol, UK*

BRITTANY L. CORREIA • *Department of Anatomy & Neurobiology, University of Tennessee Health Science Center, Memphis, TN, USA*

CHRIS I. DE ZEEUW • *Netherlands Institute for Neuroscience, Amsterdam, The Netherlands; Department of Neuroscience, Erasmus MC, Rotterdam, The Netherlands*

JENNIFER DEGER • *Jan and Dan Duncan Neurological Research Institute at Texas Children's Hospital, Houston, TX, USA; Department of Neuroscience, Baylor College of Medicine, Houston, TX, USA*

BEN DEVERETT • *Department of Anesthesiology, Perioperative and Pain Medicine, Stanford University College of Medicine, Stanford, CA, USA*

DAVID DIGREGORIO • *Laboratory of Synapse and Circuit Dynamics, Institut Pasteur, Paris Cedex, France*

KEVIN DORGANS • *Neuronal Rhythms in Movement Unit, Okinawa Institute of Science and Technology Graduate University, Okinawa, Japan; Institut des Neurosciences Cellulaires et Intégratives, CNRS, Université de Strasbourg, Strasbourg, France*

GUILLAUME DUGUÉ • *Neurophysiology of Brain Circuits Team, Institut de Biologie de l'Ecole Normale Supérieure (IBENS), Ecole Normale Supérieure, CNRS, INSERM, PSL Research University, Paris, France*

NICO A. FLIERMAN • *Netherlands Institute for Neuroscience, Amsterdam, The Netherlands; Department of Neuroscience, Erasmus MC, Rotterdam, The Netherlands*

MIA B. FOX • *Department of Anatomy & Neurobiology, University of Tennessee Health Science Center, Memphis, TN, USA*

NARUMI FUKUDA • *Graduate School of Science and Engineering, Saitama University, Saitama, Japan*

THÉO GAGNEUX • *Institut des Neurosciences Cellulaires et Intégratives, CNRS, Université de Strasbourg, Strasbourg, France*

JASON S. GILL • *Department of Pediatrics, Division of Neurology, Baylor College of Medicine, Houston, TX, USA; Jan and Dan Duncan Neurological Research Institute at Texas Children's Hospital, Houston, TX, USA*

DETLEF H. HECK • *Department of Anatomy & Neurobiology, University of Tennessee Health Science Center, Memphis, TN, USA*

MASAHIKO HIBI • *Division of Biological Science, Graduate School of Science, Nagoya University, Nagoya, Aichi, Japan*

KANAE HIYOSHI • *Graduate School of Science and Engineering, Saitama University, Saitama, Japan*

COURT HULL • *Department of Neurobiology, Duke University School of Medicine, Durham, NC, USA*

PHILIPPE ISOPE • *Institut des Neurosciences Cellulaires et Intégratives, CNRS, Université de Strasbourg, Strasbourg, France*

JOHN JACOB • *Nuffield Department of Clinical Neurosciences, University of Oxford, Oxford, UK*

AKIRA KATOH • *Department of Physiology, Tokai University School of Medicine, Isehara-shi, Kanagawa, Japan*

ANDREI KHILKEVICH • *Sainsbury Wellcome Centre for Neural Circuits and Behaviour, University College London, London, UK*

MIKHAIL KISLIN • *Princeton Neuroscience Institute, Princeton University, Princeton, NJ, USA*

YOSHIKO KOJIMA • *Washington National Primate Research Center, University of Washington, Seattle, WA, USA; Department of Otolaryngology-Head and Neck Surgery, University of Washington, Seattle, WA, USA*

BERND KUHN • *Optical Neuroimaging Unit, Okinawa Institute of Science and Technology Graduate University, Okinawa, Japan*

CHARLOTTE L. LAWRENSON • *School of Physiology, Pharmacology, and Neuroscience, University of Bristol, Bristol, UK*

CLÉMENT LÉNA • *Neurophysiology of Brain Circuits Team, Institut de Biologie de l'Ecole Normale Supérieure (IBENS), Ecole Normale Supérieure, CNRS, INSERM, PSL Research University, Paris, France*

YU LIU • *Department of Anatomy & Neurobiology, University of Tennessee Health Science Center, Memphis, TN, USA*

VICTOR LLOBET • *Neuronal Algorithms Team, Institut de Biologie de l'Ecole Normale Supérieure (IBENS), Ecole Normale Supérieure, CNRS, INSERM, PSL Research University, Paris, France*

SAMUEL NAYLER • *Department of Physiology, Anatomy and Genetics, University of Oxford, Oxford, UK*

ELENA PACI • *School of Physiology, Pharmacology, and Neuroscience, University of Bristol, Bristol, UK*

ZUZANNA PIWKOWSKA • *Laboratory of Synapse and Circuit Dynamics, Institut Pasteur, Paris Cedex, France*

DANIELA POPA • *Neurophysiology of Brain Circuits Team, Institut de Biologie de l'Ecole Normale Supérieure (IBENS), Ecole Normale Supérieure, CNRS, INSERM, PSL Research University, Paris, France*

TAKASHI SHIMIZU • *Division of Biological Science, Graduate School of Science, Nagoya University, Nagoya, Aichi, Japan*

ASUKA SHIRAISHI • *Graduate School of Science and Engineering, Saitama University, Saitama, Japan*

ROY V. SILLITOE • *Department of Pathology & Immunology, Baylor College of Medicine, Houston, TX, USA; Jan and Dan Duncan Neurological Research Institute at Texas Children's Hospital, Houston, TX, USA; Department of Neuroscience, Baylor College of Medicine, Houston, TX, USA; Program in Developmental Biology, Baylor College of Medicine, Houston, TX, USA; Development, Disease Models & Therapeutics Graduate Program, Baylor College of Medicine, Houston, TX, USA*

ROBIJANTO SOETEDJO • *Department of Physiology and Biophysics, University of Washington, Seattle, WA, USA; Washington National Primate Research Center, University of Washington, Seattle, WA, USA*

LUDOVIC SPAETH • *Institut des Neurosciences Cellulaires et Intégratives, CNRS, Université de Strasbourg, Strasbourg, France; Dominick P Purpura Department of Neuroscience, Albert Einstein College of Medicine, Bronx, NY, USA*

IZUMI SUGIHARA • *Department of Systems Neurophysiology, Tokyo Medical and Dental University Graduate School of Medical and Dental Sciences, Tokyo, Japan*

THIBAULT TARPIN • *Neurophysiology of Brain Circuits Team, Institut de Biologie de l'Ecole Normale Supérieure (IBENS), Ecole Normale Supérieure, CNRS, INSERM, PSL Research University, Paris, France*

SACHIKO TSUDA • *Graduate School of Science and Engineering, Saitama University, Saitama, Japan; Integrative Research Center for Life Sciences and Biotechnology, Saitama University, Saitama, Japan*

MARYLKA YOE UUSISAARI • *Neuronal Rhythms in Movement Unit, Okinawa Institute of Science and Technology Graduate University, Okinawa, Japan*

ANTOINE VALERA • *Institut des Neurosciences Cellulaires et Intégratives, CNRS, Université de Strasbourg, Strasbourg, France; Department of Neuroscience, Physiology and Pharmacology, University College London, London, UK*

PAMELA VALNEGRI • *Department of Neurobiology, Northwestern University, Evanston, IL, USA*

MEIKE E. VAN DER HEIJDEN • *Department of Pathology & Immunology, Baylor College of Medicine, Houston, TX, USA; Jan and Dan Duncan Neurological Research Institute at Texas Children's Hospital, Houston, TX, USA*

MAX J. VAN ESSEN • *Nuffield Department of Clinical Neurosciences, University of Oxford, Oxford, UK*

ANDRÉS P. VARANI • *Neurophysiology of Brain Circuits Team, Institut de Biologie de l'Ecole Normale Supérieure (IBENS), Ecole Normale Supérieure, CNRS, INSERM, PSL Research University, Paris, France*

SAMUEL S. -H. WANG • *Princeton Neuroscience Institute, Princeton University, Princeton, NJ, USA*

TOMOKO YAMADA • *Department of Neurobiology, Northwestern University, Evanston, IL, USA*

YUE YANG • *Department of Neurobiology, Northwestern University, Evanston, IL, USA*

Chapter 1

Cerebellar Modelling Using Human Induced Pluripotent Stem Cells

Max J. van Essen, Samuel Nayler, Elizabeth J. Apsley, John Jacob, and Esther B. E. Becker

Abstract

Human induced pluripotent stem cells (hiPSCs) have changed the way human development and disease are studied. It is now increasingly possible to create specific cell types and organ-like structures from an embryonic stem cell state, with improved recapitulation of physiological conditions. Here, we describe a robust and reproducible method to differentiate hiPSCs into cerebellar neurons with several options for downstream applications to address different experimental questions. In addition to conventional disso-ciated culture conditions, we also present a method for long-term three-dimensional culture of cerebellar organoids for advanced in vitro modelling of the cerebellar microenvironment.

Key words Cerebellum, Development, hiPSCs, Neural differentiation, Self-organization, Three-dimensional cell culture, Organoids, Stem Cells

1 Introduction

The ground-breaking technology to reprogram fully differentiated cells back into a pluripotent stem cell state has changed the way human development and disease is studied [1, 2]. The advent of hiPSC technology has accelerated the development of protocols to create specific human cell types and organ-like structures from an embryonic stem cell state in vitro. As a result, less reliance is placed on cell lines that fall short on recapitulating physiological conditions or animal models that might not accurately mimic human biology [3, 4].

Building on experience with mouse and human embryonic stem cells [5–8], several groups have recently begun using hiPSCs to study the cerebellum [9–11]. Initial studies induced cerebellar progenitor formation by addition of signalling molecules and

Max J. van Essen and Samuel Nayler contributed equally to this work.

Roy V. Sillitoe (ed.), *Measuring Cerebellar Function*, Neuromethods, vol. 177, https://doi.org/10.1007/978-1-0716-2026-7_1,
© Springer Science+Business Media, LLC, part of Springer Nature 2022

growth factors at distinct timepoints to mimic in vivo patterning [5, 8]. Subsequently, refined methods were developed that provide just a few initial key factors to recapitulate self-inductive endogenous signalling processes during cerebellar patterning [6, 7, 12]. Cerebellar hiPSC-derived cultures are now being used to study diseases of the cerebellum including cerebellar ataxia [9] and autism spectrum disorder [10]. Moreover, the pediatric brain tumor medulloblastoma has recently been modelled through the overexpression of certain oncogenes in cerebellar organoids [13].

Differentiation of pluripotent stem cells into cerebellar neurons takes them through several progenitor stages before they adopt a post-mitotic state from whereon they further mature [14–16]. In vitro cerebellar fate specification is dependent on fibroblast growth factor (FGF) signalling that is initiated in vivo by the isthmic organizer (IsO) at the midbrain-hindbrain boundary (MHB). In turn, the IsO is established by the reciprocal expression of the transcription factors orthodenticle homeobox 2 (OTX2) and gastrulation brain homeobox 2 (GBX2) [15]. Later, two main germinal zones are established from where cerebellar neuron progenitors arise. The ventricular zone (VZ) generates GABAergic neurons including Purkinje cells, interneurons as well as astroglia. The upper rhombic lip (uRL) produces glutamatergic neuron progenitors, which later develop into granule cells, large deep cerebellar nuclei (DCN) projection neurons and unipolar brush cells [15].

The complex interplay between multiple signalling hubs within the fetal cerebellum that is necessary to form highly specialized neurons such as Purkinje cells has been difficult to recapitulate in dissociated cultures in vitro. As a consequence, increasing interest has focused on utilizing three-dimensional culture systems that rely on the intrinsic self-organizing capacity of the developing cerebellum [6, 7]. The three-dimensional configuration not only mimics the microenvironment that is needed for cerebellar development, but also forms an improved model for physiological cell–cell signalling. Furthermore, studies comparing monolayer with three-dimensional differentiation of neural progenitor cells show enrichment of brain-specific extracellular matrix (ECM) components, increased neuronal maturation, and improved disease-specific ECM remodelling [17–19].

This chapter describes a robust and reproducible method for the first 35 days of cerebellar differentiation induction, followed by options to proceed with either dissociated or three-dimensional maturation of cerebellar progenitors (Fig. 1). First, plating of hiPSCs in ultra-low-attachment cell culture plates initiates the formation of embryoid bodies (EBs). Within EBs, inhibition of mesendodermal differentiation and induction of the neuroectoderm is achieved through the addition of a TGF-β receptor inhibitor [20], while induction of the MHB cell fate is promoted by key growth factors insulin and FGF2 [6]. hiPSC-derived hindbrain

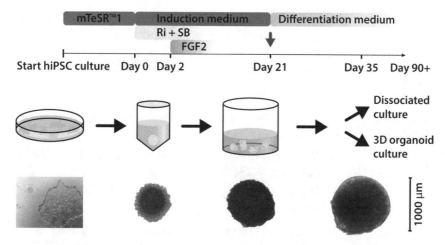

Fig. 1 Schematic overview of the cerebellar differentiation protocol. After culture in mTeSR, human induced pluripotent stem cells (hiPSCs) are seeded into sterile ultra-low-attachment V-bottom 96-well plates in Induction Medium containing Rho kinase inhibitor Y-27632 (Ri) and TGF-β signalling inhibitor SB-431542 (SB). On Day 2, fibroblast growth factor 2 (FGF2) is added to induce cerebellar lineage formation. Neural aggregates are transferred into low-attachment 24-well plates on Day 21 and cultured in Differentiation Medium. The red arrow indicates the timing of optional Matrigel® embedding. After 35 days, aggregates are either dissociated and plated down or continued as three-dimensional organoid cultures. Representative brightfield images show an hiPSCs colony, Day 1 embryoid body (EB), Day 21 neural aggregate, and Day 35 organoid

precursors express transcription factors engrailed homeobox 1 (EN1) and GBX2 and develop into cerebellar neuron progenitors that express markers of both germinal zones: uRL-specific marker atonal bHLH transcription factor 1 (ATOH1) and VZ marker kirre-like nephrin family adhesion molecule 2 (KIRREL2). After three weeks, neuralised EBs are transferred into a new medium that promotes further differentiation and maturation of progenitors into cerebellar neurons. Following another two weeks of culture, neural aggregates can be dissociated and plated down. Alternatively, the three-dimensional configuration of the organoids can be maintained to culture the cells in a more physiological microenvironment. Dissociated cultures may be better suited to study single cell types and facilitates patch clamping and high-resolution imaging, whereas three-dimensional organoids can be used to study signalling between compartmentalized tissues, neuroectodermal polarization, as well as neuronal and axon migration. Long-term culture of dissociated human cerebellar progenitors co-cultured with mouse cerebellar progenitors generates immature Purkinje cells expressing Calbindin 1 (CALB1), Forkhead box P2 (FOXP2), and Purkinje cell protein 4 (PCP4) [12]. Single-cell RNA sequencing of cerebellar organoids at 90 days of differentiation revealed transcriptionally discrete populations encompassing major cerebellar neuronal cell types including progenitors of RL, granule cells, Purkinje cells, glutamatergic DCN, Bergmann glia, roof plate, choroid plexus, and ependymal cells [21].

2   Materials

2.1   Human Induced Pluripotent Cell Culture

1. Sterile tissue culture-treated six-well plates (Corning®, Cat# 3516 or similar).

2. Matrigel® matrix, hESC-qualified (stored at −20 °C) (Corning®, Cat# 354277).

3. Ham's F-12 Nutrient Mix, GlutaMAX™ (stored at 4 °C) (Gibco™, Cat# 31765027).

4. mTESR1 medium (after constitution stored in aliquots at −20 °C, should only be thawed once) (Stemcell Technologies, Cat# 85850).

   Optional: Penicillin/Streptomycin (10.000 U/mL stock, aliquots stored at −20 °C) (Gibco™, Cat# 15140122).

5. 0.5 mM EDTA in PBS (stored at room temperature (RT)).

6. Y-27632 (Rho Kinase inhibitor) (100 mM stock in PBS, aliquots stored at −20 °C) (Abcam, Cat# ab120129).

2.2   Cerebellar Lineage Induction and Differentiation

1. Sterile ultra-low-attachment tissue culture-treated V-bottom 96-well plates (S-Bio, Cat# MS-9096VZ or similar).

2. Sterile low-attachment sterile tissue culture-treated 24-well plates (Corning®, Cat# 3473 or similar).

3. TrypLE™ Express Enzyme (1×), no phenol red (Gibco™, Cat# 12604013).

4. Induction Medium: 50% Iscove's Modified Dulbecco's Medium (Gibco™, Cat# 31980022) with 50% Ham's F-12 Nutrient Mix (Gibco™, Cat# 31765027), 7 μg/mL insulin (10 mg/mL stock in acidic water, stored in aliquots at −20 °C) (Sigma, Cat# I1882) (see Note 1), 5 mg/mL BSA (250 mg/mL stock in distilled water, stored in aliquots at −20 °C) (Sigma, Cat# A3156-5G), 1% Chemically Defined Lipid Concentrate (100× stock, stored in aliquots at −20 °C) (Gibco™, Cat# 11905031), 450 μM 1-thioglycerol (stored at 4 °C) (Sigma, Cat# M-6145), 15 μg/mL apo-transferrin (10 mg/mL stock in distilled water, aliquots stored at −20 °C) (Sigma, Cat# T1147), 1% Penicillin/Streptomycin (10.000 U/mL stock, aliquots stored at −20 °C) (Gibco™, Cat# 15140122).

5. Y-27632 (Rho Kinase inhibitor) (100 mM stock in PBS, stored in aliquots at −20 °C) (Abcam, Cat# ab120129).

6. SB-431542 (transforming growth factor-β (TGF-β) type I receptor/ALK5, ALK4 and ALK7 inhibitor) (100 mM stock

in DMSO stored in aliquots at −20 °C) (Tocris, Cat# 1614/10).

7. Recombinant Human FGF Basic/FGF2/bFGF (200 ng/μL stock in PBS with 0.1% BSA, stored in aliquots at −20 °C) (R&D Systems, Cat# 4114-TC-01 M).

8. Differentiation Medium: Neurobasal® medium (Gibco™, Cat# 21103049) with 1% GlutaMAX™ (stored at 4 °C) (Gibco™, Cat# 35050061), 1% N2 supplement (100× stock, aliquots stored at −20 °C) (Gibco™, Cat# 17502048), and 1% Penicillin/Streptomycin (10.000 U/mL stock, aliquots stored at −20 °C) (Gibco™, Cat# 15140122).

9. Growth factor-reduced Matrigel® matrix (stored in aliquots at −20 °C) (Corning®, Cat# 354230).

   Optional: Recombinant Human FGF19 (100 μg/mL stock in PBS with 0.1% BSA, stored in aliquots at −20 °C) (R&D Systems, Cat# 969-FG-025/CF), Recombinant Human CXCL12/SDF1a (100 μg/mL stock in PBS with 0.1% BSA, stored in aliquots at −20 °C) (R&D Systems, Cat# 350-NS-010/CF).

2.3 Dissociated Culture

1. Sterile tissue culture-treated 24-well plates (Corning®, Cat# 3526 or similar).

2. Sterile cover slips: Glass (Bellco RD, Cat# 1943-10012A) or plastic (Sumitomo Bakelite Cell Desk LF, Cat# MS-92132).

3. Poly-L-ornithine hydrobromine (10 mg/mL stock in distilled water, aliquots stored at −20 °C) (Sigma, Cat# P3655).

4. Laminin (1.2 mg/mL stock in distilled water, aliquots stored at −20 °C) (Gibco™, Cat# 23017015).

5. HHGN dissection solution: 10% HBSS (10× stock, stored at RT) (Invitrogen, Cat# 14180046) with 2.5 mM HEPES pH 7.3–7.5 (1 M stock, stored at 4 °C) (Invitrogen, Cat# 15630049), 35 mM Glucose (1 M stock, stored at 4 °C), and 4 mM NaHCO₃ (1 M stock, stored at RT).

6. Neuron Isolation Enzyme (with papain) (ThermoFisher, Cat #88285).

7. Base Medium: DMEM/F12 with Glutamine (Invitrogen, Cat# 11320074), 1% N2 Supplement (100× stock, aliquots stored at −20 °C) (Gibco™, Cat# 17502048), additional 1.4 mM L-Glutamine (200 mM stock) (aliquots stored at −20 °C) (Invitrogen, Cat# 25030–081), 5 μg/mL insulin (10 mg/mL stock in acidic water, aliquots stored at −20 °C) (Sigma, Cat# I1882), 1% Penicillin/Streptomycin (10.000 U/mL stock, aliquots stored at −20 °C) (Gibco™, Cat# 15140122).

8. Seeding Medium: Base medium with 10% US-defined, heat-inactivated fetal bovine serum (FBS) (GE Healthcare Life Sciences HyClone, Cat# 12359792).

9. Culture Medium: Base medium supplemented with 0.5 ng/mL Tri-Iodothyronine (T3) (20 µg/mL stock in 20 mM NaOH, aliquots stored at −20 °C) (Sigma, Cat# T6397) (*see* **Note 2**), 100 µg/mL BSA (250 mg/mL stock in distilled water, aliquots stored at −20 °C) (250 mg/mL stock, aliquots stored at −20 °C) (Sigma, Cat# A3156), 50 ng/mL BDNF (100 µg/mL stock in sterile PBS, aliquots stored at −20 °C) (R&D Systems, Cat# 248-BD-005/CF), 50 ng/mL NT3 (50 µg/mL stock in sterile PBS, aliquots stored at −20 °C) (R&D Systems, Cat# 267-N3-005/CF).

10. Feeding Medium: Culture medium with 4 µM Cytosine β-D-arabinofuranoside (Ara-C) (400 µM stock in distilled water, aliquots stored at −20 °C) (Sigma, Cat# C1768) (*see* **Note 3**), 5–10 µg/mL Laminin (1.2 mg/mL stock in distilled water, aliquots stored at −20 °C) (Gibco™, Cat# 23017015).

2.4 Three-Dimensional Organoid Culture

1. Sterile tissue culture-treated six-well plates (Corning®, Cat# 3516 or similar).

2. Differentiation Medium.

3. Cell culture transwell inserts (PTFE 0.4 µm pore size) (Millicel, Cat# PICM0RG50).

4. Sterile forceps.

2.5 Sample Harvest

1. 4% Paraformaldehyde (PFA).

2. 20% Sucrose in PBS.

3. Cryosectioning medium such as Optimal Cutting Temperature (OCT) compound.

4. Cryomolds.

5. Accumax (Sigma, Cat# A7089) or Neuronal Isolation Enzyme (with papain) (ThermoFisher, Cat #88285).

3 Methods

3.1 Human iPSC Culture and Maintenance

3.1.1 Matrigel Coating

hiPSCs require an extracellular matrix (ECM) when grown in feeder-free conditions. This ECM contains growth factors such as basic FGF and TGF-β to promote pluripotent stem cell survival and proliferation [22, 23]. In this protocol, Matrigel® matrix is used, an extract of Engelbreth-Holm-Swarm (EHS) mouse sarcoma cell line in the form of a hydrogel that solidifies at temperatures between 20 and 40 °C. To prepare this, hESC-qualified Matrigel® matrix is diluted in a basal medium of choice (i.e., DMEM/F-12) using the dilution factor given in the certificate of analysis. One mL of basement-membrane matrix will coat one well in a six-well plate (10 cm^2). As Matrigel® rapidly solidifies at room temperature, it is important to work quickly. Pipette tips should be cooled at −20 °C prior to use, and medium should be kept on ice. After adding the Matrigel® solution to the required number of wells, plates are incubated either for 30–60 min at 37 °C or overnight at 4 °C. Parafilm-sealed plates coated with Matrigel® can be stored at 4 °C for up to one week. The Matrigel® solution is removed just prior to adding the cell culture medium.

3.1.2   Thawing hiPSCs

Matrigel®-coated wells are prepared as described above. The required amount of mTeSR1 medium (2 mL per well in a six-well plate) is supplemented with 10 µM Y-27632 and warmed to 37 °C, and a 15-mL centrifuge tube is prepared with 9 mL of PBS or mTeSR1 medium. The hiPSC-containing cryovial is defrosted by swirling in a pre-warmed water bath until most of the liquid has thawed, making sure the lid stays dry (*see* **Note 4**). The cell suspension is transferred into the prepared centrifuge tube with PBS or mTeSR1 and centrifuged at 400 × *g* for 5 min. The supernatant is aspirated, before the cell pellet is loosened by tapping the tube and the cells are resuspended by adding liquid in a dropwise manner in 1 mL mTeSR1 medium with Y-27632. It is recommended to split cells across at least two wells in different densities (e.g., a 20/80 split) in case one of the densities becomes too confluent or passaging fails. After 24 h, medium is changed to mTeSR1 without Y-27632.

3.1.3   hiPSC Maintenance

A full medium change with mTeSR1 is required daily. It is optional to add 1% Penicillin/Streptomycin to mTeSR1 to reduce risk of infection. For daily feeding, mTeSR1 should only be warmed to room temperature as FGF2 present in this medium quickly degrades at higher temperatures. hiPSCs are best passaged at 70–80% confluency. Culture medium is aspirated, and cells are washed once with PBS, then incubated for 3–5 min with 0.5 mM EDTA in PBS at 37 °C. Once cells start to detach, the EDTA solution is should be removed carefully and cells are resuspended in mTeSR1 supplemented with 10 µM Y-27632 Rho Kinase inhibitor. Depending on the specific stem cell line growth rate, splits should range from 1:4 to 1:12. Culture medium is supplemented with 10 µM Y-27632 for the first 24 h after passaging to increase viability and attachment of the cells. hiPSCs should grow in monolayer colonies consisting of a homogeneous cell population (Fig. 1). Spontaneous differentiation of hiPSCs can be identified as colonies that do not have a clearly defined edge, neural rosette formation within the colony, or clusters of large (fibroblast-like) cells. Differentiated areas are usually macroscopically visible as faint white specks and can be removed mechanically by aspiration using a P10/P200 pipette tip prior to passaging (*see* **Note 5**).

3.2 Base Protocol for the First 35 Days of Differentiation

In this section, the first 35 days of the differentiation protocol are described resulting in generation of cerebellar neuron progenitors (Fig. 1). After 35 days, different methods for long-term culture maintenance are chosen based on downstream application. These are described in Subheadings 3.3 and 3.4. hiPSC-derived aggregates are referred to as 'embryoid bodies (EBs)' during the first 21 days of the protocol to distinguish them from the more developed 'neural aggregates' (Day 21–35 of differentiation) and 'organoids' (>35 days of differentiation).

3.2.1 Embryoid Body Formation

In preparation, V-bottom 96-well plates as well as appropriate volumes of TrypLE™ Express Enzyme (1×) and Induction Medium are warmed to 37 °C. Materials for cell counting are arranged (*see* **Note 6**). Two to three wells of a six-well plate with hiPSCs are selected, which will provide sufficient cell numbers for up to four 96-well plates. EB formation is optimal when hiPSCs are at 60–70% confluence (*see* **Note 7**).

hiPSC culture medium is aspirated off selected wells, and cells are washed once with PBS. Then, 1 mL TrypLE™ Express Enzyme (1×) is added to each well, and cells are incubated at 37 °C (*see* **Notes 8** and **9**). hiPSCs usually detach after 3–5 min of incubation, depending on the used stem cell line. After detaching, cells are collected in a sterile conical centrifuge tube. PBS is used to wash any residual cells from the wells, which are then added to the centrifuge tube. Together, 9 mL PBS should be added for every mL TrypLE™ enzyme in the tube to stop the enzymatic reaction. Using spent hiPSC culture medium instead of PBS may enhance viability (*see* **Note 8**). After gentle mixing, the suspension is centrifuged for 5 min at $400 \times g$ at room temperature. During centrifugation, pre-warmed Induction Medium is supplemented with Y-27632 (Rock inhibitor, Ri) to a final concentration of 50 µM and SB-431542 (SB) to a final concentration of 10 µM. Supernatant is aspirated, and the tube is gently tapped to loosen the cell pellet. Cells are diluted in 0.5–1 mL of Induction Medium + Ri + SB and counted. For each 96-well plate, $1.0–1.2 \times 10^6$ cells are added to 10 mL of Induction Medium + Ri + SB (*see* **Note 10**). Using a multichannel pipette, 100 µL of cell suspension are added to each well of a low-attachment V-bottom 96-well plate to allow re-aggregation and EB formation ($1.0–1.2 \times 10^4$ cells per well). Plates are gently placed in an incubator at 37 °C with 5% CO_2 (*see* **Notes 11** and **12**).

3.2.2 Cerebellar Lineage Induction

Two days after re-aggregation, cerebellar lineage commitment is induced through the addition of FGF2. 20 µL of Induction Medium + Ri + SB are removed from each well and replaced with

20 μL Induction Medium + Ri + SB + 250 ng/mL FGF2 to give a final concentration of 50 ng/mL FGF2 in each well. On Day 7, 30 μL Induction Medium + Ri + SB are removed from each well and replaced with 30 μL Induction Medium without Ri or SB. On Day 14, EBs require a full medium change with fresh Induction Medium, taking care not to disturb them. It is optional to increase the culture volume to 120–150 μL, which may improve cell survival.

Optional: To stimulate polarization of cerebellar plate neuroepithelium, FGF19 is added to the Induction Medium in a final concentration of 100 ng/mL on Day 14. If FGF19 is added, the medium should also be supplemented with 300 ng/mL CXCL12/SDF-1 on Day 28 [24].

3.2.3 Cerebellar Progenitor Maturation up to Day 35

On Day 21, neural aggregates are transferred into sterile low-attachment tissue culture-treated 24-well plates with Differentiation Medium. Plates containing 500 μL medium are warmed to 37 °C. Aggregates are picked up and placed into the new medium using a wide bore pipette tip. Up to four aggregates are placed per well (*see* **Notes 13** and **14**).

Neural aggregates have a tendency to fuse when placed within the same well. If this is unsuitable for the final application, aggregates can be kept apart in smaller well formats or can be embedded in undiluted growth factor-reduced Matrigel® matrix. In addition to preventing fusion of organoids, Matrigel® embedding can used to simulate basement-membrane signalling and alter growth dynamics. However, greater inter-organoid variability in generating cerebellar cell types is seen upon Matrigel® embedding with a lineage bias towards rhombic lip derivatives [21]. These factors need to be taken into account when considering Matrigel® embedding.

If Matrigel® embedding is desired, a box of sterile P200 pipette tips is cooled at −20 °C for at least 30 min prior to embedding. The appropriate amount of Matrigel® is thawed on ice; typically, 10 μL are sufficient to embed one neural aggregate. Two methods can be used for the embedding. The first method embeds neural aggregates on a Parafilm® sheet, similar to the embedding of cerebral organoids described by Lancaster et al. [25]. The second method makes use of the geometry of the 96-well plate and embeds the organoids in the well in which they have been previously cultured.

Method 1: Matrigel® Embedding Using Parafilm® Sheet

A sheet of Parafilm® is placed on top of an empty P20 pipette tip box insert, sterilized using 70% ethanol, and exposed to UV light for at least 30 min. Subsequently, dimples are made in the Parafilm® sheet by pressing down with a fingertip on the empty spaces in the pipette tip box insert under sterile conditions [25]. Afterwards, the sheet is cut into two-by-two sections to help handling of the embedded aggregates in further steps. At this stage, neural aggregates are macroscopically visible and can be picked up using a wide bore tip and placed into each dimple. All residual medium is then removed using a P10 pipette tip, leaving the aggregate bare. Using chilled P20 pipette tips, 10 μL of Matrigel® is added to each aggregate and aggregates are gently moved to the middle of the droplet using a pipette tip. After all aggregates have been embedded, the Parafilm® sheet is transferred into a sterile container such as a large petri dish and placed in an incubator at 37 °C for 30 min. Similar to the method described earlier for unembedded aggregates, a 24-well plate with 500 μL of Differentiation Medium per well is warmed to 37 °C. After incubation, each two-by-two Parafilm® grid is placed into a single well using sterile forceps. The Matrigel® embedded neural aggregates are then washed off with a P1000 pipette using the medium present in the well.

Method 2: Matrigel® Embedding in a V-Bottom 96-Well Plate

Medium is carefully aspirated from the neural aggregates that are to be embedded. Subsequently, using pre-chilled P200 pipette tips, 10 μL Matrigel® are added to each aggregate making sure they are placed in the middle of the Matrigel® droplet. Aggregates can be carefully moved using a pipette tip. The plate is returned to an incubator at 37 °C and incubated for 30 min. Similar to Subheading "Method 1: Matrigel® Embedding Using Parafilm® Sheet", the appropriate number of wells in a 24-well plate is prepared by adding 500 μL of pre-warmed Differentiation Medium to each well. After 30 min, both the 96-well plate and the 24-well plate are transferred into a flow hood. To lift the Matrigel® embedded aggregates, 100 μL of pre-warmed Differentiation Medium is carefully jetted underneath the Matrigel® droplet. Typically, Matrigel® embedded aggregates will float to the surface of the liquid and can then be transferred into the 24-well plate using a wide bore tip.

3.2.4 Neural Aggregate Culture Between Day 21 and 35

Day 21 marks one of the earliest time points to validate cerebellar lineage induction by the expression of hindbrain precursor markers, e.g., GBX2 and EN1. Therefore, it is recommended to harvest several aggregates for analysis at this timepoint. Suggested methods for validation are immunostaining and RT-qPCR (Fig. 2a–c). A full medium change is performed on Day 28 of differentiation. On Day 35, subpopulations of cells within neural aggregates express markers of the two germinal zones of the developing cerebellum; VZ marker KIRREL2 and uRL progenitor marker ATOH1. Expression of these markers is demonstrated using immunostaining and RT-qPCR (Fig. 2d–f).

3.3 Dissociated Culture of hiPSC-Derived Cerebellar Neurons

Monolayer culture of hiPSC-derived cerebellar neurons is performed on either glass or plastic coverslips, depending on preference. Coverslips are prepared by coating with poly-L-ornithine (0.5 mg/mL) and incubation overnight at 37 °C. On the day of the dissociation, poly-L-ornithine is aspirated and coverslips are

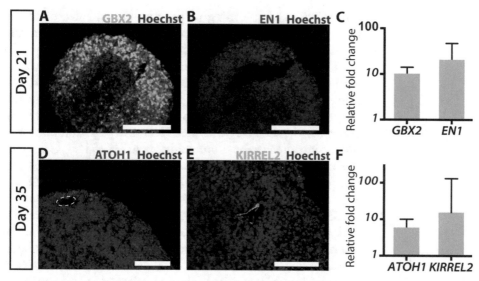

Fig. 2 Expression of markers specific to hindbrain and cerebellar progenitors. (**a, b**) Immunofluorescence staining of Day 21 neural aggregates with antibodies specific to the hindbrain markers GBX2 (green) and EN1 (red). Nuclei are visualized in blue by Hoechst staining. Scale bars 150 μm. (**c**) Fold change in mRNA expression of GBX2 and EN1 in Day 21 aggregates relative to hiPSCs and normalized to GAPDH. (**d–f**) Immunostaining of Day 35 organoids with antibodies specific to markers of rhombic lip precursors, ATOH1 (red; **d**), and ventricular zone precursors, KIRREL2 (green; **e**). Nuclei are visualized in blue by Hoechst staining. Dashed line indicates an example of a neuroepithelial lumen (rosette). Scale bars 150 μm (**a, b, d**) and 100 μm (**e**). (**f**) Fold change in mRNA expression of markers in Day 35 organoids relative to hiPSCs and normalized to GAPDH. (Panel **f** adapted under CC-BY license from Watson et al. [12])

washed once with sterile PBS, before coating with natural mouse laminin (10 mg/mL) for 2–3 h before use (*see* **Note 15**). Coverslips are then washed once with PBS and allowed to air dry in the flow hood (*see* **Note 16**).

On Day 35 of differentiation, up to ten aggregates are carefully transferred per one microcentrifuge tube using a wide bore tip and washed twice with 1 mL HHGN dissection solution. Ten aggregates will contain between 0.5 and 1×10^6 cells and between five and ten tubes are prepared in total. Subsequently, 200 μL Neuronal Isolation Enzyme with Papain are added to each tube, and tubes are incubated for 20–30 min at 37 °C with agitation every 3–5 min. It is normal for the edges of the aggregates to become fuzzy. After incubation, the enzyme solution is carefully removed from the microcentrifuge tubes and cells are gently washed three times with 1 mL HHGN. Aggregates are further dissociated into a single-cell suspension by carefully triturating 20–25 times in 500 μL of Seeding Medium, avoiding formation of bubbles (*see* **Note 17**). Subsequently, dissociated cells are pooled, resuspended in 5 mL of Seeding Medium and centrifuged at $185 \times g$ for 5 min at room temperature. The supernatant is then aspirated and the cell pellet is resuspended in 250–500 μL Seeding Medium (this volume can be adjusted according to the number and size of the aggregates) (*see* **Note 18**). Cells are counted and the cell concentration is adjusted to 8×10^6 cells/mL in Seeding Medium.

Co-culture of the dissociated human cerebellar progenitors with mouse cerebellar progenitors enhances viability and maturation. In parallel to neural aggregate dissociation, mouse cerebellar progenitors are prepared from embryonic day (E) 18.5 mouse embryos. Cerebella are dissected from the pups and pooled into a microcentrifuge tube containing cold HHGN. Cerebella are washed twice in HHGN before incubation in TrypLE™ Express Enzyme ($1\times$) for 15 min at 37 °C with agitation every 3–5 min. Afterwards, cerebella are carefully washed three times with 1 mL HHGN before 500 μL Seeding Medium is added. Cerebella are then dissociated into a single-cell suspension by triturating 20–25 times while avoiding the formation of bubbles. The cell suspension is transferred into a 15-mL conical tube and 4.5 mL of Seeding Medium are added. Cells are then centrifuged at $185 \times g$ for 5 min, after which the supernatant is removed, and the cell pellet is resuspended in 250 μL Seeding Medium. Cells are counted, and the cell concentration is adjusted to 8×10^6 cells/mL in Seeding Medium. Human cerebellar neuron progenitors are mixed with mouse cerebellar neuron progenitors at a 1:10 ratio. This cell mixture is seeded in droplets of 55 μL on prepared coverslips and incubated for 3 h at 37 °C with 5% CO_2. Subsequently, 500 μL of culture medium are added to each well, and plates are returned to the incubator. This results in a final FBS concentration of 1%. The co-cultures are fed every 3–4 days by replacing half of the medium with Feeding Medium. By Day 70 of differentiation, human Purkinje cells expressing the markers CALB1, FOXP2, and PCP4 can be visualized using immuno staining (Fig. 3). Cultures can be maintained for at least 90 total days of differentiation.

Human-only monolayer cerebellar progenitor culture is possible, however, with much decreased long-term viability compared to

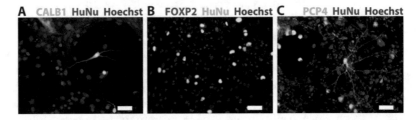

Fig. 3 Expression of Purkinje cell markers in hiPSC-derived cerebellar progenitors in murine co-culture. Immunostaining of hiPSC-derived cerebellar neurons in co-culture with murine cerebellar progenitors after 90 total days of differentiation with antibodies specific to early Purkinje cell markers CALB1 (green) (**a**), FOXP2 (red) (**b**), and PCP4 (green) (**c**). All nuclei are visualized in blue by Hoechst staining. Human nuclei are visualized by HuNu staining (red; **a** and **c**, green; **b**). Scale bars 50 μm. (Figure adapted under CC-BY license from Watson et al. [12])

co-culture with mouse progenitor cells. If human-only culture is desired, hiPSC-derived cerebellar progenitors should be prepared as described above. Subsequently, 55 µL droplets of human iPSC derived cerebellar progenitor cells are seeded on prepared coverslips and incubated for 3 h at 37 °C with 5% CO_2. After incubation, 500 µL of Culture Medium are gently added to each well, and the plates are returned to the incubator. This results in a final FBS concentration of 1%. Half of the medium is changed every 2–3 days with fresh Feeding Medium. Monolayer cultures may be maintained up to 50 days post-seeding, however, viability will decrease over time and experiments should be planned accordingly.

3.4 Cerebellar Organoid Culture

As an alternative to dissociated cultures, neural aggregates can also be cultured as three-dimensional organotypic structures, commonly referred to as organoids. Organoids can be cultured either floating in suspension or at the air–liquid interface (ALI) on transwell membranes. The latter culture method has been shown to improve oxygen availability and thereby survival and maturation in cerebral organoids [26]. However, this requires an additional transfer step, and the organoids can be slightly more difficult to process later on. Both methods should therefore be taken into consideration when designing experiments.

3.4.1 Organoid Culture in Suspension

Cerebellar organoids can be kept floating in suspension for several weeks. Medium is changed at least once a week, but the frequency may be increased depending on the size and number of organoids in each well. Organoids may also be transferred into larger size dishes to meet nutrient demands. Long-term maintenance of organoids floating in suspension may be achieved by utilizing strategies that are used for the cultivation of cerebral organoids, including the use of an orbital shaker or a spinning bioreactor that promote enhanced nutrient uptake and oxygenation [27].

3.4.2 Organoid Culture at Air-Liquid Interface (ALI)

For long-term organoid maintenance, it may be advisable to transfer cerebellar organoids onto transwell inserts to improve oxygen availability. One 30-mm PTFE transwell membrane insert (0.4 µm pore size) can hold six to twelve organoids, depending on their size and the duration of the experiment (Fig. 4).

In preparation for ALI culture of cerebellar organoids, 1 mL of Differentiation Medium is added per well of a six-well plate. While holding the plate at a 45° angle, the transwell insert is gently placed into the plate with the membrane perpendicular to the bottom of the plate using sterile forceps. If necessary, the transwell insert is gently moved around until all air bubbles are relieved. The culture plate with transwells is then placed into an incubator at 37 °C with 5% CO_2 to equilibrate.

Individual organoids are placed onto the equilibrated transwell membranes in a small drop of medium using a wide bore tip (*see*

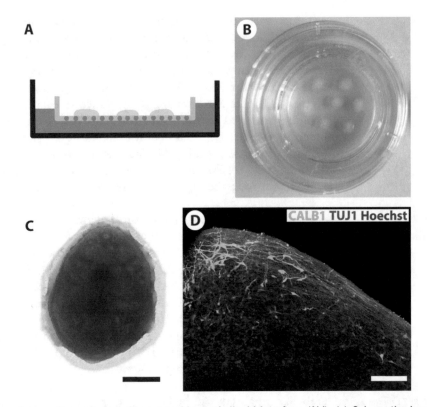

Fig. 4 Long-term culture of cerebellar organoids at air-liquid interface (ALI). (**a**) Schematic drawing of ALI culture of cerebellar organoids. Organoids are resting on the membrane of the transwell insert in a six-well plate above a layer of culture medium. Due to the large pore size of the membrane, all constituents of the culture medium can pass freely to the upper compartment. Exposure to air allows for good oxygenation of the organoids. (**b**) Photograph of a 30-mm transwell membrane with seven organoids placed at ALI. (**c**) Brightfield image of an organoid at Day 50 of differentiation. Scale bar 1000 μm. (**d**) Whole organoid immunostaining of a Day 50 hiPSC-derived cerebellar organoid that was cultured on a transwells with antibodies specific to early Purkinje cell marker CALB1 (green) and pan-neuronal marker TUJ1 (red). Nuclei are visualized in blue by Hoechst staining. Scale bar 50 μm

Notes **19** and **20**). Thereafter, all medium on the apical side of the transwell is removed. A full medium change is performed every 1–3 days, making sure no air bubbles are trapped underneath the transwell insert that could deprive the organoids of nutrients (*see* Note **21**). Organoids can be maintained in excess of 130 days on ALI culture.

3.5 Analysis

Sample preparation of the hiPSC-derived cerebellar cultures varies depending on whether pooled or single-cell analysis is required.

3.5.1 Sample
Preparation for Pooled
Analysis

Preparation for Reverse
Transcriptase-Polymerase
Chain Reaction

Adherent cells can either be lysed directly on a culture plate or coverslip and transferred into a microcentrifuge tube for snap freezing, or detached using enzymatic digestions, transferred to a microcentrifuge tube and pelleted by centrifugation at a maximum of $400 \times g$ for 4 to 5 min before snap freezing. A 24-well plate well at ~80% confluence should yield approximately 500 ng of RNA. Neural aggregates and organoids are harvested using a wide bore tip and pooled into a microcentrifuge tube. Typically, between 40 and 50 aggregates are pooled on Day 21, whereas on Day 35 around 30–40 will be adequate. RNA yields for pooled organoids range between 10 µg and 25 µg. Organoids from 50 days of differentiation onwards can be processed individually and RNA yield can range between 500 ng and 5 µg. Matrigel® should be removed prior to RNA extraction. While incubating organoids in ice cold PBS for 30 min at 4 °C with intermittent gentle trituration removes most Matrigel®, digestion solutions such as Cell Recovery Solution (Corning, Cat. #354253) may yield better results [28]. Samples are centrifuged at a maximum of $400 \times g$ for 4–5 min, and after the removal of supernatant they are ready to be processed immediately. Cell pellets can also be snap-frozen and stored at −80 °C to allow for flexibility in later processing.

Preparation
for Immunostaining

Cells growing on glass or plastic coverslips are fixed with 4% Paraformaldehyde (PFA) solution for 20 min at room temperature and processed immediately for immunofluorescence microscopy. Human cells can be distinguished from mouse cells by co-staining with an antibody against human anti-nuclear antigen (HuNu) (Fig. 3).

Floating samples are harvested using a wide bore tip and placed into 4% PFA solution for 20 min at room temperature for fixation. Subsequently, samples are gently rinsed three times with PBS and cryoprotected overnight in 20% sucrose in PBS. Samples are collected using a wide bore tip and placed in cryomolds. Any residual sucrose solution is removed as this will negatively impact the embedding and subsequent cryosectioning. Samples are embedded in optimal cutting temperature (OCT) compound and frozen on dry ice before storage at −80 °C or immediate processing.

Organoids that have been cultured on transwell membranes require some additional steps to the above-mentioned method. For optimal processing, a dissection microscope is used. Each individual organoid is cut out from the transwell membrane using a razor blade or scalpel. Two parallel lines are cut in the transwell membrane around each individual organoid, then the plate is rotated by 90°, and two parallel lines are cut perpendicular to the other two lines, eventually cutting out a square shape. The cut-out organoids including the attached small piece of transwell membrane are transferred into individual tubes filled with PBS and processed as described above. Organoids can be cryosectioned while still on the transwell membrane, however, separation of the organoid from the membrane will result in higher quality sections.

Alternatively, organoids can be stained as a whole to capture the expression of markers in full across an entire organoid following the method published by Dekkers et al. [29]. In brief, organoids that are cultured on transwells are gently washed off with PBS using a wide bore pipette tip. Washing and incubation steps are extended to allow for sufficient penetration of antibodies and subsequent antibody removal. Organoids are then cleared using a home-made fructose-glycerol solution to enable high-resolution imaging with minimal light scattering. Imaging can be performed using standard confocal laser scanning microscopy or more advanced modalities including spinning-disk confocal microscopy or light-sheet fluorescence microscopy.

Preparation for Flow Cytometry Analysis

Once appropriately harvested, cells from monolayer or suspension culture can be prepared for either live or fixed fluorescence-activated cell sorting (FACS). Cells are washed once in PBS at room temperature and enzymatically digested using Accumax or Neuron Isolation Enzyme. Optimal digestion time per cell type and experiment may vary (*see* **Notes 22** and **23**). Following digestion, all steps should be performed at 4 °C unless otherwise noted. To proceed with staining of fixed samples, cells are gently pelleted by centrifugation for 5 min at 300–500 × g before removal of supernatant. Cells are then fixed in 4% PFA for 15–20 min (*see* **Note 24**). Subsequently, cells are centrifuged for a further 5 min at 300–500 × g and washed three times before proceeding to either permeabilization if intracellular staining is desired (0.3% Triton X-100 in PBS for 15 min at room temperature), or directly to blocking and immunostaining (*see* **Note 25**). Samples may be filtered through a 70-μm filter or cell strainer prior to analysis and should be kept in the dark on ice and processed as soon as possible to prevent deterioration. For processing of live samples, aseptic technique is important when planning to recover cells for further growth. Cells should be harvested and processed as described above but leaving out the fixation and permeabilization steps. Samples are labelled in either native culture medium or suitable blocking buffer (e.g., PBS with 10% FCS) and relevant primary antibodies for 30–60 min at 4 °C, before washing and application of secondary antibody under the same conditions.

3.5.2 Sample Preparation for Single-Cell Transcriptomics

Organoids are placed into microcentrifuge tubes containing PBS either as single organoids per tube or pooled. PBS is carefully removed with a P1000 tip, and 200 μL of pre-warmed TrypLE™ enzyme (1×) are added. Tubes are placed in a pre-warmed 37 °C water bath for 5 min to digest the samples. Afterwards, the tubes are gently agitated and visually inspected to check if the samples have started to dissociate. Larger organoids may take longer to dissociate and might require another 5–15 min of incubation. To acquire a single-cell suspension, samples are further dissociated by gentle trituration in the TrypLE enzyme solution. Then, 400 μL chilled Culture Medium together with 300 μL of chilled Dulbecco's PBS (DPBS) are added to the microtube to dilute the TrypLE™ enzyme. Cells are pelleted by centrifugation at $500 \times g$ for 5 min at 4 °C. All subsequent steps should be performed on ice. The supernatant is carefully removed, and cells are gently resuspended in 200 μL DPBS using a wide bore pipette tip. An appropriate volume for cell counting is transferred to another tube, and samples are then harvested for the final application. If proceeding to fixation, 300 μL MeOH are added immediately to the cell suspension in a dropwise manner, and samples are incubated for 30 min at −20 °C. It is recommended to use this time to perform cell counts and to record total cell count, concentration, and viability. After fixation, cells can be used immediately or stored in MeOH at −20 °C for one to six weeks for later processing. If not proceeding with fixation, cells should be placed in chilled PBS in 0.1% BSA and processed immediately under the relevant guidelines for sample processing according to the chosen single-cell sequencing workflow.

4   Notes

1. Acidify sterile water to pH 2.0 with hydrochloric acid. Dissolve 100 mg insulin in 10 mL of acidic water.

2. To prepare 20 μg/mL stock solution first at 1 N NaOH per mg 3.3′, 5-triiodo-L-thyronine and swirl gently to dissolve. Then, add 49 mL of sterile medium per 1 mL of 1 N NaOH to dilute.

3. Ara-C, also known as Cytarabine, is an anti-mitotic agent that incorporates into the DNA and causes DNA fragmentation. Considering this cytotoxic effect, it is strongly advised to handle this reagent with care and prevent exposure of the skin and eyes.

4. The last remnants of ice will thaw in the time it takes to transfer the vial to the flow hood. In this way, the time that hiPSCs are in DMSO-containing freezing medium is minimized.

5. Microscopic evaluation combined with marking the underside of the tissue culture plate with a marker pen may be advantageous to carefully remove differentiated colonies.

6. An automated cell counter can help speed up the process and improve consistency in cell counting.

7. Good-quality undifferentiated hiPSCs are vital, and this is considered one of the most important conditions for successful

generation of EBs. The use of hiPSC cultures with areas of spontaneous differentiation should be avoided.

8. It is optional to collect the culture medium for subsequent use as a diluent for TrypLE enzyme.

9. Once cells are detached with TrypLE, work should proceed swiftly to minimize the time before aggregation. The ability of hiPSCs to attach to each other as well as their viability decrease quickly, and prolonged periods in floating suspension negatively impacts EB formation.

10. It is recommended to include extra dead volume of 5–10% to compensate for pipetting loss.

11. During the first 24 h after plating, the EB formation process is very fragile. It is recommended to avoid repeated opening and closing of incubator doors as this may disrupt the aggregation of hiPSCs and results in lower cell survival.

12. It is normal to see a halo of dead cells surrounding the EB during the first week of differentiation (Fig. 1). This halo will disintegrate over time until only the neural aggregate remains.

13. Wide bore tips can easily be made by cutting of the end of a P200 tip with a razor blade. Alternatively, a sterile plastic Pasteur pipette can be used.

14. To reduce variability, carry-over of Induction Medium should be avoided as much as possible as this may result in unintended transfer of growth factors.

15. Laminin should be slowly thawed on ice to avoid solidification.

16. 0.5 mL will cover one well of a 24-well plate.

17. Setting the pipette volume to 400 μL will help avoiding bubbles.

18. For 30 large neural aggregates, 1 mL of Seeding Medium is used.

19. Depending on the size, one can choose to use a wide bore P200 or P1000 tip.

20. After picking up the organoid with the wide bore tip, as much liquid as possible should be expelled, leaving the organoid in a low volume in the tip. Allowing the organoid to settle by gravity in the wide bore tip can help further to reduce the amount of liquid that is transferred with the organoid onto the transwell.

21. To avoid air bubbles underneath the membrane, sterile forceps can be used to pick up the transwell before aspirating the medium and again when adding medium back. This is best done at a 45° angle to further reduce the risk of air bubble formation.

22. Generally, non-enzymatic methods are preferred as proteolytic enzymes can cleave surface antigens [30].

23. It is recommended to start with an excess of cells to allow for at least 50,000 cells for end-point analysis. This number may vary if isolating rare live populations.

24. Optimization of the duration of fixation, blocking step, and antibody incubation is recommended for each sample type and antibody.

25. 0.1–10 μg/mL primary antibody is a suitable starting point. Antibodies should be titrated against relevant negative controls.

5 Conclusions

The method described in this chapter allows for the reproducible generation of cerebellar neurons from hiPSCs. After 21 days of aggregated culture, GBX2- and EN1-positive hindbrain neural stem cells are identified, roughly corresponding to developmental stage E9.5 in the developing mouse brain. By Day 35, these cells give rise to ATOH1-positive uRL progenitors and KIRREL2-positive VZ progenitors, mimicking the mediolateral expansion of the developing cerebellum that is observed in the murine brain around E13. Further maturation allows for the generation of major cell types of the developing cerebellum including Purkinje cells and granule neurons. The flexibility in culture options such as monolayer co-culture with mouse cerebellar progenitors or three-dimensional organoid culture makes this method versatile and suitable for many different research questions.

The current protocol has some shortcomings that need to be taken into consideration. Monolayer co-culture with mouse progenitor cells, as well as being xenogenic, may confound some downstream analysis strategies. Three-dimensional organoid models obviate the need for mouse co-cultures but come with increased complexity and variability. Another limitation of the current protocol is the relative immaturity of the produced cerebellar neurons. During human cerebellar development, maturation of cerebellar neurons extends into the first years of childhood and recapitulating this protracted process in vitro is difficult. Additionally, cerebellar organoids produced by the current protocol lack several significant factors such as vascularization, immune cells, stroma, and elements of dorsoventral patterning. Future studies should therefore consider the use of microfluidic culture systems, co-culture with immune cells, and transplantation into mice, as has been done successfully for other organoid models [31–35]. In spite of these limitations, the differentiation of hiPSCs into cerebellar neurons and organoids provides unique opportunities to study complexities of early embryogenesis, tissue-specific microenvironments, and human diseases. Especially for rare genetic disorders that often lack accurate in vitro models, patient-derived hiPSCs are groundbreaking experimental tools [9, 10, 36]. Although the reprogramming of patient-derived cells still needs to be optimized in terms of efficiency and speed, hiPSCs hold promising potential for advanced personalized and precision medicine. In the future, it is likely that patient-derived hiPSCs, which faithfully recapitulate the patient's genetic background, will be used to test a variety of different therapies to inform the most optimal treatment strategy for an individual patient. Additionally, modelling tissue-specific microenvironments may enhance our current understanding of cell–cell interactions that play important roles in many diseases. Future improvement in cerebellar differentiation protocols and their applications will undoubtedly increase our understanding of the many cerebellar disorders and transform their treatment options.

Acknowledgments

This work was supported by funds from the Rosetrees Trust, Action for A-T, BrAshA-T, and the John Fell Fund. M. J. v. E. was supported by a Clinical Research Training Fellowship from the Cancer Research UK Oxford Centre. S.N. was the recipient of a fellowship from the Oxford Nuffield Medical Trust. E. J. A. was supported by an iCASE studentship from the Biotechnology and Biological Sciences Research Council (UKRI-BBSRC).

References

1. Takahashi K, Tanabe K, Ohnuki M et al (2007) Induction of pluripotent stem cells from adult human fibroblasts by defined factors. Cell 131: 861–872

2. Takahashi K, Yamanaka S (2006) Induction of pluripotent stem cells from mouse embryonic and adult fibroblast cultures by defined factors. Cell 126:663–676

3. Fatehullah A, Tan SH, Barker N (2016) Organoids as an in vitro model of human development and disease. Nat Cell Biol 18:246–254

4. Rowe RG, Daley GQ (2019) Induced pluripotent stem cells in disease modelling and drug discovery. Nat Rev Genet 20:377–388

5. Su HL, Muguruma K, Matsuo-Takasaki M et al (2006) Generation of cerebellar neuron precursors from embryonic stem cells. Dev Biol 290:287–296

6. Muguruma K, Nishiyama A, Ono Y et al (2010) Ontogeny-recapitulating generation and tissue integration of ES cell-derived Purkinje cells. Nat Neurosci 13:1171–1180

7. Muguruma K, Nishiyama A, Kawakami H et al (2015) Self-organization of polarized cerebellar tissue in 3D culture of human pluripotent stem cells. Cell Rep 10:537–550

8. Salero E, Hatten ME (2007) Differentiation of ES cells into cerebellar neurons. Proc Natl Acad Sci U S A 104:2997–3002

9. Ishida Y, Kawakami H, Kitajima H et al (2016) Vulnerability of Purkinje cells generated from spinocerebellar ataxia type 6 patient-derived iPSCs. Cell Rep 17:1482–1490

10. Sundberg M, Tochitsky I, Buchholz DE et al (2018) Purkinje cells derived from TSC patients display hypoexcitability and synaptic deficits associated with reduced FMRP levels and reversed by rapamycin. Mol Psychiatry 23:2167–2183

11. Silva TP, Bekman EP, Fernandes TG et al (2020) Maturation of human pluripotent

stem cell-derived cerebellar neurons in the absence of co-culture. Front Bioeng Biotechnol 8:1–17

12. Watson LM, Wong MMK, Vowles J et al (2018) A simplified method for generating purkinje cells from human-induced pluripotent stem cells. Cerebellum 17:419–427

13. Ballabio C, Anderle M, Gianesello M et al (2020) Modeling medulloblastoma in vivo and with human cerebellar organoids. Nat Commun 11:1–18

14. Liu A, Joyner AL (2001) Early anterior/posterior patterning of the midbrain and cerebellum. Annu Rev Neurosci 24:869–896

15. Leto K, Arancillo M, Becker EBE et al (2016) Consensus paper: cerebellar development. Cerebellum 15:789–828

16. van Essen MJ, Nayler S, Becker EBE et al (2020) Deconstructing cerebellar development cell by cell. PLoS Genet 16:e1008630

17. Simão D, Silva MM, Terrasso AP et al (2018) Recapitulation of human neural microenvironment signatures in iPSC-derived NPC 3D differentiation. Stem Cell Rep 11:552–564

18. Yan Y, Song L, Bejoy J et al (2018) Modeling neurodegenerative microenvironment using cortical organoids derived from human stem cells. Tissue Eng Part A 24:1125–1137

19. Lee HK, Velazquez Sanchez C, Chen M et al (2016) Three dimensional human neurospheroid model of Alzheimer's disease based on differentiated induced pluripotent stem cells. PLoS One 11:e0163072

20. Chambers SM, Fasano CA, Papapetrou EP et al (2009) Highly efficient neural conversion of human ES and iPS cells by dual inhibition of SMAD signaling. Nat Biotechnol 27:275–280

21. Nayler S, Agarwal D, Curion F et al (2021) High-resolution transcriptional landscape of xeno-free human induced pluripotent stem cell-derived cerebellar organoids. Sci Rep 11:12959

22. Rosler ES, Fisk GJ, Ares X et al (2004) Long-term culture of human embryonic stem cells in feeder-free conditions. Dev Dyn 229:259–274

23. Yao S, Chen S, Clark J et al (2006) Long-term self-renewal and directed differentiation of human embryonic stem cells in chemically defined conditions. Proc Natl Acad Sci U S A 103:6907–6912

24. Muguruma K (2017) 3D culture for self-formation of the cerebellum from human pluripotent stem cells through induction of the isthmic organizer. In: Methods in molecular biology. Humana Press, New York, NY, pp 31–41

25. Sutcliffe M, Lancaster MA (2019) A simple method of generating 3D brain organoids using standard laboratory equipment. Methods Mol Biol 1576:1–12

26. Giandomenico SL, Mierau SB, Gibbons GM et al (2019) Cerebral organoids at the air–liquid interface generate diverse nerve tracts with functional output. Nat Neurosci 22:669–679

27. Lancaster MA, Knoblich JA (2014) Generation of cerebral organoids from human pluripotent stem cells. Nat Protoc 9:2329–2340

28. Bagley JA, Reumann D, Bian S et al (2017) Fused cerebral organoids model interactions between brain regions. Nat Methods 14:743–751

29. Dekkers JF, Alieva M, Wellens LM et al (2019) High-resolution 3D imaging of fixed and cleared organoids. Nat Protoc 14:1756–1771

30. Wang S, Wang B, Pan N et al (2015) Differentiation of human induced pluripotent stem cells to mature functional Purkinje neurons. Sci Rep 5:9232

31. Smith Q, Gerecht S (2014) Going with the flow: microfluidic platforms in vascular tissue engineering. Curr Opin Chem Eng 3:42–50

32. Homan KA, Gupta N, Kroll KT et al (2019) Flow-enhanced vascularization and maturation of kidney organoids in vitro. Nat Methods 16:255–262

33. Tsai S, McOlash L, Palen K et al (2018) Development of primary human pancreatic cancer organoids, matched stromal and immune cells and 3D tumor microenvironment models. BMC Cancer 18:335

34. Mansour AA, Gonçalves JT, Bloyd CW et al (2018) An in vivo model of functional and vascularized human brain organoids. Nat Biotechnol 36:432–441

35. Yin X, Mead BE, Safaee H et al (2016) Engineering stem cell organoids. Cell Stem Cell 18:25–38

36. Wong MMK, Watson LM, Becker EBE (2017) Recent advances in modelling of cerebellar ataxia using induced pluripotent stem cells. J Neurol Neuromed 2:11–15

Chapter 2

From Cerebellar Genes to Behaviors in Zebrafish

Masahiko Hibi and Takashi Shimizu

Abstract

The cerebellum is involved in some forms of motor coordination and learning, and also in non-motor cognitive and emotional functions. These functions of the cerebellum rely on the structure of the neural circuit which is generally conserved among vertebrates. As the zebrafish became a versatile animal model to study the development and function of the vertebrate brain, its cerebellum was studied over the last decade. Although the zebrafish cerebellum has a small and simple structure, it has layer and neural circuit structures as well as motor and non-motor functions similar to the mammalian cerebellum, and the same genetic program. Therefore, the zebrafish cerebellum is a good model for understanding the development and function of the cerebellum of mammals, including humans. In this review, we outline the structure and development of the zebrafish cerebellum, and explain the molecular and transgenic tools that are used to study it. Finally, we discuss how studies on the zebrafish cerebellar circuits contribute to our understanding the vertebrate cerebellum and human diseases caused by cerebellar defects.

Key words Calcium imaging, Cerebellum, CRISPR/Cas9, Fear conditioning, Mutants, Transgenic zebrafish

1 Introduction

The cerebellum is not only involved in motor coordination and learning, but also in cognitive and emotional functions, such as fear conditioning and reward expectations [1–10]. However, little is still known about the mechanisms regulating how cerebellar neural circuits elicit these functions. The development and function of the cerebellum have been studied mainly by using mammals. However, as the mammalian cerebellum is large and complex, it is difficult to study the neuronal activity of the entire cerebellum in living animals. Therefore, alternative models could assist the study of the vertebrate cerebellum.

The zebrafish brain has become a versatile model for studying the function and development of the vertebrate brain since it is smaller than the mammalian brain, the zebrafish body is transparent during early larval stages (and also at late larval stages in a pigmentless mutant background), and the development of its brain is

Roy V. Sillitoe (ed.), *Measuring Cerebellar Function*, Neuromethods, vol. 177, https://doi.org/10.1007/978-1-0716-2026-7_2,
© Springer Science+Business Media, LLC, part of Springer Nature 2022

relatively fast. There are many tools and methods available to study the development and function of the zebrafish brain. These include transgenic (Tg) zebrafish and mutants having defects in the structure and/or function of the brain (Subheading 2). In addition, genome editing with the CRISP/Cas9 system can modify genes of interest. For functional analyses of the zebrafish brain, Ca^{2+} and voltage live imaging of neuronal/glial activity, and optogenetic manipulation can be used to study zebrafish neural circuits (Subheading 2). Therefore, studies with zebrafish can provide valuable information that is not available for other vertebrates. In the past decade, evidence has accumulated showing that the zebrafish cerebellum is a good model for understanding the vertebrate cerebellum.

1.1 Anatomy of Zebrafish Cerebellar Circuits

Organization of the zebrafish cerebellum has been studied in detail [11, 12] (reviewed in [13–16]). It has three domains: the rostralmost region is called the valvular cerebelli (Va), the middle region is called the corpus cerebelli (CCe), and the lateral and caudal regions are functionally linked and called the eminentia granularis (EG) and lobus caudalis cerebelli (LCa), respectively (Fig. 1a, b). Like the mammalian cerebellum, the Va and CCe are composed of three layers, called the molecular layer (ML), the Purkinje cell layer (PCL), and the granular layer (GL), progressing from superficial to deep layers. The EG and LCa have a distinct layer structure in which the GL is located at their surface (Fig. 1b–d). There is no clear white matter layer in the zebrafish cerebellum. The zebrafish cerebellum has several types of neurons and glial cells, similar to the mammalian cerebellum, including granule cells (GCs), Purkinje cells (PCs), Golgi cells (GoCs), stellate cells (SCs), and eurydendroid cells (ECs) (Fig. 1c, d). ECs are teleost-specific excitatory projection neurons (efferent neurons) with connections similar to those of neurons in the deep cerebellar nuclei (DCNs) in mammals (Fig. 1). Although GCs, PCs, GoCs, and SCs are present in the zebrafish cerebellum, GABAergic basket cells, Lugaro cells, candelabrum cells, and glutamatergic unipolar brush cells have not been reported for zebrafish. In addition to neurons, Bergmann glial cells (BGs) are present but it is not clear whether the zebrafish cerebellum has non-BG astrocytes. The position of cerebellar neurons and glial cells in the zebrafish cerebellum is essentially the same as in the mouse cerebellum, with some differences.

As in the mammalian cerebellum, the zebrafish cerebellum receives two types of afferent fibers: climbing fibers (CFs) and mossy fibers (MFs). The CFs and MFs form synapses with PCs and GCs, respectively (Fig. 1c, d). CFs are axons of neurons in the inferior olivary nuclei (IOs), whereas MFs are axons of the precerebellar nuclei (PNs) located in various regions of the brain. Information from the MFs is conveyed by the axons of GCs (called parallel fibers, PFs) to the dendrites of PCs. The information from

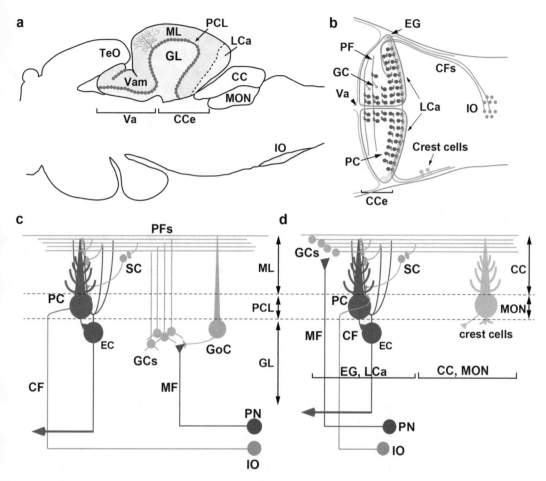

Fig. 1 Structure of zebrafish cerebellum and cerebellar neural circuits. (**a**) Structure of zebrafish adult cerebellum. Sagittal section image. (**b**) Structure of zebrafish larval cerebellar neural circuitry. Dorsal view image. (**c, d**) Schematic representation of the cerebellar neural circuits in the Va and CCe (**c**) or EG/LCa and CC/MON (**d**). *CC* crista cerebellaris, *CCe* corpus cerebelli, *CF* climbing fiber, *EC* eurydendroid cell, *EG* eminentia granularis, *GC* granule cells, *GL* granular layer, *GoC* golgi cell, *IO* inferior olivary nucleus, *LCa* lobus caudalis cerebelli, *MF* mossy fiber, *MON* medial octavolateralis nucleus, *PC* Purkinje cell, *PCL* Purkinje cell layer, *PF* parallel fiber, *PN* precerebellar nuclei (except IO), *SC* stellate cell, *TeO* tectum opticum, *Va* valvular cerebelli, *Vam* medial division of valvular cerebelli. ((**a** and **b**) were modified from [18] (**c**) was modified from [13–16])

CFs and MFs is integrated in PCs. In mammals, PCs send output to the outside of the cerebellum via neurons in the DCNs, which are located far from the PCL. In contrast, ECs, located near the PCL, receive inputs from axons of nearby PCs and probably also from PFs, and send outputs to the outside of the cerebellum (Fig. 1c, d). Information from the CF and MF may be integrated in the ECs. Similar to mammals, PCs also directly project to the vestibular nuclei (the descending octaval nucleus [DON] in zebrafish) [11, 17, 18]. Axons of GCs in the rostral and medial regions

(Va and CCe) form synapses with dendrites of PCs or stellate cells in the ML, as in mammals (Fig. 1c). Whereas axons of GCs in the lateral and caudal regions (EG and LCa) form synapses with PCs in the cerebellum, they further extend caudally to form synapses with dendrites of crest cells, which are PC-like cells located in the medial octavolateral nucleus (MON) of the dorsal hindbrain [11, 18] (Fig. 1d). Although this caudo-lateral circuit is specific to the teleost cerebellum, it is likely involved in the vestibular and lateral line system, and is related to the neural circuits in the flocculonodular lobe in the mammalian cerebellum [13–16].

1.2 Development of Zebrafish Cerebellar Circuits

1.2.1 Establishment of the Cerebellum Domain

The cerebellar domain is established in the rostral-most hindbrain as a result of signals from the isthmic organizer, which is located at the mid-hindbrain boundary (MHB) [19–22]. Molecular mechanisms for the establishment of the cerebellum domain are essentially conserved among vertebrates (Fig. 2a). In amniotes, two homeodomain-containing transcription factors, Otx2 and Gbx2, are expressed in a largely complementary manner in the rostral and caudal regions of the neuroectoderm, and determine the position of the isthmic organizer [23–28]. In zebrafish, *gbx1* and *gbx2* are expressed in overlapping regions of the hindbrain [29, 30]. Gbx1 and Gbx2 function redundantly to control formation of the isthmic organizer and cerebellum development, by repressing *otx* genes [31]. After the isthmic organizer forms, a set of genes, fibroblast growth factor 8a (*fgf8a*), *wnt* genes (*wnt1/3a/10b*), *pax2a*, and *lmx1b* are expressed in the MHB, and restrict and maintain the isthmic organizer and control the formation of the cerebellum [31–34] (Fig. 2a). Among them, *wnt* genes and *fgf8a* are expressed in the midbrain and hindbrain regions, respectively, and play major roles in morphogenesis of the cerebellum. The Wnt proteins prevent apoptosis and induce cell proliferation [32]. Fgf8a is required for the isthmic organizer and cerebellum formation in zebrafish [33]. Fgf8a controls cerebellar fate by repressing *otx* genes [35]. Fgf signaling also functions downstream of Gbx1/2 for the isthmic organizer program and cerebellar differentiation in zebrafish [31].

1.2.2 Differentiation of Neurons

After the establishment of the cerebellum domain, cerebellar neurons are generated from two neural progenitor domains: the upper rhombic lip (URL, also called the cerebellar rhombic lip) and the cerebellar ventricular zone (VZ) [36–38] (Fig. 2b, c). The URL is located at the dorso-caudal edge of the cerebellum primordium. Like in mammals [39–45], the neural progenitors express the basic helix-loop-helix (bHLH)-type proneural *atoh1* genes (*atoh1a/b/c*) and give rise to glutamatergic neurons such as GCs in zebrafish [46, 47]. In amniotes, the *Atoh1*+ neural progenitors migrate tangentially from the URL to form the external granular (or germinal) layer (EGL), which covers the entire dorsal surface of the

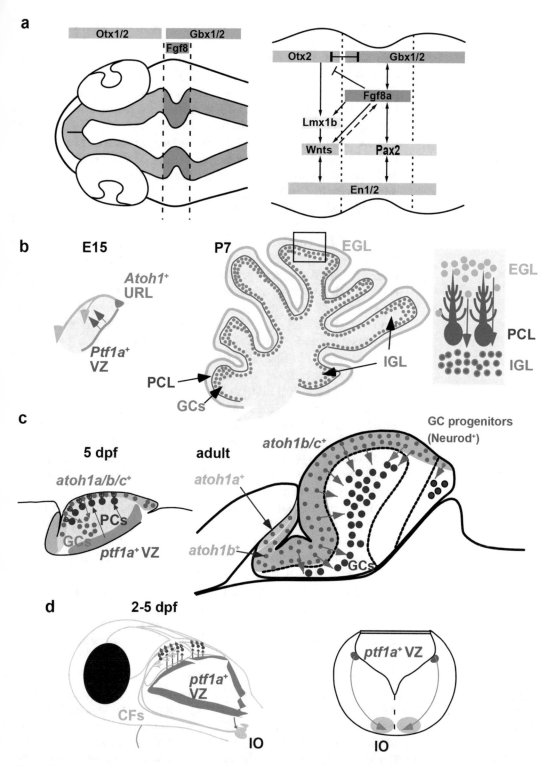

Fig. 2 Development of cerebellar neural circuitry in zebrafish. (**a**) Establishment of the cerebellar domains. Schematic representation of the early rostro-caudal patterning of the neuroectoderm that determines the cerebellum domain (left panel), and the genetic program that establishes and maintains the isthmic organizer (right panel). (**b**) Mouse cerebellum development on embryonic day 15 (E15, left panel) and postnatal day

cerebellum [37] (Fig. 2b). In zebrafish, the *atoh1*+ neural progenitor domains near the midline extend rostrally and those at the caudo-lateral edge extend laterally, forming continuous medial and caudal *atoh1*+ domains [46, 47] (Fig. 2c). The *atoh1*+ cells are proliferative, and a portion of these cells start to express another bHLH factor Neurod1 and differentiate into post-mitotic GCs in zebrafish [47]. They migrate ventrally and stay at the GL in the Va and CCe, whereas they stay in the surface domain in the EG and LCa. The VZ is located at the rostral roof of the fourth ventricle and ventral to the URL. As in mammals [48, 49], the neural progenitors in the VZ express the bHLH proneural gene *ptf1a* and give rise to GABAergic neurons, including PCs and stellate cells [47]. They migrate dorsally from the VZ and form the PCL [47]. *Atoh1*+ neural progenitors give rise to projection neurons in the DCNs in mice [42, 43, 45], whereas the origin of ECs is not entirely clear. At least a portion of ECs express the proneural gene *olig2* and the majority of the *olig2*+ cells are derived from the *ptf1a*-expressing VZ cells; they also migrate dorsally from the VZ and stop near the PCL [11, 47]. Reelin secreted from GCs is involved in the dorsal migration of PCs, ECs, and possibly also BGs, which are likely generated in the VZ [50].

In mice, the *Atoh1*+ neural progenitors in the dorsal hindbrain (lower rhombic lip [LRL]) give rise to neurons in the PNs that project the MFs [42, 43, 51]. Although cells derived from the *atoh1*+ neural progenitors migrate ventrally in the zebrafish hindbrain [47], it has not yet been reported whether they contribute to precerebellar neurons of MFs in zebrafish. As in mammals, *ptf1a*+ neural progenitors in the caudal hindbrain give rise to the IO neurons that project the CFs [47] (Fig. 2d).

Differentiation of the cerebellar neurons and the IO neurons is observed from the early larval stage (3 days post-fertilization [dpf]). A simple cerebellar circuit structure is observed at 5 dpf. Neurogenesis takes place for long periods: PCs are generated during the first month (larval stage) and GCs are continuously generated throughout the life [47, 52, 53] (Fig. 2c). RNA sequencing (RNA-seq) with zebrafish Tg larvae identified genes expressed in GCs, PCs, ECs, BGs, or IO neurons during the formation of the cerebellar neural circuits [54]. Future studies will identify genes

Fig. 2 (continued) 7 (P7, middle panel). Sagittal section views. Right panel is a detailed illustration of the box in the middle panel. (**c**) Zebrafish cerebellum development at larval (5 dpf, left panel) and adult (middle panel) stages. Sagittal section views. (**d**) Development of Purkinje cells (PCs) and neurons in the inferior olivary nuclei (IOs). The *ptf1a*-expressing neural progenitors give rise to PCs and IO neurons in the rostral and caudal hindbrain, respectively. During development, PCs migrate dorsally and IO neurons migrate ventrally from the ventricular zone. *EGL* external granular layer, *IGL* internal granular layer, *URL* upper rhombic lip, *VZ* ventricular zone. The other abbreviations are described in the legend of Fig. 1. ((**a**) was modified from [117, 118]. (**b, c**, and **d**) were modified from [13, 16])

that are involved in neuronal differentiation and neural circuit formation of the cerebellum.

1.3 Function of Zebrafish Cerebellar Circuits

1.3.1 Eyeblink Conditioning, Vestibular-Ocular Reflex, and Optokinetic Response

As cerebellum-dependent learning paradigms, eyeblink conditioning, and adaptation of the vestibular-ocular reflex (VOR) of teleost species have been studied. The goldfish reportedly learns an eyeblink-like conditioned response, and cerebellar lesions prevent this learning [55, 56]. Cerebellar lesions in goldfish also affect VOR adaptation [57, 58]. In zebrafish, the VOR starts at 3 dpf [59]. However, the VOR at this stage might not depend on the cerebellar neural circuitry. The cerebellum may be involved in the adaptation of the VOR, and this occurs at a later developmental stage. In addition to the VOR, zebrafish larvae express an optokinetic response (OKR) [60–64]. The cerebellum is involved in the OKR in zebrafish larvae [17].

1.3.2 Motor Adaptation

The zebrafish cerebellum is also involved in motor adaptation. Zebrafish modify their motor output during swimming according to visual information (visual feedback) that is obtained from their surroundings. When a zebrafish larva is paralyzed with a muscle relaxant agent, as shown in an artificially moving environment, the fish cannot swim but try to adapt motor output (activities of motoneurons) to changes in the virtual environment in the brain (fictive swimming or fictive optomotor response [OMR]). The cerebellar neural circuits are activated during the fictive swimming adaptation [65, 66]. Lesions of the CFs prevent the adaptation of fictive swimming [65]. Optogenetic activation of PCs affects the OMR [17]. These reports suggest that cerebellar neural circuits are involved in the visual feedback-dependent motor adaptation in zebrafish.

1.3.3 Fear Conditioning

The role of the cerebellum in classical conditioning has been studied. The repeated pairing of a conditioned stimulus (CS) (e.g., a tone or light) and an aversive unconditioned stimulus (US) (e.g. an electric shock) can induce CS-evoked avoidance responses in unrestrained animals (adaptive avoidance learning). However, if animals are restrained, the CS induces freezing behaviors, such as bradycardia (classical fear conditioning). Cerebellar lesions or drug-mediated silencing of the cerebellar function in goldfish result in an impaired conditioned bradycardia response [4, 67]. Zebrafish are also able to acquire classical fear conditioning and active avoidance learning from larval stages, although the timing of the robust responses depends on the experimental conditions [68, 69]. We previously showed that zebrafish acquire the conditioned bradycardia responses from the late larval stage (around 20 dpf, Subheading 3.2) [8] (Figs. 4 and 5). Botulinum-toxin-mediated inhibition of GCs in the CCe does not suppress CS-evoked bradycardia, but rather prolongs the response [8]. We found cerebellar neurons

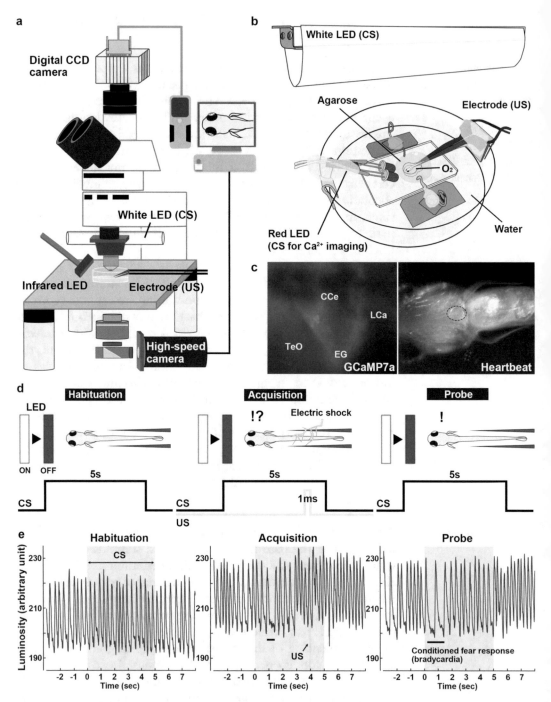

Fig. 4 Method for studying classical fear conditioning in zebrafish. (**a**, **b**) Schematic representation of the experimental setup for classical fear conditioning using about 20-dpf zebrafish larvae. **c** Imaging of Ca^{2+} signals of cerebellar neurons (left panel); the heartbeats (right panel) in *Tg(elavl3:Gal4-VP16);Tg(UAS: GCaMP7a)* larvae. Images showing the movement of the heart were taken from below with an infrared digital camera. (**d**) Experimental paradigm for delayed classical fear conditioning. (**e**) A typical heartbeat pattern in the classical fear conditioning, showing the heart movement in the 15th trial of the habituation session, the 16th trial of the acquisition session, and the eighth trial of the probe session, and the timing of CS (gray boxes) and US (yellow line). The *x*-axis shows time (sec). The *y*-axis shows the heartbeats as monitored by luminosity

associated with the classical fear conditioning, and these are likely GCs [8]. These data suggest the roles of the cerebellar neural circuits in controlling recovery from the conditioned fear responses in zebrafish. The zebrafish cerebellum is also involved in operant conditioning. Cerebellar neurons are activated prior to movements in the operant conditioning and lesions in the cerebellum affect decision-making in operant conditioning [70].

2   Materials

2.1   Molecular Markers

RNA-seq analysis with Tg fish expressing a fluorescent protein in the cerebellar neurons (Subheading 2.2) revealed that orthologues of mammalian genes that are known to be expressed in cerebellar neurons are mostly expressed in the same type of cerebellar neurons in zebrafish [54]. Therefore, in situ hybridization with zebrafish orthologue genes can be used to identify cell types in the zebrafish cerebellum (Table 1). However, there are some differences between mammalian and zebrafish cerebella. In mammalian cerebella, many PC genes, including zebrin II (encoded by *aldolase C* in mice and *aldolase Ca* [*aldoca*] in zebrafish), are known to display stripe expression patterns [13, 71]. So far, none of the PC genes have shown a stripe expression pattern in the zebrafish cerebellum [11, 72]. There are many parvalbumin genes present in the zebrafish genome. Although parvalbumin is expressed in both PCs and other GABAergic interneurons (SCs and GoCs) in the mouse cerebellum, parvalbumin 7 is specifically expressed in PCs in the zebrafish cerebellum [11]. Although the origin of ECs might be different from that of projection neurons in the DCN, ECs express *vglut2a* (*vglut2a*, *slc17a6b* in zebrafish) more than *vglut1* (*slc17a7*), similar to the DCN neurons. In contrast, *olig2* is expressed in ECs in the zebrafish cerebellum [11, 47, 73], but it is expressed in the progenitors of PCs in the mammalian cerebellum [74]. While gene markers identify cell types, in situ hybridization with them is technically cumbersome, and it is also difficult to mark neurite structures. Instead, protein markers, which can be detected by immunohistochemistry with antibodies, are useful to identify cell types as well as the sub-cellular structure of neurons and glia (Table 2). For instance, although *vglut1* transcripts were detected in GC somata, Vglut1 protein was detected in the axonal termini of GCs [11].

Fig. 4 (continued) (arbitrary unit) at an appropriate region of the heart in the video. Each peak represents one heartbeat. In the acquisition trial, the US elicited the bradycardia response. In the late acquisition and probe sessions, bradycardia, denoted by horizontal bars, occurred after the start of the CS presentation. The abbreviations are described in the legend of Fig. 1. (The figure was modified from [8])

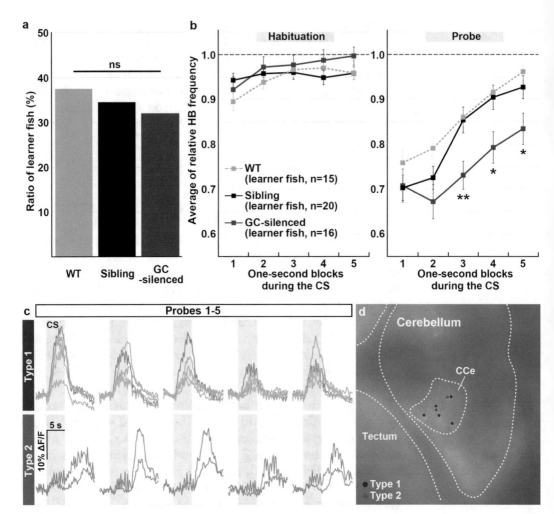

Fig. 5 Cerebellar neural circuitry is involved in classical fear conditioning in zebrafish. (**a, b**) Effect of inhibition of granule cell activity on classical fear conditioning. Tg larvae with inhibited GC activity (GC-silenced larvae) were obtained by crossing the GC-specific Gal4 line gSA2AzGFF152B and the *Tg(UAS:BoTxBLC-GFP)* line, which expresses a fusion protein of GFP and the light chain of botulinum toxin light B in a Gal4-dependent manner. (**a**) Percentage of fish showing CS-dependent bradycardia (learner fish). There was no significant difference (ns) in the learner rate among wild-type, GC-silenced and sibling groups ($p = 0.865$, Fisher's exact test). (**b**) CS-evoked bradycardia responses in wild-type, GC-silenced, and their sibling learner fish. Relative HB frequency during 1 s period of the 5 s CS presentation (average of each one-second block) is shown. Average of the data from 10 trials in the habituation and probe sessions was calculated and plotted in graphs. The graphs show the average and standard errors (SE). The CS-evoked bradycardia responses during the probe session differed between the GC-silenced and the sibling learner groups (*$p < 0.05$, **$p < 0.01$, two-way repeated measures ANOVA with Bonferroni's post-hoc test). The bradycardia response of the GC-silenced larvae was prolonged in the probe session. (**c, d**) Conditioning-associated neurons. Two types of conditioning-associated neurons were observed. Type I neurons responded quickly to the CS, while type II neurons were activated after a delay. (**c**) Neuronal activity ($\Delta F/F$) of type I and II neurons of *Tg(elavl3:Gal4-VP16);Tg(UAS:GCaMP7a)* larvae during the probe session (1st–fifth trials), data shown from five type I neurons and two type II neurons. Gray boxes indicate the timing of the CS presentation. (**d**) Location of type I (red dots) and II neurons (blue dots) in the CCe. (The figure was modified from [8])

Table 1
Genetic markers for zebrafish cerebellar circuits

Cell types	Gene symbol [ref.]	Product
Purkinje cells	*aldoca* [11, 54]	Aldolase Ca
	pvalb7 [11, 54]	Parvalbumin 7
	ca8 [11]	Carbonic anhydrase VIII
	gad1b [11]	Glutamate decarboxylase 1b
	gad2 [11]	Glutamate decarboxylase 2
	lhx1a [11]	LIM homeobox 1a
Granule cells	*vglut1* (*slc17a7*) [11]	Vesicular glutamate transporter 1
	cbln12 [54]	Cerebellin 12
	neurod1 [47, 94]	Neurogenic differentiation
	Reln [11, 50, 54]	Reelin
	barhl1b [11, 54]	BarH-like homeobox 1b
	pax6a [11, 54]	Paired box 6a
	eomesa [11, 54]	Eomesodermin homolog a
	zic1 [11, 54]	Zic family member 1
	fat2 [54]	FAT atypical cadherin 2
GABAergic interneurons	*gad1b* [11, 54]	Glutamate decarboxylase 1b
(Golgi and stellate cells)	*gad2* [11, 54]	Glutamate decarboxylase 2
Eurydendroid cells	*vglut2a* (*slc17a6b*) [11, 54]	Vesicular glutamate transporter 2b
	olig2 [11, 54]	Oligodendrocyte lineage transcription factor 2
Inferior olivary nuclei	*pou4f1* [11]	POU class 4 homeobox 1
	grm5a [111]	Glutamate receptor, metabotropic 5a
	foxp2 [a]	Forkhead box P2
Crest cells	*pvalb7* [11, 54]	Parvalbumin 7
	grid2ipa [54]	Grid2-interacting protein, a
Bergmann glial cells	*fabp7a* (*blbp*) [11]	Fatty acid binding protein 7, brain, a
	glastb (*slc1a3b*) [11]	Solute carrier family 1 member 3b
	gdf10a [54]	Growth differentiation factor 10a
	gdf10b [54]	Growth differentiation factor 10b
	grik2 [54]	Glutamate receptor, ionotropic, kainate 2

Only representative markers are shown. Please *see* [11, 54] for details and other genetic markers
[a] Unpublished result. Note: It has recently been reported in [112]

2.2 Transgenic Fish   We and others previously reported Tg zebrafish lines that express a fluorescent protein or the modified version of transcriptional activator Gal4 (e.g. GFF) in PCs, GCs, ECs, IO neurons or BGs with the Tol2 transposon (Fig. 3, Table 2) [17, 72, 75, 76]. Many of

Table 2
Protein and transgenic markers of cerebellar neural circuits

Cell types	Markers [ref.]	Localization
Purkinje cells	Protein	
	parvalbumin 7 [11]	S, A, D
	Zebrin II (Aldoca) [11]	S, A, D
	Carbonic anhydrase 8 (Ca8) [11]	S, A, D
	glutamate receptor, ionotropic, delta2 (Grid2) [113]	D
	Transgenic fish	
	Tg(aldoca:gap43-Venus) [72, 76]	
	Tg(aldoca:gap43-mCherry) [76]	
	Tg(aldoca:GFF) [76]	
Granule cells	Protein	
	Neurod1 [47]	N
	Vglut1 [11]	T
	Transgenic fish	
	Tg (cbln12:Venus) [75]	
	Tg (cbln12:GFF) [*1]	
	TgBAC (neurod1:EGFP) [114]	
	vglut1[GFP] [*1]	
	gSA2AzGFF152B (Va, CCe > LCa, EG)[*2] [76]	
	hspGFF57A [76]	
	hspGFFDMC90A [76]	
	gSAIGFF23C (LCa, EG)[*3] [76]	
Eurydendroid cells	Transgenic fish	
	Tg(olig2:EGFP)[*4] [11, 73]	
	hspzGFFgDMC156A [76]	
GABAergic interneurons	Protein	
(Golgi and stellate cells)	Pax2 [53]	N
Inferior olivary nuclei	Protein	
(CFs)	Vglut2a [11]	T
	Transgenic fish	
	hspGFFDMC28C [76]	
	TgBAC(mnx2b:GFF)[*]	
CFs and MFs	Transgenic fish	
	TgBAC(vglut2a:EGFP) [11]	

(continued)

Table 2
(continued)

Cell types	Markers [ref.]	Localization
Crest cells	Protein	
	parvalbumin 7 [11]	S, A, D
	glutamate receptor, ionotropic, delta2 (Grid2) [113]	D
Bergmann glial cells	Protein	
	Fabp7a [11]	C
	Transgenic fish	
	SAGFF(LF)251A [76]	
	SAGFF(LF)226B (only larval stage) [76]	
Oligodendrocytes	Protein	
	Myelin basic protein (MBP) [11]	M
	Transgenic fish	
	Tg(olig2:EGFP) [11]	

The proteins in this list can be detected by immunohistochemistry with antibodies. For the Gal4 (GFF) transgenic lines, fluorescent proteins are detected in somata and neurites when crossed with UAS:GFP or UAS:RFP lines. Please *see* [11, 18, 47] for details. [*1] Unpublished results. [*2] Expression is detected mainly in the Va and CCe. [*3] Expression is detected only in the LCa and EG. [*4] Expression is detected in ECs and oligodendrocytes in the zebrafish cerebellum. *A* axons, *C* cytosol, *D* dendrites, *M* myelin, *N* nucleus, *S* soma, *T* axon terminal

them were obtained from gene and/or enhancer trap screening, while the rest of them were established by using enhancer and promoter elements of genes specifically expressed in PCs or GCs, such as *aldoca*, *ca8* (for PCs), and *cbln12* [17, 72, 75, 76] (Tables 2 and 3). By combining the Gal4 lines with Tg lines carrying a reporter or effector located downstream of Gal4 binding sequences (upstream activating sequence: UAS), any transgenes can be expressed in the zebrafish cerebellar neural circuits (Gal4-UAS system). For instance, Tg fish expressing a fluorescent protein mark specific types of neurons and can be used to find targets for electrophysiological recordings [66, 77–82]. The Ca^{2+} indicator GCaMP or voltage sensors can be expressed to monitor neuronal activity of cerebellar neurons [8, 17, 83, 84]. Neurotoxin (e.g. tetanus toxin, botulinum toxin), cell death-inducing tools (e.g. nitroreductase, KillerRed), and optogenetic tools can be expressed to manipulate functions of the cerebellar neural circuits in zebrafish [8, 85, 86]. Neuronal type-specific enhancer/promoters are also used to express these proteins. Tg fish used for functional analyses are listed in Table 3. New or improved tools, especially for sensitive neuronal activity imaging and optogenetic control of cytoplasmic signaling, will hopefully be available soon from the zebrafish research community.

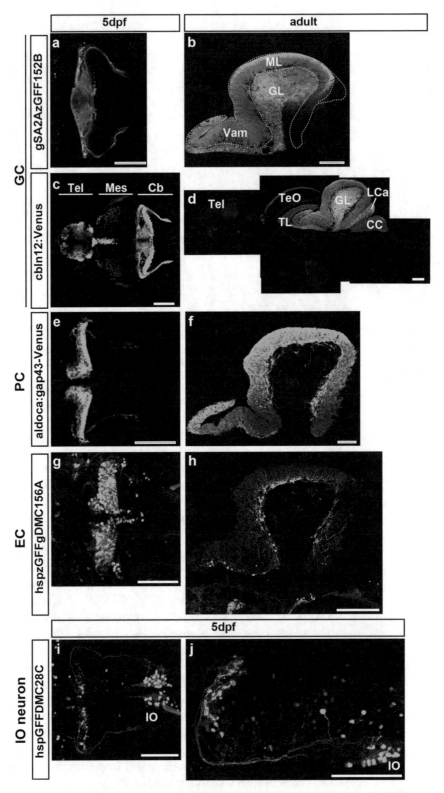

Fig. 3 Transgenic fish for zebrafish cerebellum research. Expression of GFP or Venus in gSA2AzGFF152B;*Tg (UAS:GFP)* (**a**, **b**, GCs), *Tg(cbln12:Venus)* (**c**, **d**, GCs), *Tg(aldoca:gap43-Venus)* (**e**, **f**, PCs), hspzGFFgDMC156A;

2.3 Zebrafish Mutants

Forward genetic screening of zebrafish mutants having abnormalities in the development of PCs and/or GCs was possible using the mutagen ethyl nitrourea (ENU) [11]. From the F3 screening and positional cloning of mutant loci, we previously reported that type IV collagen controls axogenesis of GCs [76] and Rab escort protein 1 (Rep1) is involved in the formation or maintenance of GCs and PCs [87]. These genes are related to two human diseases, Alport syndrome (type IV collagen) and choroideremia (Rep1), although cerebellum-related symptoms have been reported for these human diseases. Nevertheless, zebrafish mutants isolated from forward genetics might help to understand the pathogenesis of these diseases.

Genome editing with the CRISPR/Cas9 system has been used to generate zebrafish mutants. Injection of Cas9 mRNA (or protein) with single guided (sg) RNA or crRNA:tracrRNA (double guided [dg]RNA) recognizing target DNA in fertilized eggs introduces indel mutations in the target [88–91]. It took more than 3 months (one generation) to establish mutant lines and homozygous mutant fish. However, a recent CRISP/Cas9 system using Cas9 ribonucleoproteins (RNPs) containing synthetic Alt-R-modified dgRNA (from Integrated DNA Technologies [IDT]) induced an indel mutation with extremely high efficiency and generated null mutants at the F0 generation [92]. Genome editing with CRISPR/Cas9 and TALENs (transcription activator-like effector nucleases) methods generated zebrafish (and medaka) mutants of genes involved in the development and/or function of the cerebellum, including *reelin* and *contactin1* [50, 93, 94]. With the improved CRISPR/Cas9 method, the functions of genes involved in the development and function of the cerebellum will surely be revealed. The CRISPR/Cas9 method also enables loci to be targeted by knock-in transgenes (e.g. GFP and Gal4) [88, 95, 96].

3 Methods

3.1 Histology, Mutagenesis, Transgenesis, and Electrophysiology

General methods of histology (in situ hybridization and immunohistochemistry), transcriptome, and generation of mutants and transgenic fish for zebrafish cerebellum research have been reported [11, 18, 47, 50, 54]. Electrophysiological analyses of the zebrafish

Fig. 3 (continued) Tg(UAS:GFP) (**g**, **h**, ECs), and hspGFFDMC28C;*Tg(UAS:GFP)* (**i**, **j**, IO neurons). Expression in 5-dpf (**a**, **c**, **e**, **g**, **i**, **j**) or adult brain (**b**, **d**, **f**, **h**). Dorsal views (**a**, **c**, **e**, **g**, **i**), lateral view (**j**), and sagittal sections (**b**, **d**, **f**, **h**). Samples were immunostained with anti-GFP antibody (green) and anti-parvalbumin 7 (magenta, for PCs). *Cb* cerebellum, *Mes* mesencephalon, *Tel* telencephalon. The other abbreviations are described in the legend of Figs. 1 and 2. Scale bars: (**a**, **c**, **e**, **g**, **i**, **j**) 100 μm, (**b**, **d**, **f**, **h**) 200 μm. ((**a**, **b**, **e**, **f**, **g**, **h**, **i**, and **j**) are modified from [18]. (**c** and **d**) were modified from [75])

Table 3
Transgenic fish can be used for functional analyses

Enhancer/ promoter	Reporter or effectors	Function of reporters/effectors
aldoca (Purkinje cells)	*Tg(aldoca:GCaMP6s)*[a]	Ca^{2+} indicator
	Tg(aldoca:ChRWR-EGFP)[a]	Channel rhodopsin, depolarization
	Tg(aldoca:ArchT-GFP)[a]	Arch rhodopsin, hyperpolarization
	Tg(aldoca:BoTxBLC-GFP)[a]	Botulinum toxin
	Tg(aldoca:NTR-TagRFP)[a]	Nitroreductase
UAS (or UAS-hsp70l)	*Tg(UAS:GCaMP6s)* (or GCaMP7a) [8]	Ca^{2+} indicator
	Tg(UAS:ChRWR-EGFP) [116]	Channel rhodopsin, depolarization
	Tg(UAS:BoTxBLC-GFP) [86]	Botulinum toxin
	Tg(UAS:NTR-TagRFP) [116]	Nitroreductase
	Tg(UAS:KillerRed) [85]	Reactive oxygen species (ROS) generator
	Tg(UAS:AcGFP-P2A-WGA) [18]	Wheat germ agglutinin (WGA), forward neuronal tracing

[a]Unpublished. A reporter or effectors are expressed by using the UAS lines, cell-type specific Gal4 lines in Table 2 (e.g. *Tg (aldoca:GFF)*, *Tg(cbln12:GFF)*, and Gal4 gene/enhancer trap lines) or pan-neuronal Gal4 lines such as *Tg(elavl3:Gal4-VP16)*. There are many UAS:effector Tg lines that are available from the zebrafish research community

cerebellum have also been reported [66, 77–82]. Neural circuit tracing with the recombinant rabies virus for afferent connections in the zebrafish cerebellum was also reported [75]. There are many reports describing behavior analyses related to the zebrafish cerebellum [17, 68, 70, 79, 83, 94, 97]. Readers should refer to those papers for details of these analyses.

3.2 Classical Fear Conditioning

As an example of functional analyses of the zebrafish cerebellum, we studied the role of the zebrafish cerebellum in classical fear conditioning [8]. We used a delayed fear conditioning paradigm with about 20-dpf zebrafish larvae (Fig. 4). The larvae were restrained in 4% agarose gel and O$_2$ gas was provided from below to support respiration (Fig. 4a, b). When a white or red light-emitting diode (LED) was extinguished, this served as the CS, and the electric shock served as the US. Classical fear conditioning consists of three sessions: habituation, acquisition, and probe. In the habituation session, the larvae were exposed to 5-s CS for 10–15 trials. In the acquisition session, the CS was paired with the US (1 ms, delivered 4 s after the onset of CS) for 20 trials (Fig. 4d). In the probe session, the larvae were exposed to the CS alone for 10 trials. In a control experiment (backward conditioning), the US was delivered 2 s before the CS. A 50-s interval was provided between trials or sessions. We monitored the larva's heartbeat (HB) to assess the conditioned response (Fig. 4c). A typical larva exhibited arousal bradycardia response to the first CS in the habituation session. The arousal response disappeared after 1–2 trials, and the HB was unaffected by the CS thereafter. At the beginning of the acquisition session, the bradycardia responses were elicited only by the US. After around the 10th–15th trial in the acquisition session, the conditioned bradycardia responses occurred shortly after the onset of CS and HB recovered during the presentation of CS. The conditioned response persisted in the probe session where no US was applied. We defined "learners" as zebrafish larvae whose HB frequency was unchanged during the late habituation session but was significantly reduced (bradycardia response) during the probe session. About 40% of around 20-dpf wild-type zebrafish larvae were learners (Fig. 5a). No learners were observed when the larvae were subjected to the backward conditioning.

We inhibited neurotransmitter release in GCs by expressing the botulinum toxin B by crossing the GC-specific Gal4 line gSA2AzGFF152B [18] and the *Tg(UAS:BoTxBLC-GFP)* line [86]. In the resultant (GC-silenced) larvae, about half of the GCs in the CCe expressed BoTxBLC-GFP, whereas only a minor population (about 1.5%) of the GCs in the EG and LCa expressed it. There was no apparent difference in swimming behavior between the GC-silenced larvae and their siblings. When subjected to the classical fear conditioning, there was no significance difference in learning rates among wild-type, GC-silenced, and their sibling larvae (Fig. 5a). However, the CS-evoked bradycardia response in the probe session was prolonged in the GC-silenced larvae (Fig. 5b), suggesting that GCs play a role in recovery from the conditioned bradycardia response.

To monitor neuronal activity in the zebrafish cerebellum during the classical fear conditioning, we expressed GCaMP by crossing the pan-neuronal Gal4-driven *Tg(elavl3:Gal4-VP16)* [98] and a reporter *Tg(UAS:GCaMP7a)* line [99] (Fig. 4c). To detect GCaMP fluorescence, Tg fish on the *casper (mitfa;roy)* background, which lack melanophores and iridophores, were used. To excite GCaMP, the brain needs to be illuminated with a blue light with a spectrum that overlaps the white LED that was used for the CS. To avoid this overlap, a red LED was used, and an extinguished red LED was used as the CS. The paired-associated learning with the red LED also induced a CS-evoked bradycardia response. The cerebellum of the larva embedded in agarose was observed with a water-immersion objective lens equipped with a cooled CCD camera (ORCA-R2, Hamamatsu Photonics, Japan). The change in GCaMP7a fluorescence intensity ($\Delta F/F$) was calculated to estimate neuronal activity. In the probe session, the CS evoked an increase in $\Delta F/F$ in response to the CS in some cerebellar neurons, which we termed "conditioning-associated neurons." A substantial population of the larvae that were subjected to the conditioning displayed more than four conditioning-associated neurons. These neurons were located in the CCe. We found that the CS-evoked activity in conditioning-associated neurons gradually increased during the acquisition session and decreased during the probe session. The bradycardia responses occurred when conditioning-associated neurons appeared in the acquisition and probe sessions. These data suggest that memory forms progressively during conditioning and gradually disappears with the repeated presentation of the unpaired CS. We found two types (I and II) of conditioning-associated neurons in the zebrafish cerebellum (Fig. 5c). Type I neurons were activated immediately upon the presentation of CS, whereas type II neurons were activated after a delay. Type I and II neurons were located close to each other in the CCe (Fig. 5d). Histological analysis implicated that the conditioning-associated neurons were GCs. These data indicate that GCs control the recovery from the conditioned fear responses in zebrafish. The detailed experimental procedure is described in [8].

There are many unanswered questions. Although integration of the CS and US occurs at, or upstream of, the level of inputs to the GCs in the classical fear conditioning, it remains elusive how the GCs receive and integrate the inputs. Furthermore, the contribution of other neural circuit components, including synapses between PFs and PCs and between CFs and PCs, is unknown. Furthermore, Ca^{2+} imaging with GCaMP only detects strong neuronal activities (e.g. complex spikes in PCs) but cannot detect weak activities that are also involved in learning processes (e.g. simple spikes in PCs). Voltage imaging and an electrophysiological approach are required to clarify this issue. Moreover, the roles of the cerebellum in active avoidance learning were not fully elucidated[1].

4   Discussion and Prospective

Accumulating evidence suggests that the zebrafish cerebellum is involved in not only motor control/learning but also in higher-order brain functions. Functional imaging and optogenetic manipulation of target components in zebrafish will reveal the mechanisms by which the cerebellum elicits higher-order brain functions. The knowledge obtained from zebrafish cerebellum research could also provide important clues for understanding human diseases related to the development of cerebellar neural circuits, including degenerative diseases (e.g. spinocerebellar ataxia [SCA]), neuropsychiatric diseases (e.g. autism spectrum disorders [ASDs]), brain

---

[1] The role of the zebrafish cerebellum in active avoidance has recently been reported in [100].

injury, and malignant tumors (medulloblastoma). For instance, transgenic expression of the SCA-causative mutant SCA13 gene in PCs mimicked the pathology and symptoms seen in human SCA13 patients [101]. The CRISPR/Cas9 method can readily generate zebrafish disease models by generating mutants of genes for human cerebellum-related genetic diseases. ASDs are developmental disorders characterized by social impairment and repetitive, restricted social behaviors. Anatomical and functional imaging of the brains of ASD patients suggest that abnormalities of the cerebellum structure are often linked to ASDs [6, 102–106]. Mouse mutants having a disrupted PC function recapitulated ASD-related social and repetitive behaviors [107]. Thus, abnormal development of cerebellar neural circuits may be implicated in at least some forms of ASDs. Zebrafish mutants having a mutation in a high-risk gene of ASDs showed abnormal social behavior [108]. Social behavior assays have employed zebrafish [109, 110]. Therefore, studies with the zebrafish cerebellum may contribute to understanding roles of the cerebellum in social behaviors and the pathogenesis of ASDs caused by cerebellar defects. Furthermore, once zebrafish models for human cerebellum-related diseases are established, screening of chemicals that ameliorate the abnormalities could be feasible and may lead to the development of therapeutic strategies against the diseases.

In conclusion, studies with zebrafish cerebellar neural circuits will lead to a better understanding of the development and function of the cerebellum, and may provide alternative views for understanding the pathology and etiology of human cerebellum diseases and developing therapeutic strategies against them.

Acknowledgements

The authors thank past and current members of the Hibi Laboratory for their contribution to the works cited here.

References

1. Ito M (2006) Cerebellar circuitry as a neuronal machine. Prog Neurobiol 78 (3–5):272–303

2. Ito M (2008) Control of mental activities by internal models in the cerebellum. Nat Rev Neurosci 9(4):304–313

3. Raymond JL, Lisberger SG, Mauk MD (1996) The cerebellum: a neuronal learning machine? Science 272(5265):1126–1131

4. Yoshida M, Okamura I, Uematsu K (2004) Involvement of the cerebellum in classical fear conditioning in goldfish. Behav Brain Res 153(1):143–148

5. Glickstein M (2007) What does the cerebellum really do? Curr Biol 17(19):R824–R827

6. Strata P (2015) The emotional cerebellum. Cerebellum 14(5):570–577

7. Adamaszek M, D'Agata F, Ferrucci R, Habas C, Keulen S, Kirkby KC, Leggio M, Marien P, Molinari M, Moulton E, Orsi L, Van Overwalle F, Papadelis C, Priori A, Sacchetti B, Schutter DJ, Styliadis C,

Verhoeven J (2017) Consensus paper: cerebellum and emotion. Cerebellum 16(2):552–576

8. Matsuda K, Yoshida M, Kawakami K, Hibi M, Shimizu T (2017) Granule cells control recovery from classical conditioned fear responses in the zebrafish cerebellum. Sci Rep 7(1):11865

9. Wagner MJ, Kim TH, Savall J, Schnitzer MJ, Luo L (2017) Cerebellar granule cells encode the expectation of reward. Nature 544(7648):96–100

10. Schmahmann JD (2019) The cerebellum and cognition. Neurosci Lett 688:62–75

11. Bae YK, Kani S, Shimizu T, Tanabe K, Nojima H, Kimura Y, Higashijima S, Hibi M (2009) Anatomy of zebrafish cerebellum and screen for mutations affecting its development. Dev Biol 330(2):406–426

12. Wullimann MF, Rupp B, Reichert H (1996) Neuroanatomy of the zebrafish brain: a topological atla. Birkhäuser Verlag

13. Hashimoto M, Hibi M (2012) Development and evolution of cerebellar neural circuits. Develop Growth Differ 54(3):373–389

14. Hibi M, Matsuda K, Takeuchi M, Shimizu T, Murakami Y (2017) Evolutionary mechanisms that generate morphology and neural-circuit diversity of the cerebellum. Develop Growth Differ 59(4):228–243

15. Hibi M, Shimizu T (2012) Development of the cerebellum and cerebellar neural circuits. Dev Neurobiol 72(3):282–301

16. Hibi M, Shimizu T (2014) Deciphering cerebellar neural circuitry involved in higher order functions using the zebrafish model. In: Kondo H, Kuroiwa A (eds) New principles in developmental processes. Springer, New York, pp 161–184

17. Matsui H, Namikawa K, Babaryka A, Koster RW (2014) Functional regionalization of the teleost cerebellum analyzed in vivo. Proc Natl Acad Sci U S A 111(32):11846–11851

18. Takeuchi M, Matsuda K, Yamaguchi S, Asakawa K, Miyasaka N, Lal P, Yoshihara Y, Koga A, Kawakami K, Shimizu T, Hibi M (2015) Establishment of Gal4 transgenic zebrafish lines for analysis of development of cerebellar neural circuitry. Dev Biol 397(1):1–17

19. Hidalgo-Sanchez M, Millet S, Simeone A, Alvarado-Mallart RM (1999) Comparative analysis of Otx2, Gbx2, Pax2, Fgf8 and Wnt1 gene expressions during the formation of the chick midbrain/hindbrain domain. Mech Dev 81(1–2):175–178

20. Joyner AL, Liu A, Millet S (2000) Otx2, Gbx2 and Fgf8 interact to position and maintain a mid-hindbrain organizer. Curr Opin Cell Biol 12 (6):736–741

21. Simeone A (2000) Positioning the isthmic organizer where Otx2 and Gbx2 meet. Trends Genet 16(6):237–240

22. Wurst W, Bally-Cuif L (2001) Neural plate patterning: upstream and downstream of the isthmic organizer. Nat Rev Neurosci 2(2):99–108

23. Acampora D, Avantaggiato V, Tuorto F, Simeone A (1997) Genetic control of brain morphogenesis through Otx gene dosage requirement. Development 124(18):3639–3650

24. Broccoli V, Boncinelli E, Wurst W (1999) The caudal limit of Otx2 expression positions the isthmic organizer. Nature 401(6749):164–168

25. Millet S, Campbell K, Epstein DJ, Losos K, Harris E, Joyner AL (1999) A role for Gbx2 in repression of Otx2 and positioning the mid/hindbrain organizer. Nature 401(6749):161–164

26. Simeone A, Acampora D, Gulisano M, Stornaiuolo A, Boncinelli E (1992) Nested expression domains of four homeobox genes in developing rostral brain. Nature 358(6388):687–690

27. Suda Y, Matsuo I, Aizawa S (1997) Cooperation between Otx1 and Otx2 genes in developmental patterning of rostral brain. Mech Dev 69(1–2):125–141

28. Wassarman KM, Lewandoski M, Campbell K, Joyner AL, Rubenstein JL, Martinez S, Martin GR (1997) Specification of the anterior hindbrain and establishment of a normal mid/hindbrain organizer is dependent on Gbx2 gene function. Development 124(15):2923–2934

29. Kikuta H, Kanai M, Ito Y, Yamasu K (2003) gbx2 Homeobox gene is required for the maintenance of the isthmic region in the zebrafish embryonic brain. Dev Dyn 228 (3):433–450

30. Rhinn M, Lun K, Ahrendt R, Geffarth M, Brand M (2009) Zebrafish gbx1 refines the midbrain-hindbrain boundary border and mediates the Wnt8 posteriorization signal. Neural Dev 4:12

31. Su CY, Kemp HA, Moens CB (2014) Cerebellar development in the absence of Gbx function in zebrafish. Dev Biol 386(1):181–190

32. Buckles GR, Thorpe CJ, Ramel MC, Lekven AC (2004) Combinatorial Wnt control of zebrafish midbrain-hindbrain boundary formation. Mech Dev 121(5):437–447

33. Reifers F, Bohli H, Walsh EC, Crossley PH, Stainier DY, Brand M (1998) Fgf8 is mutated in zebrafish *acerebellar* (*ace*) mutants and is required for maintenance of midbrain-hindbrain boundary development and somitogenesis. Development 125(13):2381–2395

34. Brand M, Heisenberg CP, Jiang YJ, Beuchle D, Lun K, Furutani-Seiki M, Granato M, Haffter P, Hammerschmidt M, Kane DA, Kelsh RN, Mullins MC, Odenthal J, van Eeden FJ, Nusslein-Volhard C (1996) Mutations in zebrafish genes affecting the formation of the boundary between midbrain and hindbrain. Development 123: 179–190

35. Foucher I, Mione M, Simeone A, Acampora D, Bally-Cuif L, Houart C (2006) Differentiation of cerebellar cell identities in absence of Fgf signalling in zebrafish Otx morphants. Development 133(10):1891–1900

36. Wingate RJ (2001) The rhombic lip and early cerebellar development. Curr Opin Neurobiol 11(1):82–88

37. Wingate RJ, Hatten ME (1999) The role of the rhombic lip in avian cerebellum development. Development 126(20):4395–4404

38. Zervas M, Millet S, Ahn S, Joyner AL (2004) Cell behaviors and genetic lineages of the mesencephalon and rhombomere 1. Neuron 43(3):345–357

39. Abdelfattah AS, Kawashima T, Singh A, Novak O, Liu H, Shuai Y, Huang YC, Campagnola L, Seeman SC, Yu J, Zheng J, Grimm JB, Patel R, Friedrich J, Mensh BD, Paninski L, Macklin JJ, Murphy GJ, Podgorski K, Lin BJ, Chen TW, Turner GC, Liu Z, Koyama M, Svoboda K, Ahrens MB, Lavis LD, Schreiter ER (2019) Bright and photostable chemigenetic indicators for extended in vivo voltage imaging. Science 365(6454):699–704

40. Alder J, Cho NK, Hatten ME (1996) Embryonic precursor cells from the rhombic lip are specified to a cerebellar granule neuron identity. Neuron 17(3):389–399

41. Ben-Arie N, Bellen HJ, Armstrong DL, McCall AE, Gordadze PR, Guo Q, Matzuk MM, Zoghbi HY (1997) *Math1* is essential for genesis of cerebellar granule neurons. Nature 390 (6656):169–172

42. Machold R, Fishell G (2005) *Math1* is expressed in temporally discrete pools of cerebellar rhombic-lip neural progenitors. Neuron 48 (1):17–24

43. Wang VY, Rose MF, Zoghbi HY (2005) *Math1* expression redefines the rhombic lip derivatives and reveals novel lineages within the brainstem and cerebellum. Neuron 48 (1):31–43

44. Wilson LJ, Wingate RJ (2006) Temporal identity transition in the avian cerebellar rhombic lip. Dev Biol 297(2):508–521

45. Wingate R (2005) Math-map(ic)s. Neuron 48(1):1–4

46. Chaplin N, Tendeng C, Wingate RJ (2010) Absence of an external germinal layer in zebrafish and shark reveals a distinct, anamniote ground plan of cerebellum development. J Neurosci 30(8):3048–3057

47. Kani S, Bae YK, Shimizu T, Tanabe K, Satou C, Parsons MJ, Scott E, Higashijima S, Hibi M (2010) Proneural gene-linked neurogenesis in zebrafish cerebellum. Dev Biol 343(1–2):1–17

48. Hoshino M (2006) Molecular machinery governing GABAergic neuron specification in the cerebellum. Cerebellum 5(3):193–198

49. Hoshino M, Nakamura S, Mori K, Kawauchi T, Terao M, Nishimura YV, Fukuda A, Fuse T, Matsuo N, Sone M, Watanabe M, Bito H, Terashima T, Wright CV, Kawaguchi Y, Nakao K, Nabeshima Y (2005) Ptf1a, a bHLH transcriptional gene, defines GABAergic neuronal fates in cerebellum. Neuron 47(2):201–213

50. Nimura T, Itoh T, Hagio H, Hayashi T, Di Donato V, Takeuchi M, Itoh T, Inoguchi F, Sato Y, Yamamoto N, Katsuyama Y, Del Bene F, Shimizu T, Hibi M (2019) Role of Reelin in cell positioning in the cerebellum and the cerebellum-like structure in zebrafish. Dev Biol 455(2):393–408

51. Landsberg RL, Awatramani RB, Hunter NL, Farago AF, DiPietrantonio HJ, Rodriguez CI, Dymecki SM (2005) Hindbrain rhombic lip is comprised of discrete progenitor cell populations allocated by Pax6. Neuron 48(6):933–947

52. Kaslin J, Kroehne V, Benato F, Argenton F, Brand M (2013) Development and specification of cerebellar stem and progenitor cells in zebrafish: from embryo to adult. Neural Dev 8:9

53. Kaslin J, Kroehne V, Ganz J, Hans S, Brand M (2017) Distinct roles of neuroepithelial-like and radial glia-like progenitor cells in cerebellar regeneration. Development 144(8):1462–1471

54. Takeuchi M, Yamaguchi S, Sakakibara Y, Hayashi T, Matsuda K, Hara Y, Tanegashima C, Shimizu T, Kuraku S, Hibi M (2017) Gene expression profiling of granule cells and Purkinje cells in the zebrafish

cerebellum. J Comp Neurol 525(7):1558–1585

55. Gomez A, Duran E, Salas C, Rodriguez F (2010) Cerebellum lesion impairs eyeblink-like classical conditioning in goldfish. Neuroscience 166(1):49–60

56. Rodriguez F, Duran E, Gomez A, Ocana FM, Alvarez E, Jimenez-Moya F, Broglio C, Salas C (2005) Cognitive and emotional functions of the teleost fish cerebellum. Brain Res Bull 66(4–6):365–370

57. Pastor AM, de la Cruz RR, Baker R (1994) Cerebellar role in adaptation of the goldfish vestibuloocular reflex. J Neurophysiol 72(3):1383–1394

58. Pastor AM, De la Cruz RR, Baker R (1997) Characterization of Purkinje cells in the goldfish cerebellum during eye movement and adaptive modification of the vestibulo-ocular reflex. Prog Brain Res 114:359–381

59. Mo W, Chen F, Nechiporuk A, Nicolson T (2010) Quantification of vestibular-induced eye movements in zebrafish larvae. BMC Neurosci 11:110

60. Beck JC, Gilland E, Tank DW, Baker R (2004) Quantifying the ontogeny of optokinetic and vestibuloocular behaviors in zebrafish, medaka, and goldfish. J Neurophysiol 92(6):3546–3561

61. Brockerhoff SE, Hurley JB, Janssen-Bienhold U, Neuhauss SC, Driever W, Dowling JE (1995) A behavioral screen for isolating zebrafish mutants with visual system defects. Proc Natl Acad Sci U S A 92(23):10545–10549

62. Easter SS Jr, Nicola GN (1996) The development of vision in the zebrafish (Danio rerio). Dev Biol 180(2):646–663

63. Marsh E, Baker R (1997) Normal and adapted visuooculomotor reflexes in goldfish. J Neurophysiol 77(3):1099–1118

64. Portugues R, Engert F (2009) The neural basis of visual behaviors in the larval zebrafish. Curr Opin Neurobiol 19(6):644–647

65. Ahrens MB, Li JM, Orger MB, Robson DN, Schier AF, Engert F, Portugues R (2012) Brain-wide neuronal dynamics during motor adaptation in zebrafish. Nature 485(7399):471–477

66. Scalise K, Shimizu T, Hibi M, Sawtell NB (2016) Responses of cerebellar Purkinje cells during fictive optomotor behavior in larval zebrafish. J Neurophysiol 116(5):2067–2080

67. Yoshida M, Hirano R (2010) Effects of local anesthesia of the cerebellum on classical fear conditioning in goldfish. Behav Brain Funct 6:20

68. Aizenberg M, Schuman EM (2011) Cerebellar-dependent learning in larval zebrafish. J Neurosci 31(24):8708–8712

69. Valente A, Huang KH, Portugues R, Engert F (2012) Ontogeny of classical and operant learning behaviors in zebrafish. Learn Mem 19(4):170–177

70. Lin Q, Manley J, Helmreich M, Schlumm F, Li JM, Robson DN, Engert F, Schier A, Nobauer T, Vaziri A (2020) Cerebellar Neurodynamics predict decision timing and outcome on the single-trial level. Cell 180(3):536–551. e517

71. Apps R, Hawkes R (2009) Cerebellar cortical organization: a one-map hypothesis. Nat Rev Neurosci 10(9):670–681

72. Tanabe K, Kani S, Shimizu T, Bae YK, Abe T, Hibi M (2010) Atypical protein kinase C regulates primary dendrite specification of cerebellar Purkinje cells by localizing Golgi apparatus. J Neurosci 30(50):16983–16992

73. McFarland KA, Topczewska JM, Weidinger G, Dorsky RI, Appel B (2008) Hh and Wnt signaling regulate formation of olig2+ neurons in the zebrafish cerebellum. Dev Biol 318(1):162–171

74. Seto Y, Nakatani T, Masuyama N, Taya S, Kumai M, Minaki Y, Hamaguchi A, Inoue YU, Inoue T, Miyashita S, Fujiyama T, Yamada M, Chapman H, Campbell K, Magnuson MA, Wright CV, Kawaguchi Y, Ikenaka K, Takebayashi H, Ishiwata S, Ono Y, Hoshino M (2014) Temporal identity transition from Purkinje cell progenitors to GABAergic interneuron progenitors in the cerebellum. Nat Commun 5:3337

75. Dohaku R, Yamaguchi M, Yamamoto N, Shimizu T, Osakada F, Hibi M (2019) Tracing of afferent connections in the zebrafish cerebellum using recombinant rabies virus. Front Neural Circuits 13:30

76. Takeuchi M, Yamaguchi S, Yonemura S, Kakiguchi K, Sato Y, Higashiyama T, Shimizu T, Hibi M (2015) Type IV collagen controls the Axogenesis of cerebellar granule cells by regulating basement membrane integrity in zebrafish. PLoS Genet 11(10):e1005587

77. Hsieh JY, Ulrich B, Issa FA, Wan J, Papazian DM (2014) Rapid development of Purkinje cell excitability, functional cerebellar circuit, and afferent sensory input to cerebellum in zebrafish. Front Neural Circuits 8:147

78. Chang W, Pedroni A, Hohendorf V, Giacomello S, Hibi M, Koster RW, Ampatzis K (2020) Functionally distinct Purkinje cell types show temporal precision in encoding

locomotion. Proc Natl Acad Sci U S A 117(29):17330–17337

79. Knogler LD, Kist AM, Portugues R (2019) Motor context dominates output from purkinje cell functional regions during reflexive visuomotor behaviours. Elife 8:e42138

80. Harmon TC, Magaram U, McLean DL, Raman IM (2017) Distinct responses of Purkinje neurons and roles of simple spikes during associative motor learning in larval zebrafish. Elife 6:e22537

81. Harmon TC, McLean DL, Raman IM (2020) Integration of swimming-related synaptic excitation and inhibition by olig2(+) Eurydendroid neurons in larval zebrafish cerebellum. J Neurosci 40(15):3063–3074

82. Sengupta M, Thirumalai V (2015) AMPA receptor mediated synaptic excitation drives state-dependent bursting in Purkinje neurons of zebrafish larvae. Elife 4:e09158

83. Knogler LD, Markov DA, Dragomir EI, Stih V, Portugues R (2017) Sensorimotor representations in cerebellar granule cells in larval zebrafish are dense, spatially organized, and non-temporally patterned. Curr Biol 27(9):1288–1302

84. Miyazawa H, Okumura K, Hiyoshi K, Maruyama K, Kakinuma H, Amo R, Okamoto H, Yamasu K, Tsuda S (2018) Optical interrogation of neuronal circuitry in zebrafish using genetically encoded voltage indicators. Sci Rep 8(1):6048

85. Ehrlich DE, Schoppik D (2019) A primal role for the vestibular sense in the development of coordinated locomotion. Elife 8:e45839

86. Sternberg JR, Severi KE, Fidelin K, Gomez J, Ihara H, Alcheikh Y, Hubbard JM, Kawakami K, Suster M, Wyart C (2016) Optimization of a neurotoxin to investigate the contribution of excitatory interneurons to speed modulation in vivo. Curr Biol 26(17):2319–2328

87. Song KH, Woo SR, Chung JY, Lee HJ, Oh SJ, Hong SO, Shim J, Kim YN, Rho SB, Hong SM, Cho H, Hibi M, Bae DJ, Kim SY, Kim MG, Kim TW, Bae YK (2017) REP1 inhibits FOXO3-mediated apoptosis to promote cancer cell survival. Cell Death Dis 8(1):e2536

88. Auer TO, Duroure K, De Cian A, Concordet JP, Del Bene F (2014) Highly efficient CRISPR/Cas9-mediated knock-in in zebrafish by homology-independent DNA repair. Genome Res 24(1):142–153

89. Hwang WY, Fu Y, Reyon D, Maeder ML, Tsai SQ, Sander JD, Peterson RT, Yeh JR, Joung JK (2013) Efficient genome editing in zebrafish using a CRISPR-Cas system. Nat Biotechnol 31(3):227–229

90. Jao LE, Wente SR, Chen W (2013) Efficient multiplex biallelic zebrafish genome editing using a CRISPR nuclease system. Proc Natl Acad Sci U S A 110(34):13904–13909

91. Ota S, Hisano Y, Ikawa Y, Kawahara A (2014) Multiple genome modifications by the CRISPR/Cas9 system in zebrafish. Genes Cells 19(7):555–564

92. Hoshijima K, Jurynec MJ, Klatt Shaw D, Jacobi AM, Behlke MA, Grunwald DJ (2019) Highly efficient CRISPR-Cas9-based methods for generating deletion mutations and F0 embryos that lack gene function in zebrafish. Dev Cell 51(5):645–657. e644

93. Di Donato V, De Santis F, Albadri S, Auer TO, Duroure K, Charpentier M, Concordet JP, Gebhardt C, Del Bene F (2018) An attractive Reelin gradient establishes synaptic lamination in the vertebrate visual system. Neuron 97(5):1049–1062. e1046

94. Takeuchi M, Inoue C, Goshima A, Nagao Y, Shimizu K, Miyamoto H, Shimizu T, Hashimoto H, Yonemura S, Kawahara A, Hirata Y, Yoshida M, Hibi M (2017) Medaka and zebrafish contactin1 mutants as a model for understanding neural circuits for motor coordination. Genes Cells 22(8):723–741

95. Kimura Y, Hisano Y, Kawahara A, Higashijima S (2014) Efficient generation of knock-in transgenic zebrafish carrying reporter/driver genes by CRISPR/Cas9-mediated genome engineering. Sci Rep 4:6545

96. Ota S, Taimatsu K, Yanagi K, Namiki T, Ohga R, Higashijima SI, Kawahara A (2016) Functional visualization and disruption of targeted genes using CRISPR/Cas9-mediated eGFP reporter integration in zebrafish. Sci Rep 6:34991

97. Cong L, Wang Z, Chai Y, Hang W, Shang C, Yang W, Bai L, Du J, Wang K, Wen Q (2017) Rapid whole brain imaging of neural activity in freely behaving larval zebrafish (Danio rerio). Elife 6:e28158

98. Kimura Y, Satou C, Fujioka S, Shoji W, Umeda K, Ishizuka T, Yawo H, Higashijima S (2013) Hindbrain V2a neurons in the excitation of spinal locomotor circuits during zebrafish swimming. Curr Biol 23(10):843–849

99. Muto A, Ohkura M, Abe G, Nakai J, Kawakami K (2013) Real-time visualization of neuronal activity during perception. Curr Biol 23(4):307–311

100. Koyama W, Hosomi R, Matsuda R, Kawakami K, Hibi M, Shimizu T (2021) Invovement of cerebellar neural circuits in active avoidanc

conditioning in zebrafish. eNeuro 8(3): ENEURO.0507-20.2021

101. Namikawa K, Dorigo A, Zagrebelsky M, Russo G, Kirmann T, Fahr W, Dubel S, Korte M, Koster RW (2019) Modeling neurodegenerative spinocerebellar ataxia type 13 in zebrafish using a Purkinje neuron specific tunable Coexpression system. J Neurosci 39(20):3948–3969

102. Courchesne E, Yeung-Courchesne R, Press GA, Hesselink JR, Jernigan TL (1988) Hypoplasia of cerebellar vermal lobules VI and VII in autism. N Engl J Med 318(21):1349–1354

103. Fatemi SH, Halt AR, Realmuto G, Earle J, Kist DA, Thuras P, Merz A (2002) Purkinje cell size is reduced in cerebellum of patients with autism. Cell Mol Neurobiol 22(2):171–175

104. Wang SS, Kloth AD, Badura A (2014) The cerebellum, sensitive periods, and autism. Neuron 83(3):518–532

105. Skefos J, Cummings C, Enzer K, Holiday J, Weed K, Levy E, Yuce T, Kemper T, Bauman M (2014) Regional alterations in purkinje cell density in patients with autism. PLoS One 9(2):e81255

106. Steinlin M (2008) Cerebellar disorders in childhood: cognitive problems. Cerebellum 7(4):607–610

107. Tsai PT, Hull C, Chu Y, Greene-Colozzi E, Sadowski AR, Leech JM, Steinberg J, Crawley JN, Regehr WG, Sahin M (2012) Autistic-like behaviour and cerebellar dysfunction in Purkinje cell Tsc1 mutant mice. Nature 488(7413):647–651

108. Ruzzo EK, Perez-Cano L, Jung JY, Wang LK, Kashef-Haghighi D, Hartl C, Singh C, Xu J, Hoekstra JN, Leventhal O, Leppa VM, Gandal MJ, Paskov K, Stockham N, Polioudakis D, Lowe JK, Prober DA, Geschwind DH, Wall DP (2019) Inherited and de novo genetic risk for autism impacts shared networks. Cell 178(4):850–866.e26

109. Stednitz SJ, McDermott EM, Ncube D, Tallafuss A, Eisen JS, Washbourne P (2018) Forebrain control of behaviorally driven social orienting in zebrafish. Curr Biol 28(15):2445–2451.e3

110. Dreosti E, Lopes G, Kampff AR, Wilson SW (2015) Development of social behavior in young zebrafish. Front Neural Circuits 9:39

111. Haug MF, Gesemann M, Mueller T, Neuhauss SC (2013) Phylogeny and expression divergence of metabotropic glutamate receptor genes in the brain of zebrafish (Danio rerio). J Comp Neurol 521(7):1533–1560

112. Itoh T, Takeuchi M, Sakagami M, Asakawa K, Sumiyama K, Kawakami K, Shimizu T, Hibi M (2020) Gsx2 is required for specification of neurons in the inferior olivary nuclei from Ptf1a-expressing neural progenitors in zebrafish. Development 147(19):dev190603

113. Mikami Y, Yoshida T, Matsuda N, Mishina M (2004) Expression of zebrafish glutamate receptor delta2 in neurons with cerebellum-like wiring. Biochem Biophys Res Commun 322(1):168–176

114. Obholzer N, Wolfson S, Trapani JG, Mo W, Nechiporuk A, Busch-Nentwich E, Seiler C, Sidi S, Sollner C, Duncan RN, Boehland A, Nicolson T (2008) Vesicular glutamate transporter 3 is required for synaptic transmission in zebrafish hair cells. J Neurosci 28(9):2110–2118

115. Umeda K, Shoji W, Sakai S, Muto A, Kawakami K, Ishizuka T, Yawo H (2013) Targeted expression of a chimeric channelrhodopsin in zebrafish under regulation of Gal4-UAS system. Neurosci Res 75(1):69–75

116. Tabor KM, Bergeron SA, Horstick EJ, Jordan DC, Aho V, Porkka-Heiskanen T, Haspel G, Burgess HA (2014) Direct activation of the Mauthner cell by electric field pulses drives ultrarapid escape responses. J Neurophysiol 112(4):834–844

117. Harada H, Sato T, Nakamura H (2016) Fgf8 signaling for development of the midbrain and hindbrain. Develop Growth Differ 58(5):437–445

118. Martinez S, Andreu A, Mecklenburg N, Echevarria D (2013) Cellular and molecular basis of cerebellar development. Front Neuroanat 7:18

Chapter 3

Silencing the Output of Cerebellar Neurons Using Cell Type-Specific Genetic Deletion of Vesicular GABA and Glutamate Transporters

Meike E. van der Heijden, Amanda M. Brown, and Roy V. Sillitoe

Abstract

Neurons are distinguished from other cell types by their inherent properties such as electrical excitability and their ability to rapidly communicate signals at chemical synapses. Oftentimes, researchers manipulate these properties to study the contribution of neurons to circuit computation and behavior. However, many of the preferred techniques lack cell type-specificity, rely on invasive surgeries that may damage the circuit, or are compensated for by other molecular mechanisms. It has been especially difficult to precisely eliminate synaptic transmission in expansive circuits, such as those in the cerebellum, because inhibitory drugs and light sources for optogenetics do not diffuse over large enough areas. Here, we describe how neurons can be silenced by impairing neurotransmitter release through conditional deletion of synaptic vesicular transporter genes. This approach is non-invasive, cell type-specific, and regionally unbiased, making it desirable for understanding the unique contributions of specific cerebellar neurons to circuit function in vivo.

Key words Cerebellum, Mouse, Conditional genetics, Conditional silencing, Cre-LoxP, Vesicular neurotransmitter transporters, *In vivo* electrophysiology

---

1 Introduction

The cerebellar circuit consists of repetitive computational units that are further organized in micro-units, which are arranged into medial-lateral zones; zones typically extend across multiple folia in the rostral-caudal axis [1, 2]. These cerebellar units are largely comprised of the same anatomical components across the entire cerebellum, although some differences are present in specific local circuits. At the center of the cerebellar computational unit is the large inhibitory Purkinje cell (PC) that forms the sole output of the cerebellar cortex. PCs synapse primarily on cerebellar nuclei cells. Each PC receives a number of different excitatory and inhibitory inputs that help modulate the frequency and pattern of its intrinsically generated action potentials. There are two types of excitatory

Roy V. Sillitoe (ed.), *Measuring Cerebellar Function*, Neuromethods, vol. 177, https://doi.org/10.1007/978-1-0716-2026-7_3,

inputs that project onto the PC: climbing fibers from the inferior olive and parallel fibers from cerebellar granule cells (GCs). GCs themselves receive excitatory input from mossy fibers originating from a large number of nuclei in the midbrain, brainstem, and spinal cord. PCs also receive an abundance of inhibitory inputs from different types of inhibitory interneurons, predominantly from stellate cells and basket cells, but also from Lugaro cells and candelabrum cells [3]. There are two types of cerebellar interneurons that do not directly synapse on PCs but instead terminate on GCs, thereby serving to modulate PC activity indirectly. These are the excitatory unipolar brush cells [4] and the inhibitory Golgi cells [3]. All neurons in the cerebellum and their respective synapses are highly organized within a layered circuit, suggesting that each of the components in the circuit contribute to circuit computations in a unique way (Fig. 1). Testing this hypothesis warrants precise, cell type-specific manipulations of cellular function.

To understand how each of these cerebellar cell types contributes to cerebellar-guided behaviors, one must understand not only the anatomy, but also the function of each neuron. Neuroscientists often investigate neural function by eliminating the neurons' means of communicating with its downstream targets, which occurs through the conversion of electrical activity into chemical signals that depend on neurotransmitter release at the synapse. Therefore, the function of specific cerebellar cell types can be investigated by modulating neurotransmitter release across the entire circuit in a cell type-specific, preferably non-invasive, manner. Many methods for perturbing neural function have focused on impairing one or more of the processes that lead to neurotransmitter release, but each one has its limitations [5]. For example, classical methods of hyperpolarizing neurons using GABA receptor agonists [6, 7], glutamate receptor antagonists [8], or preventing synaptic vehicle fusion using tetanus toxin (TTX) [9] lack cell type specificity, can only be applied to affect relatively small regions, and the spread of the drug is hard to control and quantify in vivo. Newer techniques such as optogenetics can be cell type-specific [10, 11], but they are invasive for in vivo applications because of the need to surgically implant an optic fiber, they have spatial limitations due to the spreading of light in the tissue, and they can have unexpected/paradoxical opposing effects on neurotransmitter release due to local differences in ion concentrations at the synapse [12] and when depolarization block occurs [13]. There are several non-invasive options for silencing neurons, including conditional expression of designer receptors that are exclusively activated by designer drugs (DREADDs, [14, 15]) or using a tetanus toxin allele [16, 17]. However, these options are often limited by the availability of highly specific genetic alleles [18–20], and the in vivo cellular effects of silencing are often hard to track, control for, and define [21, 22].

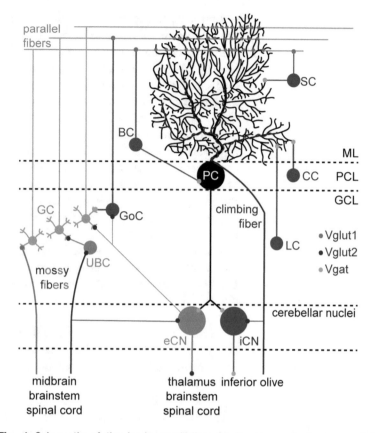

Fig. 1 Schematic of the basic cerebellar circuit using rodent as a model. Abbreviations of different cell types: *S* stellate cell, *B* basket cell, *P* Purkinje cell, *C* candelabrum cell, *Lu* Lugaro cell, *G* granule cell, *Go* Golgi cell, *U* unipolar brush cell, *CN* cerebellar nuclei cell. *ML* molecular layer, *PC* Purkinje cell layer, *GC* granule cell layer. Note: for simplicity, this schematic does not include glycinergic cerebellar nuclei cells, albeit they do exist. There are both Vglut1 and Vglut2-expressing mossy fibers that project to the cerebellar cortex where they terminate in alternating sagittal stripes [46]. Mossy fibers and climbing fibers send collaterals to both excitatory and inhibitory cerebellar nuclei cells. The CN projection to G and Go cells only exists in certain lobules and only from the interposed CN [47]. Also not shown are the direct Purkinje cell to granule cell projections in the posterior cerebellum [48]

Deletion of synaptic vesicular transporters is an alternative, non-invasive way to effectively impair neurotransmitter release, without affecting neural excitability [23–25]. Removing vesicular transporters prevents the transport (uptake) of neurotransmitter into the presynaptic vesicle. This results in the release of vesicles that are devoid of the specific neurotransmitter typically carried by that vesicle and, therefore, the elimination of fast neurotransmission and communication with the postsynaptic cell when the targeted cell fires an action potential. Conditional deletion of vesicular transporters has the benefit that it relies on the genetic Cre-LoxP system and, therefore, is completely non-invasive and only affects the intersectional domain of cells that express both the Cre and a manipulated allele of the specific vesicular transporter (Fig. 2) [7, 10, 24–26]. This adds in one extra layer of cell type-specificity over Cre-LoxP systems with "effector" molecules (*see* Chapter 7 for a summary on opto- and chemogenetics) because cerebellar neurons frequently express just one type of vesicular transporter, and not all neurons expressing the Cre allele will express the same transporter molecule (*see* Tables 1 and 2). Furthermore, vesicular transporter type switching is unlikely to occur because transporter "fate" is determined early during development [23], and because the manipulation is independent of neural excitability, molecular compensation through differential expression of ion channels does not tend to counteract the loss of transporter molecules. Additionally, only two genetic mouse lines are usually needed for deletion (a Cre allele and a flox allele) and the effectiveness of the approach can be validated using straightforward methods including immunofluorescence staining and in situ hybridization [7, 10, 26] (**Note 1**).

The Cre-LoxP approach is a binary system inspired by the phage-derived enzyme "Cre" that recognizes target sequences in the genome called "LoxP sites," neither of which naturally occur in the mammalian genome [27]. LoxP sites are 34 bp long, have an asymmetric 8 bp center, with 13 bp sequences on each site that are palindromic to each other. The Cre enzyme recognizes these sequences and performs site-directed recombination at the LoxP sites, causing excision or inversion of the DNA sequence between the two LoxP sites, depending on whether the two sites are orientated in the same or opposing directions, respectively. In this chapter, we describe the use of engineered mice that express the Cre allele under the control of a region-specific promotor inserted into the mouse genome, but note that the Cre can also be delivered with viral methods.

The second required genetic component of the Cre-LoxP system is the presence of a flanked-by-LoxP-sites ("floxed") target gene. Usually, the LoxP sites are inserted in the introns of the target gene so that the engineered sequence alone is less likely to interfere with the endogenous expression of the gene or affect proper protein translation from that gene. A conditional knockout mouse can either be heterozygous or homozygous for the Cre allele and must be homozygous for the flox allele. Homozygosity for the flox allele is key, as one copy of the wild-type gene is usually sufficient to compensate for the loss of the other copy. However, even though a conditional knockout mouse carries the Cre gene and floxed gene in all its cells, not all cells in a given region are affected. The floxed gene should theoretically be deleted in all cells that express the Cre protein, but because not all Cre-expressing cells use the same vesicular transporter; some cells are thus left

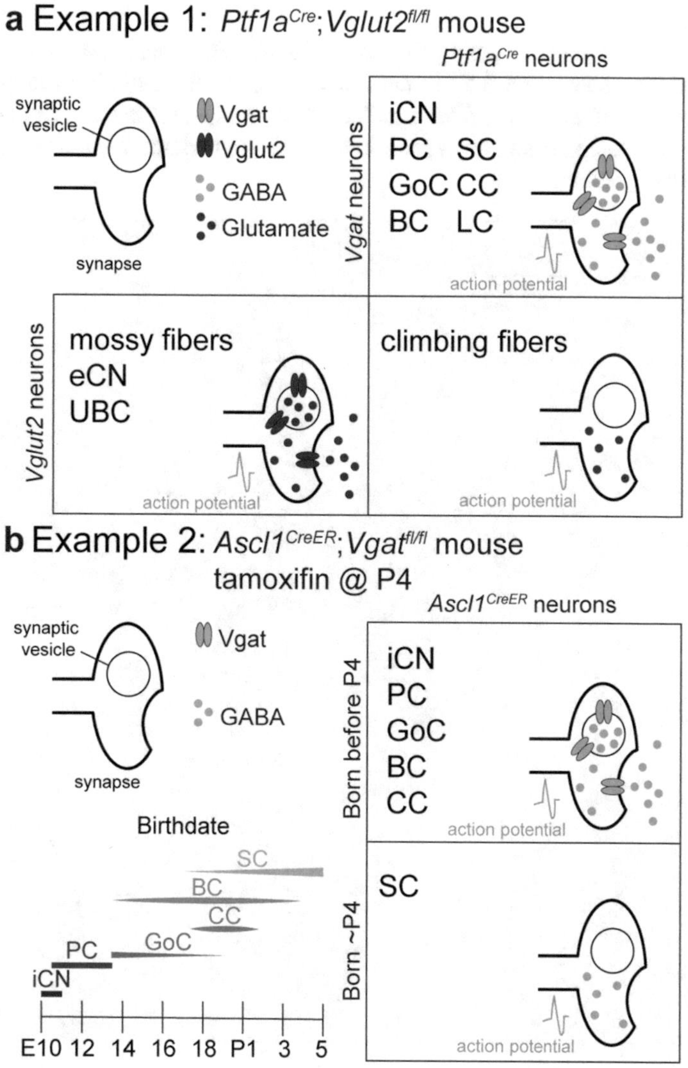

Fig. 2 Schematic for how conditional deletions of vesicular GABA and glutamate transporters work when using the Cre-LoxP system. (**a**) Example of conditional silencing of climbing fibers in *Ptf1a^Cre^;Vglut2^fl/fl^* mice. While many excitatory neurons in the cerebellum express *Vglut2* and most inhibitory neurons express *Ptf1a-Cre*, only the climbing fibers express both. Therefore, only climbing fibers are silenced in this specific conditional knockout paradigm. (**b**) Example of conditional silencing of stellate cells using *Ascl1^CreER^;Vgat^fl/fl^* mice. Most inhibitory neurons in the cerebellum express *Ascl1^Cre^*, although all of those inhibitory neurons are born before P4, except for the majority of stellate cells (SC) that are born during and after this time. Therefore, injecting tamoxifen at P4 targets and silences a population of SC's. Timeline of birthdates of key cerebellar inhibitory cell types based on genetic studies performed by Sudarov et al. [28]. *BC* basket cell, *CC* candelabrum cell, *eCN* excitatory nuclei cell, *iCN* inhibitory cerebellar nuclei cell, *GC* granule cell, *GoC* Golgi cell, *LC* Lugaro cell, *PC* Purkinje cell, *SC* stellate cell, *UBC* unipolar brush cell. Birthdating: *E* embryonic, *P* postnatal

Table 1
Cre lines used for targeting the cerebellar circuit. Engineered mice indicated as *"Gene^Cre"* have knock-in alleles with the Cre sequences inserted directly after the gene-driving promoter in their original genomic locus. Alleles indicated as *"Gene-Cre"* contain Cre sequences that are driven under a gene-specific promoter, which is all contained within a sequence that is often randomly inserted into the genome

Cre line	Cell type	Time onset
$Atoh1^{Cre}$ [49] $Atoh1^{CreER}$ [50] *also known as Math1*	All excitatory neurons: · Excitatory nuclei cells · Unipolar brush cells · Granule cells (and their precursors)	E9.5 E15.5 E12.5
$Ptf1a^{Cre}$ [51]	All inhibitory neurons: · Inhibitory nuclei cells. · Purkinje cells · Golgi cells · Molecular layer interneurons Inferior Olive neurons	
$Ascl1$-Cre [38] $Ascl1^{CreER}$ [30] *also known as Mash1*	Inhibitory neurons: · Inhibitory nuclei cells · Purkinje cells · Golgi cells · Basket cells · Candelabrum · Stellate cells	E10.5 E10.5–12.5 E13.5 E17.5 E17.5 P0
$PCP2$-Cre [52] *also known as L7*	Purkinje cells	
$GABARa6^{Cre}$ [53]	Granule cells	
$Ntsr1$-Cre [54]	Excitatory nuclei cells	
$Ucn3$-Cre [55]	Subset of excitatory nuclei cells (IPN)	

Table 2
LoxP-flanked vesicular transporter alleles

Flox allele	Cell type
$Vglut1^{flox}$ [56]	Granule cells
$Vglut2^{flox}$ [25]	Inferior olive, UBCs, excitatory CN
$Vgat^{flox}$ [24]	All inhibitory neurons in cerebellar cortex GABAergic nuclei cells

functionally intact. Therefore, only those cells that express the floxed transporter are functionally affected by this genetic manipulation. Figure 2a exemplifies how conditional genetics can be used for regional specificity based on the dual components of the Cre-LoxP system.

It is important to mention that conditional deletion using LoxP-flanked conditional alleles and genetically expressed Cre lines is typically irreversible and therefore the history of a cell's expression of the Cre protein is equally important as the ongoing expression of the Cre allele. An ideal example of this concept in the cerebellum is the expression of CreER under control of the *Ascl1* promotor [28, 29]. Even though the *Ascl1* promotor is not active in any mature cerebellar neurons, it is expressed in the progenitor cells of many different cell types in the cerebellar cortex and is important for the development of PC and other inhibitory inter-neurons (*see* Table 1). But, because of the timing of when the $Ascl1^{CreER}$ allele turns on, it can be used to selectively target specific inhibitory neurons as they differentiate from the cerebellar ventric-ular zone. Therefore, one can use the *timing* of endogenous *Ascl1* expression and the inducible $Ascl1^{CreER}$ mouse line to narrow down the genetic manipulation to a cell type with preferred identity [28, 30]. In this mouse line, the Cre gene is fused to a mutated estrogen receptor (ER) sequence. When the *CreER* fusion gene is expressed, the Cre protein is sequestered to the cell membrane due to the signaling sequences on the ER protein [27]. This prevents the enzymatic activity of the Cre protein from affecting the LoxP sites in the nucleus, as DNA recombination requires direct contact between the DNA and the active Cre molecules. Upon binding of tamoxifen to ER (because of the engineered mutation in the ER, tamoxifen has a powerful affinity for the ER compared to the normal biological scenario where estrogen strongly interacts with the wild-type ER), the CreER protein is then shuttled to the nucleus where the Cre protein can now execute the recombination event. Thus, for a CreER protein to effectively interact with a floxed sequence, the CreER has to already be present in the cells of interest, during the time window that tamoxifen is delivered to the animal. As a result, CreER-LoxP systems are advantageous because they provide precise temporal (based on the timing of when tamoxifen is provided plus when the cells start to express the CreER) and spatial (based on cell type expression of the pro-motor) control over gene function.

In this chapter, we describe how the Cre-LoxP recombination system can be used to generate mutant mice that lack vesicular transporters in specific cerebellar cell types to investigate how tar-geted populations of cerebellar neurons contribute to circuit func-tion and animal behavior.

2 Materials

2.1 Mouse Lines

There are a number of Cre lines that are commercially available or shared by individual investigators that can be used to target specific cellular components of the cerebellar circuit. Table 1 gives a list of useful Cre lines that are commonly used to target the cerebellar circuit, especially in regard to our recent efforts in assessing the functional consequences of manipulating different cell types. There are also several inducible (CreER) transgenic/knock-in mouse lines available for spatiotemporal manipulation, which utilize the expression of specific gene regulatory elements that are active in the developing cerebellum (also *see* Table 1 for expression and onset of Cre activity in specific cell types). Similarly, there are multiple floxed alleles available that can be used for conditional deletion of vesicular transporters in the cerebellum, as summarized in Table 2 (**Note 2**).

2.2 Visualization of Molecular Expression

In situ *hybridization*: A commercial in situ hybridization (ISH) kit can be used to generate the probes for ISH, with accompanying instructions for how to perform the ISH procedure (for example, Thermo Fisher, Cat# F32956). Certainly, there are many in vitro transcriptional "homemade" assays available that can be used to generate reliable probes for detecting endogenous mRNA patterns. The following cDNA sequences can be used to generate such probes for the purpose of determining the presence of vesicular transporters using in situ hybridization [31]. Please also refer to our previous work for specific examples showing the expression of vesicular transporters in the cerebellum, as detected by in situ hybridization [7, 26].

Vglut1:

5′CAGAGCCGGAGGAGATGAGCGAGGAGAAGTGTGGC
TTTGTTGGCCACGACCAGCTGGCTGGCAGTGACGAAAG
TGAAATGGAGGACGAGGCTGAGCCCCCAGGGGCGCCCC
CCGCGCCGCCTCCGTCCTACGGGGCCACACACAGCACA
GTGCAGCCTCCGAGGCCCCCGCCCCCTGTCCGGGACTA
CTGACCACGGGCCTCCCACTGTGGGGCAGTTTCCAGGA
CTTCCACTCCATACACCTCTAGCCTGAGCGGCAGTGTCG
AGGAACCCCACTCCTCCCCTGCCTCAGGCTTAAGATGCA
AGTCCTCCCTTGTTCCCAGTGCTGTCCGACCAGCCCTCT
TTCCCTCTCAACTGCCTCCTGCGGGGGGTGAAGCTGCA
CACTAGCAGTTTCAAGGATACCCAGACTCCCCTGAAAGT
CGTTCTCCGCTTGTTTCTGCCTGTGTGGGCTCAAATCTC
CCCTTTGAGGGCTTTATTTGGAGGGACAGTTCAACCTCT
TCCTCTCTTGTGGTTTTGAGGTTTCACCCCTTCCCCCAA
GACCCCAGGGATTCTCAGGCTACCCCGAGATTATTCAGG
TGGTCCCCTACTCAGAAGACTTCATGGTCGTCCTCTAT
TAGTTTCAAGGCTCGCCTAACCAATTCTACATTTTTCC

AAGCTGGTTTAACCTAACCACCAATGCCGCCGTTCCCAG
GACTGATTCTCACCAGCGTTTCTGAGGGA3'.

Vglut2: 5'CCAAATCTTACGGTGCTACCTCACAGGAGAA
TGGAGGCTGGCCTAACGGCTGGGAGAAAAAGGAAGAATT
TGTGCAAGAAGGTGCGCAAGACGCGTACACCTATAAGGA
CCGAGATGATTATTCATAACGATGCTAGTTGCTGGATTCA
TTTGTAGTGTTTGTGAATCAATTAATTGTGATTGCACAAA
AATAATTTTAAAAATGTGGTGTGAACATGTAAACATATCAA
CCAAGCAAGTCTTGCTGTTCAAAAACAAAAACAAAAAAAT
CTGAATTCAAAACAGACCATGAGATTCCCATCAAGTGCAA
TCTGTGGCAGTTGTCACGTTATGCCGTCTTCATTCAGGC
CATTTGTCCTTTCGTTTGTGATTTAAAGGTTTCCTGTAGA
AATAAGTAGGTATTCGTTGGACCCATCACCATTTTAGAGA
GCACAACTACAACAGTTGGCACATGTCATCCTACAGAAG
TTAGGAAGCCAAAGCTACTGGATCATGCAAACTGCACTT
ATTTATTACACTGGACTGCAAACTATCCCAGGGAAAGCCT
GTCTAGAGACATAGTGGAACAGGAAAGATGGCT3'.

Vgat:
5' GCCATTCAGGGCATGTTCGTGCTGGGCCTACCCTA
CGCCATCCTCCACGGCGGCTACCTGGGGTTGTTCCTCA
TCATCTTCGCCGCAGTGGTGTGCTGCTACACCGGCAAG
ATCCTCATCGCGTGCCTGTACGAGGAGAACGAAGACGG
GGAGGTGGTGCGCGTGCGGGACTCGTATGTGGCCATAG
CTAACGCATGCTGCGCTCCTCGATTCCCCACCCTGGGC
GGCCGCGTGGTCAATGTGGCGCAGATCATCGAGCTGGT
GATGACGTGTATCTTGTACGTCGTGGTGAGCGGCAACCT
CATGTACAACAGTTTCCCGGGGCTGCCCGTGTCGCAGA
AGTCCTGGTCCATCATAGCCACAGCGGTGCTGCTGCCCT
GCGCCTTCCTGAAGAATCTCAAGGCCGTGTCCAAGTTCA
GTCTGCTGTGTACGCTGGCCCACTTCGTCATCAACATCC
TGGTCATCGCTTACTGTCTCTCTCGCGCGCGTGATTGG
GCCTGGGAGAAGGTGAAGTTCTACATCGACGTCAAGAAG
TTTCCCATCTCCATTGGCATCATCGTGTTCAGCTACACG
TCGCAGATCTTCCTGCCCTCTCTCGAAGGCAACATGCAG
CAGCCCAGCGAATTCCACTGCATGATGAACTGGACACAC
ATCGCCGCCTGCGTGCTCAAGGGTCTCTTCGCGCTCGT
CGCCTACCTCACCTGGGCCGACGAGACCAAGGAAGTCA
TCACGGATAACCTGCCCGGCTCCATCCGCGCCGTGGTC
AACCTCTTCCTGGTGGCCAAGGCGCTGCTGTCCTATCC
GTTGCCCTTCTTCGCGGCCGTCGAAGTGCTGGAGAAGT
CTCTCTTCCAGGAAGGCAGTCGCGCCTTCTTCCCCGCC
TGCTATGGAGGCGACGGTCGCCTTAAGTCCTGGGGGCT
GACGCTGCGCTGCGCGCTGGTGGTCTTCACGCTGC3'.

Immunohistochemistry: There are a variety of antibodies that are commercially available for visualizing VGLUT1 (mouse: 1:500, Synaptic Systems, Cat# 135511; rabbit: 1:1,000; Synaptic Systems, Cat# 135 302), VGLUT2 (mouse: 1:500, EMD Millipore, Cat# MAB5504; rabbit: 1:500; Synaptic Systems, Cat# 135 403;

guinea pig: 1:500, Synaptic Systems, Cat# 135 404), and VGAT (guinea pig: 1:500; Synaptic Systems, Cat# 131 004). Secondary antibodies against mouse, rabbit, or guinea pig conjugated with fluorophores (we prefer Alexa) can be purchased from Invitrogen; we routinely use them at a concentration of 1:1500. Horseradish peroxidase conjugated secondary antibodies can also be used to produce robust signals for colorimetric visualization of the signal.

2.3   In Vivo Electrophysiology

Electrophysiology can be used to confirm the successful removal of the targeted cell type's ability to signal using fast neurotransmission. The recordings can be performed in combination with an additional genetic manipulation or conducted on the background of mice that express a genetically encoded opsin. The reasoning and methods for these manipulations are described in the following. In short, however, optogenetically manipulating the targeted cell population while performing electrophysiological recordings of a downstream cell can inform about whether the excision or inversion of the vesicular transporter allele did indeed remove the targeted cells' ability to communicate with downstream partners via fast neurotransmission (Fig. 3).

Craniotomy and head-plate surgery: Electrophysiology can be performed using either awake or anesthetized *in vivo* preparations. Both approaches require a surgery [7, 10, 29, 32]. Analgesics typically used include buprenorphine (0.6 mg/kg subcutaneous (SC)) and meloxicam (4 mg/kg SC). Anesthesia can be induced using 3% isoflurane gas and the levels maintained at 2% during surgery. All surgeries are performed on a stereotaxic platform (David Kopf Instruments, Tujenga, CA, USA), which helps us accurately place the craniotomy and target specific lobules.

Optopatcher: If an opsin is expressed throughout the cell membrane of the targeted population, delivery of light with an optical fiber can be effective if the fiber is placed as close as possible to the tip of the recording electrode. This would allow for optogenetic stimulation of synaptic terminals belonging to the targeted cell population that are directly adjacent to the recorded cell. A device called an Optopatcher (ALA Scientific Instruments Inc., Farmingdale, NY, USA) allows for the placement of an optical fiber (Thorlabs, Newton, NJ, USA; #FT200UMT) within a glass recording electrode (Harvard Apparatus, Cambridge, MA, USA; #30-0057). With this configuration, we are able to stimulate the expressed opsin with the appropriate wavelength using LED's (we have successfully used channelrhodopsin in a recent in vivo study [10], which is stimulated with a 465 nm light source) [33]. Glass electrodes are pulled at the time of recording (Sutter Instrument, Novato, CA, USA; #P-1000) and filled with a 0.9% w/v NaCl solution. The optical fiber is passed through the center of the electrode and secured so that the end is as close to the tip of the electrode as possible without occluding or damaging the electrode.

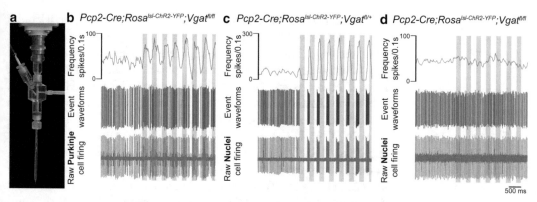

Fig. 3 Examples of electrophysiological recordings to test the effectiveness of genetic-based cell silencing. (**a**) Photo of an Optopatcher glass electrode. (**b**) Example electrophysiology trace of raw Purkinje cell firing (green) at baseline and in response to optogenetic stimulation using channelrhodopsin (light blue bars). Detected simple spikes (dark blue) and complex spikes (red) increase in firing rate (pink trace) in response to stimulation. (**c**) Example electrophysiology trace of raw cerebellar nuclei cell firing (green) in an animal expressing Cre and channelrhodopsin in Purkinje cells. Detected spikes (dark blue) decrease in firing rate (pink trace) in response to optogenetic stimulation of the nearby Purkinje cell terminals (light blue bars). (**d**) Example *in vivo* electrophysiology trace of raw cerebellar nuclei cell firing (green) in an animal expressing Cre, channelrhodopsin, and a homozygous floxed vesicular GABA transporter allele in Purkinje cells. There is no change in the firing rate (pink trace) of detected spikes (dark blue) in response to optogenetic stimulation of the nearby Purkinje cell terminals (light blue bars). This demonstrates the successful removal of fast neurotransmission between Purkinje cells and nuclei cells after genetically deleting *Vgat*

Light stimulation patterns can be programmed into and triggered by a CED Power1401 data acquisition interface (CED, Cambridge, UK) paired with the Spike2 software (CED, Cambridge, UK). Homemade optrodes are equally as effective, but care must be taken when securely attaching the fiber optic to the glass electrode of choice. In this regard, an excellent feature of the Optopatcher is that the fiber optic is weaved into the barrel of the electrode, creating an ideal configuration between the two.

Optical fiber implantation: If stimulation of the synaptic terminals of the targeted cell population is not possible or desired, optical fibers can be implanted to stimulate other regions of the cell body. In this regard, the optical fiber must be cut to the appropriate length to reach the desired cell population (s) (Thorlabs, Newton, NJ, USA; #S90R), glued into a ceramic ferrule (Thorlabs, Newton, NJ, USA; #CFLC230-10), and polished (Thorlabs, Newton, NJ, USA; #LF5P, #LF3P, #LF1P, #LF03P, #LFCF) to allow for the optimal passage of light. Ceramic mating sleeves (Thorlabs, Newton, NJ, USA; #ADAL1-5) are used to connect the implanted ferrule to an optical patch cable. The ferrule is secured to the skull using C&B Metabond Adhesive Luting Cement (Parkell, Edgewood, NY, USA), followed by dental cement for added strength (A-M Systems, Sequim, WA, USA; dental cement powder #525000 and solvent #526000).

Electrophysiological recording: Recording electrodes with an impedance of 4–13 MΩ made of either glass (described above) or tungsten (e.g. Thomas Recording, Giessen, Germany) can be used for either awake or anesthetized in vivo recordings of single neurons. Electrodes are connected to a preamplifier headstage (NPI Electronic Instruments, Tamm, Germany), which is moved by a motorized micromanipulator (MP-225; Sutter Instrument Co., Novato, CA, USA). Output from the headstage is filtered and amplified (ELC-03XS amplifier, NPI Electronic Instruments, Tamm, Germany). Amplifier output is sent to both an audio monitor (AM10, Grass Technologies, West Warwick, RI, USA) and a digitizer (CED Power 1401, CED, Cambridge, UK) before being recorded and analyzed offline using Spike2 software (CED, Cambridge, UK).

3   Methods

3.1   Generation of Conditional Vgat and Vglut Knockout Mice Using Existing Alleles

In order to generate mice with conditional deletions of either of the vesicular transport genes, one should cross a mouse positive for the preferred Cre allele (Table 1) plus heterozygous or homozygous for the preferred floxed vesicular transporter allele (Table 2) to a mouse that is homozygous for the preferred floxed allele. The genotype of the offspring can be determined using standard PCR genotyping. Only mice positive for the Cre allele *and* homozygote for the floxed allele are conditional knockout mice. All other littermates can be used as controls in the study.

To activate the Cre protein in CreER mice, one must provide the ER-ligand, tamoxifen. Tamoxifen (Sigma) is a highly hydrophobic molecule with a strong affinity for the human estrogen receptor. For tamoxifen preparation in the CreER-LoxP approach, one needs to dissolve the tamoxifen in corn or peanut oil. Dosages for tamoxifen-induced CreER activation vary depending on the biology of the target cells and the specific experimental questions, but the range is usually between 50 µg/g [28] and 200 µg/g [29]. In young postnatal mice, the tamoxifen can be injected subcutaneously (we typically use the neckfold skin), whereas for prenatal timepoints tamoxifen is delivered to the pregnant dam through oral gavage (or intraperitoneal injection) and it can be provided in combination with 50 ug/g progesterone to prevent spontaneous abortions [29] (**Note 3**).

To assess the expression of the Cre lines, one can cross these mice to a Cre reporter transgenic mouse. Two popular Cre reporter alleles are $Rosa^{lsl\text{-}tdTomato}$ (JAX: 007908) [34], which has a tdTomato sequence inserted after a LoxP-flanked STOP cassette in the universally active ROSA26 locus with a CAG promoter, and $Tau^{lsl\text{-}mGFP\text{-}nLacZ}$ (JAX: 021162) [35], which has an mGFP sequence and a nuclear-targeted β-galactosidase sequence inserted after a loxP-flanked STOP cassette in the neuronally-expressed Tau locus. In these mice, only cells with a history of Cre activity will express tdTomato or mGFP (throughout the cell) and β-galactosidase (in the nucleus), respectively. Mice carrying the Cre allele and one or two copies of the reporter allele can be used to assess Cre activity, for both its timing of expression and pattern of expression. Note that there is a plethora of other useful reporter alleles, but each Cre/reporter genetic combination must be empirically assessed for reliability of expression from generation to generation, faithfulness of the Cre to the spatiotemporal pattern of the endogenous regulatory elements, leakiness without Cre recombination, and strength of the Cre driver in each particular allele combination.

3.2 Visualization of Gene and Protein Expression in Tissue Slices

The well-structured, layered organization of the cerebellum, within which the localization of cell bodies and synaptic contacts has been described in detail (Fig. 1), facilitates the validation of Cre expression and deletion of the target allele using classical visualization techniques including in situ hybridization and immunohistochemistry. In situ hybridization allows the visualization of mRNA localization in the cytoplasm, whereas immunohistochemistry allows the visualization of the vesicular transporter proteins at the different synapses.

Detection of reporter expression: Anesthetize the mice using your approved treatment regime, flush through the heart with phosphate-buffered saline (PBS pH 7.4) to remove the blood from the tissue, and then thoroughly perfuse the animal with ice-cold 4% paraformaldehyde (PFA). Remove the brain from the skull and submerge in 4% PFA overnight at 4 °C to postfix the tissue. Cryoprotect the tissue using in a three-step sucrose gradient (10% > 20% > 30% sucrose in PBS) at 4 °C, with each step completed when the tissue sinks to bottom of the vessel. Freeze the tissue in OCT (Optimal Cutting Temperature solution; at this stage, the tissue can be stored long-term at −80 °C) and cut free-floating sections at 40 μm into PBS. Depending on the strength of the fluorescent reporter allele, sections can either be mounted immediately and the signal visualized directly or a secondary stain for the reporter (e.g. for GFP) can be performed to intensify the visualized signal as described in the following section called "*Immunohistochemistry.*"

In situ *hybridization:* Anesthetize the mice, dissect the brains from the skulls, and immediately flash-freeze the tissue by placing them in OCT-filled cryomolds that are sitting in liquid nitrogen. Cut cerebellar sections at 25 μm and immediately mount the sections onto electrostatically coated glass slides (Probe On Plus Fisher Brand; Fisher Scientific). The tissue can be probed for mRNA according to the manual of commercially available ISH kits or using the many reliable protocols. Whether using a kit or homemade reagents, every probe typically needs to be optimized for the specific tissue of interest, and also taking the age of the mice into account.

Immunohistochemistry. Perfuse the mice and prepare the tissue as described above in *"Detection of reporter expression."* For IHC using fluorescent-tagged antibodies, block the tissue in 10% normal goat or donkey serum in PBS with 0.1% TritonX (PBS-T) for 2 h at room temperature. Next, incubate the tissue in the preferred primary antibody in blocking buffer overnight at room temperature. Wash the tissue three times for 5 min in PBS-T. Next, incubate the tissue in the preferred secondary antibody diluted in the blocking solution for 2 h at room temperature. Finally, wash the tissue three times in PBS-T before mounting the sections on electrostatically coated glass slides (make sure there are no folds or curling of the tissue). All antibody and washing steps should be performed while the tissue is rocking gently.

Imaging the stained tissue sections: Immunofluorescent microscopy can be used to visualize the localization of the reporter allele, immunohistochemistry signal, or in situ hybridization signal. The raw data, usually in the form of digital microscope images, can be corrected for brightness and contrast using open-source software such as ImageJ (available at: https://imagej.nih.gov/ij/) or specialty commercial software such as Adobe Photoshop.

3.3 Electrophysiology— Genetic Crosses and Required Recording Components

It is important to test whether the action potentials of the targeted cells still have the ability to affect the firing of the downstream populations. To achieve this, we combine a Cre allele, a floxed vesicular transporter allele, and an opsin allele that is preceded by a floxed stop sequence. If the opsin is preceded by a floxed stop cassette, the Cre will allow expression of the opsin in the same cells that the Cre will also perform the excision or inversion of the targeted vesicular transporter sequence. Thus, all three alleles are expressed in the same targeted cell population. Ideally, a high-fidelity opsin such as channelrhodopsin (for example, $Rosa^{lsl-ChR2-YFP}$ mice (JAX:012569) [36]) should be used for this purpose so that strong and immediate manipulation of action potentials in the targeted cell population can be achieved in vivo. With this configuration, we use the optogenetics to stimulate the targeted cells and then use electrophysiological recordings to assess the readout of cells that are directly downstream of the targeted neurons. This approach can be accomplished by directly pairing a recording electrode with a stimulating optical fiber.

Surgery for electrophysiological recording: Our general surgical techniques have been previously described in detail [32]. Therefore, we will only briefly describe the procedure in order to highlight some key important recommendations for success. Perform all procedures using sterile surgery techniques and in accordance with your IACUC-approved protocol. First provide preemptive analgesics at least 30 min before the procedure. To implant ports used for awake recordings, induce anesthesia using 3% isoflurane and maintain anesthesia with 2% isoflurane. Otherwise, for anesthetized recordings, induce anesthesia with ketamine (80 mg/kg) admixed with xylazine (16 mg/kg) or dexmedetomidine. Firmly secure the animal's head into the stereotax. Remove the fur from the surgical site with a margin wide enough for unrestricted work, first using clippers followed by depilatory cream. Clean and sterilize the skin using at least 3 alternating applications of an alcohol and betadine scrub solution applied in a spiral motion. Cut the skin, clear tissue away from the skull, and find the stereotaxic coordinates of bregma. Calculate the position of the desired craniotomy site from bregma and perform a craniotomy in this location. Either immediately perform an anesthetized recording or secure a cannula around the craniotomy site and a head-plate on an adjacent area on the skull with Metabond. Then, secure the head apparatus with dental cement for later awake recordings. If performing the recordings later, allow the animal to recover with a continued supply of analgesics (use only according to your approved protocols) plus continue to provide other appropriate post-operative care routines, as required.

Electrophysiological recording and optogenetic stimulation: As described before, we use a device called an Optopatcher that permits the placement of an optical fiber within a glass recording electrode [33]. This allows for a very short distance between the light-stimulating fiber and the tip of a recording electrode. Therefore, recordings of a cell directly downstream of a genetically manipulated neuron can be made while optogenetically stimulating only the closest opsin-expressing synaptic terminals, many of which directly contact the recorded cell. First, pull a glass electrode, backfill it with NaCl solution so that there are no bubbles, and place it onto the preamplifier headstage. Using a micromanipulator, direct the electrode over the craniotomy site, and slowly lower it into the tissue, noting its stereotaxic coordinates relative to bregma and the surface of the brain. It is important to record and stimulate both the opsin-expressing cell population as well as the downstream cell population. This is done to ensure that the opsin-expressing cell is responding, as expected, to light stimulation and to determine optimal LED power to achieve the desired alteration in firing frequency. When a downstream cell is encountered and stable, record its electrical activity for at least 100 s before stimulating in order to determine the baseline firing properties. Light stimulation can then be sent in short (~50–500 ms) square wave pulses, triggered by the recording software, or whatever other configuration best suits your experimental objectives. Be sure to annotate in your recordings the exact moments the tissue is illuminated in order to later determine whether there is a properly timed response to stimulation. If channelrhodopsin is successfully expressed in your target cell population, a short-latency increase in firing rate should be observed in the opsin-expressing cell in response to light stimulation. If the animal is homozygous for a floxed vesicular transporter and expresses Cre, the same stimulation should have no effect on the firing rate of a downstream cell (Fig. 3). This finding should be supported by performing the same experiment in an animal that expresses channelrhodopsin, but that is either heterozygous for the floxed transporter or is homozygous for the wild-type allele. It is also valuable to conduct a set of control experiments in mice that do not have the Cre allele. In this case, the downstream cell should have a short-latency alteration in firing rate in the expected direction based on the type of neurotransmitter released by the opsin-expressing cell. For example, if an inhibitory cell expresses channelrhodopsin, then stimulating that cell should cause the firing of the downstream cell to decrease in frequency. Finally, if the opsin is not expressed in the synaptic terminals of the targeted cell (but is expressed in other compartments of the cell), first confirm the opsin-expressing cells' response to optogenetic stimulation using the Optopatcher. Then, use an implanted optical fiber to direct the light stimulation to the cell bodies of the genetically targeted cell population and repeat the single-unit recording method described above with either tungsten or glass electrodes, keeping in mind that there will be a slightly longer latency between optogenetic stimulation and the expected firing response in the downstream cell.

3.4 Phenotyping the Genetically Manipulated Mice

One goal of silencing neurons by deleting vesicular transporters is to determine the role of specific cell types during circuit function and behavior. Since the genetic silencing of cerebellar subtypes is non-invasive, this approach of neural silencing lends itself especially well to combining it with the assessment of behavior using various paradigms and circuit analyses using different recording approaches. Several chapters in this book have detailed descriptions on how to assess cerebellar function using mouse behavior and neural recordings. Uusisaari and colleagues describe how to perform slice electrophysiology, Apps and colleagues describe how to measure cellular function and behavior within specific zonal circuits, Hull and colleagues describe the development of novel behavioral tasks for cerebellar functional analysis, and Katoh and colleagues describe how to assess cerebellar learning paradigms. Furthermore, Heck and colleagues summarize single unit and local field potential recordings, Popa, Lena, and colleagues detail how to record cerebellar neurons using tetrodes and electrode arrays, all together describing how to use integrated, systems level electrophysiology approaches in the cerebellum. The genetic silencing tools that we have described can be used with all of these in vivo and in vitro functional analysis techniques (**Note 4**).

4 Notes

1. Genetic silencing using a combination of cerebellum-specific Cre (and CreER) lines and LoxP-flanked vesicular transporter alleles have several benefits over other methods of silencing cerebellar cell types: it is non-invasive, it eliminates both evoked and spontaneous neurotransmitter release, it adds an extra level to the specificity for neural targeting (due to the duality of the Cre-Lox system), it is relatively straightforward to validate, it allows for easy combination with analyses of mouse behavior and neural recordings [7, 30, 37, 38], and finally, only two mutant genotypes need to be maintained to generate conditional knockout mice. However, when using this approach for silencing neural cell types, investigators must keep several constraints in mind.

2. The approach is limited by the availability of a specific genetic intersection between a Cre line and expression of the vesicular transporter. Thus far, there are only two widely available Cre lines that target just one specific cerebellar cell type (the $Pcp2$-Cre line for Purkinje cells, and the $GABAR\alpha6^{Cre}$ line for granule cells). However, cell type specificity for other Cre lines rely on the specificity of its combination with the floxed allele or temporal activation of the Cre allele. For example, $Ptf1a^{Cre}$;$Vglut2^{fl/fl}$ only targets excitatory neurons in the inferior olive without affecting inhibitory ($Ptf1a$-positive, but $Vglut2$-negative) neurons in the cerebellar cortex [7]. And activation of the Cre protein in $Ascl1^{CreER}$ mice during postnatal time points targets the late-born stellate cells whereas activation during late prenatal time points targets basket cells and some candelabrum cells [28, 29, 39]. In addition, all in vivo genetic techniques are susceptible to a number of issues related to gene regulation, although fortunately there are some practical measures one can take to help deal with them. Because genetic drift, allele suppression, and unexpected expression of Cre have previously been reported, every researcher should validate the expression of the Cre protein and most importantly the recombination of the Cre driver using reporter alleles to exclude the possibility of leaky or germ-line expression [40]. There is also evidence that edited alleles can adapt to prevent gene silencing [41], which could warrant further testing for the expected deletion of the targeted genes using one of the methods described in this chapter.

3. Cre-mediated excision is permanent. As many of the Cre alleles are under control of developmentally expressed promotors, the expression of the Cre protein usually occurs during early development of the specific neurons. It is known that compensatory mechanisms and neural network rearrangements are more likely to occur during early development. Therefore, experimenters should consider that these mechanisms may attenuate some of the effects of silencing specific cell populations. Although we have not

observed compensation with one vesicular transporter after the loss of another [7], we recommend that in every genetic combination that is used to target a cell type, the experimenter should confirm the absence of molecular compensation in the target cell, and in the connected cells. Conversely, neuronal activity is known to play a role in the development of circuit assembly. For example, $Pcp2::Cre^{TG};Vgat^{fl/fl}$ mice that lack neurotransmission from Purkinje cells show abnormal stripe (zone) formation [26] and abnormal synapses between Purkinje cells and basket cells [39]. Furthermore, granule cell input onto Purkinje cell through parallel fibers is fundamental for the correct arrangement of climbing fibers on Purkinje cells [42–44]. As a result, loss of neurotransmitter release during developmental stages may cause secondary effects on circuit function that may not directly reflect the function of the silenced cell type during adulthood. Nevertheless, anatomical investigations of $Pcp2$-$Cre;Vgat^{fl/fl}$ [10, 26, 39], $Ascl1^{CreER};Vgat^{fl/fl}$ [29], and $Ptf1a^{Cre};Vglut2^{fl/fl}$ [7] mice have not shown large, unexpected, anatomical differences between the conditional knockout mice and control littermate mice, suggesting that these secondary effects may be minimal [45], especially in regard to their overall impact on behavior.

4. Conditional genetic deletion of vesicular transporters is an ideal, non-invasive approach for silencing large numbers of specific cerebellar cell types in mice of any age. The simplicity, reliability, and robustness of this genetic approach allows for easy combination with multiple other experimental assays for studying in vivo cerebellar function and behavior.

Acknowledgements

This work was supported by funds from Baylor College of Medicine (BCM) and Texas Children's Hospital. RVS received support from The Hamill Foundation, BCM IDDRC U54HD083092 (Intellectual and Developmental Disabilities Research Center), and the National Institutes of Neurological Disorders and Stroke (NINDS) R01NS089664 and R01NS100874. AMB received support from F31NS101891.

References

1. Sillitoe RV, Joyner AL (2007) Morphology, molecular codes, and circuitry produce the three-dimensional complexity of the cerebellum. Annu Rev Cell Dev Biol 23:549–577. https://doi.org/10.1146/annurev.cellbio.23.090506.123237

2. Beckinghausen J, Sillitoe RV (2019) Insights into cerebellar development and connectivity. Neurosci Lett 688:2–13. https://doi.org/10.1016/j.neulet.2018.05.013

3. Schilling K, Oberdick J, Rossi F, Baader SL (2008) Besides Purkinje cells and granule neurons: an appraisal of the cell biology of the interneurons of the cerebellar cortex. Histochem Cell Biol 130:601–615. https://doi.org/10.1007/s00418-008-0483-y

4. Mugnaini E, Sekerková G, Martina M (2011) The unipolar brush cell: a remarkable neuron finally receiving deserved attention. Brain Res Rev 66:220–245. https://doi.org/10.1016/j.brainresrev.2010.10.001

5. Wiegert JS, Mahn M, Prigge M et al (2017) Silencing neurons: tools, applications, and experimental constraints. Neuron 95:504–529. https://doi.org/10.1016/j.neuron.2017.06.050

6. Johnston GAR (2014) Muscimol as an ionotropic GABA receptor agonist. Neurochem Res 39:1942–1947. https://doi.org/10.1007/s11064-014-1245-y

7. White JJ, Sillitoe RV (2017) Genetic silencing of olivocerebellar synapses causes dystonia-like behaviour in mice. Nat Commun 8:14912. https://doi.org/10.1038/ncomms14912

8. Stone TW (1993) Neuropharmacology of quinolinic and kynurenic acids. Pharmacol Rev 45:309–379

9. Fremont R, Calderon DP, Maleki S, Khodakhah K (2014) Abnormal high-frequency burst firing of cerebellar neurons in rapid-onset dystonia-parkinsonism. J Neurosci 34:11723–11732. https://doi.org/10.1523/JNEUROSCI.1409-14.2014

10. Brown AM, White JJ, van der Heijden ME et al (2020) Purkinje cell misfiring generates high-amplitude action tremors that are corrected by cerebellar deep brain stimulation. Elife 9:e51928. https://doi.org/10.7554/eLife.51928

11. Prestori F, Montagna I, D'Angelo E, Mapelli L (2020) The optogenetic revolution in cerebellar investigations. Int J Mol Sci 21(7):2494. https://doi.org/10.3390/ijms21072494

12. Messier JE, Chen H, Cai Z-L, Xue M (2018) Targeting light-gated chloride channels to neuronal somatodendritic domain reduces their excitatory effect in the axon. Elife 7:e38506. https://doi.org/10.7554/eLife.38506

13. Herman AM, Huang L, Murphey DK et al (2014) Cell type-specific and time-dependent light exposure contribute to silencing in neurons expressing Channelrhodopsin-2. elife 3:e01481. https://doi.org/10.7554/eLife.01481

14. Badura A, Verpeut JL, Metzger JW et al (2018) Normal cognitive and social development require posterior cerebellar activity. Elife 7:e36401. https://doi.org/10.7554/eLife.36401

15. Stoodley CJ, D'Mello AM, Ellegood J et al (2017) Altered cerebellar connectivity in autism and cerebellar-mediated rescue of autism-related behaviors in mice. Nat Neurosci 20:1744–1751. https://doi.org/10.1038/s41593-017-0004-1

16. Zhang Y, Narayan S, Geiman E et al (2008) V3 spinal neurons establish a robust and balanced locomotor rhythm during walking. Neuron 60:84–96. https://doi.org/10.1016/j.neuron.2008.09.027

17. Park H, Kim T, Kim J et al (2019) Inputs from sequentially developed parallel fibers are required for cerebellar organization. Cell Rep 28:2939–2954.e5. https://doi.org/10.1016/j.celrep.2019.08.010

18. Vardy E, Robinson JE, Li C et al (2015) A new DREADD facilitates the multiplexed chemogenetic interrogation of behavior. Neuron 86:936–946. https://doi.org/10.1016/j.neuron.2015.03.065

19. Ray RS, Corcoran AE, Brust RD et al (2011) Impaired respiratory and body temperature control upon acute serotonergic neuron inhibition. Science 333:637–642. https://doi.org/10.1126/science.1205295

20. Kim JC, Cook MN, Carey MR et al (2009) Linking genetically defined neurons to behavior through a broadly applicable silencing allele. Neuron 63:305–315. https://doi.org/10.1016/j.neuron.2009.07.010

21. Manvich DF, Webster KA, Foster SL et al (2018) The DREADD agonist clozapine N-oxide (CNO) is reverse-metabolized to clozapine and produces clozapine-like interoceptive stimulus effects in rats and mice. Sci Rep 8:3840. https://doi.org/10.1038/s41598-018-22116-z

22. Gomez JL, Bonaventura J, Lesniak W et al (2017) Chemogenetics revealed: DREADD occupancy and activation via converted clozapine. Science 357:503–507. https://doi.org/10.1126/science.aan2475

23. Wojcik SM, Rhee JS, Herzog E et al (2004) An essential role for vesicular glutamate transporter 1 (VGLUT1) in postnatal development and control of quantal size. Proc Natl Acad Sci U S A 101:7158–7163. https://doi.org/10.1073/pnas.0401764101

24. Tong Q, Ye C-P, Jones JE et al (2008) Synaptic release of GABA by AgRP neurons is required for normal regulation of energy balance. Nat Neurosci 11:998–1000. https://doi.org/10.1038/nn.2167

25. Tong Q, Ye C, McCrimmon RJ et al (2007) Synaptic glutamate release by ventromedial hypothalamic neurons is part of the neurocircuitry that prevents hypoglycemia. Cell Metab 5:383–393. https://doi.org/10.1016/j.cmet.2007.04.001

26. White JJ, Arancillo M, Stay TL et al (2014) Cerebellar zonal patterning relies on Purkinje cell neurotransmission. J Neurosci 34: 8231–8245. https://doi.org/10.1523/JNEUROSCI.0122-14.2014

27. McLellan MA, Rosenthal NA, Pinto AR (2017) Cre-loxP-mediated recombination: general principles and experimental considerations. Curr Protoc Mouse Biol 7:1–12. https://doi.org/10.1002/cpmo.22

28. Sudarov A, Turnbull RK, Kim EJ et al (2011) Ascl1 genetics reveals insights into cerebellum local circuit assembly. J Neurosci 31: 11055–11069. https://doi.org/10.1523/JNEUROSCI.0479-11.2011

29. Brown AM, Arancillo M, Lin T et al (2019) Molecular layer interneurons shape the spike activity of cerebellar Purkinje cells. Sci Rep 9: 1742. https://doi.org/10.1038/s41598-018-38264-1

30. Kim EJ, Leung CT, Reed RR, Johnson JE (2007) In vivo analysis of Ascl1 defined progenitors reveals distinct developmental dynamics during adult neurogenesis and gliogenesis. J Neurosci 27:12764–12774. https://doi.org/10.1523/JNEUROSCI.3178-07.2007

31. Lein ES, Hawrylycz MJ, Ao N et al (2007) Genome-wide atlas of gene expression in the adult mouse brain. Nature 445:168–176. https://doi.org/10.1038/nature05453

32. White JJ, Lin T, Brown AM et al (2016) An optimized surgical approach for obtaining stable extracellular single-unit recordings from the cerebellum of head-fixed behaving mice. J Neurosci Methods 262:21–31. https://doi.org/10.1016/j.jneumeth.2016.01.010

33. Katz Y, Yizhar O, Staiger J, Lampl I (2013) Optopatcher--an electrode holder for simultaneous intracellular patch-clamp recording and optical manipulation. J Neurosci Methods 214: 113–117. https://doi.org/10.1016/j.jneumeth.2013.01.017

34. Madisen L, Zwingman TA, Sunkin SM et al (2010) A robust and high-throughput Cre reporting and characterization system for the whole mouse brain. Nat Neurosci 13: 133–140. https://doi.org/10.1038/nn.2467

35. Hippenmeyer S, Vrieseling E, Sigrist M et al (2005) A developmental switch in the response of DRG neurons to ETS transcription factor signaling. PLoS Biol 3:e159. https://doi.org/10.1371/journal.pbio.0030159

36. Madisen L, Mao T, Koch H et al (2012) A toolbox of Cre-dependent optogenetic transgenic mice for light-induced activation and silencing. Nat Neurosci 15:793–802. https://doi.org/10.1038/nn.3078

37. Kayakabe M, Kakizaki T, Kaneko R et al (2013) Motor dysfunction in cerebellar Purkinje cell-specific vesicular GABA transporter knockout mice. Front Cell Neurosci 7:286. https://doi.org/10.3389/fncel.2013.00286

38. Parras CM, Hunt C, Sugimori M et al (2007) The proneural gene Mash1 specifies an early population of telencephalic oligodendrocytes. J Neurosci 27:4233–4242. https://doi.org/10.1523/JNEUROSCI.0126-07.2007

39. Zhou J, Brown AM, Lackey EP, et al. (2020) Purkinje cell neurotransmission patterns cerebellar basket cells into zonal modules defined by distinct pinceau sizes. Elife. https://doi.org/10.7554/eLife.55569

40. Song AJ, Palmiter RD (2018) Detecting and avoiding problems when using the Cre-lox system. Trends Genet 34:333–340. https://doi.org/10.1016/j.tig.2017.12.008

41. Hosur V, Low BE, Li D, et al. (2020) Genes adapt to outsmart gene-targeting strategies in mutant mouse strains by skipping exons to reinitiate transcription and translation. Genome Biol 21:168. https://doi.org/10.1186/s13059-020-02086-0

42. Hashimoto K, Yoshida T, Sakimura K et al (2009) Influence of parallel fiber-Purkinje cell synapse formation on postnatal development of climbing fiber-Purkinje cell synapses in the cerebellum. Neuroscience 162:601–611. https://doi.org/10.1016/j.neuroscience.2008.12.037

43. Kano M, Hashimoto K (2012) Activity-dependent maturation of climbing fiber to Purkinje cell synapses during postnatal cerebellar development. Cerebellum 11:449–450. https://doi.org/10.1007/s12311-011-0337-3

44. Pan M-K, Li Y-S, Wong S-B et al (2020) Cerebellar oscillations driven by synaptic pruning deficits of cerebellar climbing fibers contribute to tremor pathophysiology. Sci Transl Med 12(526):eaay1769. https://doi.org/10.1126/scitranslmed.aay1769

45. Kobayashi S, Kim J, Yanagawa Y et al (2020) Hyper-formation of GABA and glycine co-releasing terminals in the mouse cerebellar nuclei after deprivation of GABAergic inputs from Purkinje cells. Neuroscience 426: 88–100. https://doi.org/10.1016/j.neuroscience.2019.11.030

46. Gebre SA, Reeber SL, Sillitoe RV (2012) Parasagittal compartmentation of cerebellar mossy fibers as revealed by the patterned expression of vesicular glutamate transporters VGLUT1 and VGLUT2. Brain Struct Funct 217:165–180. https://doi.org/10.1007/s00429-011-0339-4

47. Gao Z, Proietti-Onori M, Lin Z et al (2016) Excitatory cerebellar Nucleocortical circuit provides internal amplification during associative conditioning. Neuron 89:645–657. https://doi.org/10.1016/j.neuron.2016.01.008

48. Guo C, Witter L, Rudolph S et al (2016) Purkinje cells directly inhibit granule cells in specialized regions of the cerebellar cortex. Neuron 91:1330–1341. https://doi.org/10.1016/j.neuron.2016.08.011

49. Yang H, Xie X, Deng M et al (2010) Generation and characterization of Atoh1-Cre knock-in mouse line. Genesis 48:407–413. https://doi.org/10.1002/dvg.20633

50. Machold R, Fishell G (2005) Math1 is expressed in temporally discrete pools of cerebellar rhombic-lip neural progenitors. Neuron 48:17–24. https://doi.org/10.1016/j.neuron.2005.08.028

51. Hoshino M, Nakamura S, Mori K et al (2005) Ptf1a, a bHLH transcriptional gene, defines GABAergic neuronal fates in cerebellum. Neuron 47:201–213. https://doi.org/10.1016/j.neuron.2005.06.007

52. Lewis PM, Gritli-Linde A, Smeyne R et al (2004) Sonic hedgehog signaling is required for expansion of granule neuron precursors and patterning of the mouse cerebellum. Dev Biol 270:393–410. https://doi.org/10.1016/j.ydbio.2004.03.007

53. Aller MI, Jones A, Merlo D et al (2003) Cerebellar granule cell Cre recombinase expression. Genesis 36:97–103. https://doi.org/10.1002/gene.10204

54. Gong S, Doughty M, Harbaugh CR et al (2007) Targeting Cre recombinase to specific neuron populations with bacterial artificial chromosome constructs. J Neurosci 27:9817–9823. https://doi.org/10.1523/JNEUROSCI.2707-07.2007

55. Harris JA, Hirokawa KE, Sorensen SA et al (2014) Anatomical characterization of Cre driver mice for neural circuit mapping and manipulation. Front Neural Circuits 8:76. https://doi.org/10.3389/fncir.2014.00076

56. Murao N, Yokoi N, Honda K et al (2017) Essential roles of aspartate aminotransferase 1 and vesicular glutamate transporters in β-cell glutamate signaling for incretin-induced insulin secretion. PLoS One 12:e0187213. https://doi.org/10.1371/journal.pone.0187213

Chapter 4

Mapping Structure–Function Relationships within Cerebellar Circuits

Richard Apps, Charlotte L. Lawrenson, Elena Paci, and Nadia L. Cerminara

Abstract

A key organizational feature of the cerebellum is its division into a series of cerebellar modules. The cortical component of each module is termed a zone. Each zone is defined by its climbing input originating from a well-defined region of the inferior olive, which targets one or more longitudinal zones of Purkinje cells within the cerebellar cortex. In turn, Purkinje cells within each zone project to specific regions of the cerebellar and vestibular nuclei. The behavioral significance of zones remains poorly understood.

This chapter will describe approaches to physiologically and anatomically identify cerebellar cortical zones in both acute and chronic preparations (rat and cat) as well as methods to record from zones during behavior in order to determine the role that they have in cerebellar function.

Key words Cerebellar zones, Cerebellar modules, Inferior olive, Cerebellar nuclei, Rat, Cat, Tract tracing, Retrograde, Anterograde, Electrophysiology

1 Introduction

The mapping of structure–function relations is central to understanding the relationship between brain and behavior. The cerebellum is an ideal structure to examine such relationships. It has a well-defined cytoarchitecture (for reviews *see* [1–4]) that has been studied extensively across the macro-, meso-, and microscale levels.

On a meso-scale, a key feature of the cerebellum is its division into a series of modules [1, 5–8]. In brief, each module has been defined anatomically by its climbing fiber input originating from a specific subdivision of the inferior olivary nucleus [7, 9–13]. Climbing fibers with a common olivary origin target one or more rostro-caudally oriented zones of Purkinje cells within the cerebellar cortex. Zones are defined as rostro-caudally extended arrays of Purkinje cells within the cerebellar cortex with specific olivocerebellar and cortico-nuclear connections and climbing fiber input relayed through a common set of olivocerebellar pathways

Roy V. Sillitoe (ed.), *Measuring Cerebellar Function*, Neuromethods, vol. 177, https://doi.org/10.1007/978-1-0716-2026-7_4,

[1]. Therefore, individual zones not only differ in their anatomical connectivity but can also be characterized by systematic differences in their physiological characteristics, including somatic receptive field and response latency to peripheral stimulation (for reviews *see* [1, 4, 5, 14–16]). Cortical zones are about 1–2 mm in mediolateral width, but traverse cerebellar lobules in the rostro-caudal direction for more than 10 mm [15]. In turn, Purkinje cells located within each cortical zone have a highly convergent projection to specific territories within the cerebellar and vestibular nuclei [11, 12, 17–20]. These nuclei provide cerebellar output to spinal, brainstem, and thalamic nuclei to influence behavior. Individual modules are therefore defined as a longitudinal zone of Purkinje cells within the cerebellar cortex together with its olivo-cortico-nuclear connections and associated recurrent pathways ([1]; *see* Fig. 1). Modules are highly conserved across species.

It is generally agreed that these modules are an important organizational feature of the cerebellum, with each module either acting individually or in combination with others to control behaviors such as movement, and the regulation of autonomic, emotional, and cognitive behavior [21].

Our laboratory has studied the structure–function relationship of cerebellar modules by accessing their cortical zonal components (e.g. [9, 11, 12, 17, 22–26]; *see* also Table 1), focussing predominately on modules located within the paravermal cerebellum (C1, C2 and C3), but also the vermal A module [27] and the D modules [28, 29], the latter located laterally within the hemispheres. We have taken advantage of the ability to electrophysiologically identify climbing fiber input to individual cortical zones on the cerebellar surface to guide injections of tracers to map the underlying anatomical connectivity of individual modules. In studies in awake animals we have also examined how transmission in climbing fiber pathways targeting the same cortical zones is modulated during different behaviors. This guide will help provide details on how to identify cerebellar cortical zones anatomically and electrophysiologically. These methods can also be used in combination with techniques such as optogenetics or pharmacology to test the relationship between cerebellar connectivity and behavior (e.g. [27, 30]).

2 Materials

In this section we describe the materials and procedures that are specific to studying the anatomical and electrophysiological organization within the cerebellum.

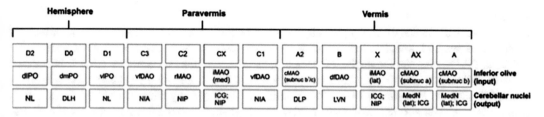

Fig. 1 Simplified block diagram of cerebellar zones and modules. Each longitudinal zone is defined by its inferior olive climbing fiber input and Purkinje cortico-nuclear output. From the medial to the lateral plane (right to left in the figure) are shown: the A, AX, X, B, and A2 zones (in the vermis), the C1, CX, C2, and C3 zones (in the paravermis), and the D1, D0, and D2 zones (in the hemisphere). Longitudinal zones in the paraflocculus and flocculus are not shown. Note that some longitudinal zones are not necessarily present in all cerebellar lobules in the adult animal (for example, the X and B zones). cMAO (subnuc a), subnucleus a of caudal medial accessory olive; cMAO (subnuc b), subnucleus b of caudal medial accessory olive; cMAO (subnuc b1/c), subnucleus b1 and c of caudal medial accessory olive; dfDAO, dorsal fold of dorsal accessory olive; DLH, dorsolateral hump; DLP, dorsolateral protuberance of medial nucleus; dlPO, dorsal lamella of the principal olive; dmPO, dorsomedial subnucleus of the principal olive; ICG, interstitial cell group; iMAO (lat), lateral part of intermediate medial accessory olive; iMAO (med), medial part of intermediate medial accessory olive; LVN, lateral vestibular nucleus; M1, primary motor cortex; MedN (lat), lateral part of medial nucleus; MedN (med), medial part of medial nucleus; NIA, nucleus interpositus anterior; NIP, nucleus interpositus posterior; NL, lateral nucleus; PML, paramedian lobule; rMAO, rostral medial accessory olive; vfDAO, ventral fold of dorsal accessory olive; vlPO, ventral lamella of the principal olive. (Reproduced from [55] with permission from Springer Nature)

2.1 Anesthetics

1. Sodium pentobarbitone salt (Sigma Aldrich).
2. 96% Ethanol.
3. Propylene glycol (Sigma Aldrich).
4. Propofol (Propoflo Plus, Abbott Animal Health, USA).
5. Sterile distilled water.

2.2 Glass Pipettes

1. Narashige electrode vertical puller.
2. Filamented capillary glass 1.5 mm OD, 0.84 ID (e.g. WPI).
3. Glass jar with lid with a piece of foam block with slits cut in and glued around top.

2.3 Tracers

1. Retrograde tracers—Red and green fluorescent beads (Lumafluor, New City, NY, USA).
2. Anterograde tracers—Fluoro-ruby (ThermoFisher, Dextran Tetramethylrhodamine 10,000 MW) and Fluoro-emerald (Dextran Fluorescien 10,000 MW).
3. Sonicator.
4. Hamilton Syringe 10 μL.

Table 1
Key references of studies examining the physiological responses and/or anatomical inputs/outputs of cerebellar zones and modules. References for the paraflocculus and flocculus have not been included. *BPN* basilar pontine nuclei, *CP* copula pyramidis, *DAO* dorsal accessory olive, *ECN* external cuneate nucleus, *IO* inferior olive, *LRN* lateral reticular nucleus, *LS* lobulus simplex, *LVN* lateral vestibular nucleus, *M1* primary motor cortex, *MAO* medial accessory olive, *n/a* not applicable, *NIA* nucleus interpositus anterior, *NIP* nucleus interpositus posterior, *NL* lateral nucleus, *NRTP* nucleus reticularis tegmenti pontis, *PML* paramedian lobule, *PO* principal olive, *RF* reticular formation, *STN* spinal trigeminal nucleus, *IV* cerebellar lobule IV, *V* cerebellar lobule V

Reference	Species	Zone/Module	Cerebellar location/ lobule	Physiological pathways	Anatomical projections
[56]	Rat	C1	LS, PML, CP	Ipsilateral forelimb and hindlimb, contralateral M1	n/a
[48]	Cat	B	Anterior vermis	Bilateral forelimbs and hindlimbs	n/a
[17]	Cat	C1, C3	V	Ipsilateral forelimb	Retrograde—IO (DAO) Anterograde—NIA
[57]	Cat	C1, C3	V	Ipsilateral forelimb	n/a
[22]	Cat	C1, C3	V, PML	Ipsilateral forelimb	n/a
[25]	Cat	C2	V, PML	Bilateral forelimbs	n/a
[58]	Cat	C1, C2	PML	Ipsilateral forelimb	Retrograde—IO (MAO, DAO); LRN
[59]	Cat	X, C1	Anterior lobe	Ipsilateral hindlimb and trunk, contralateral forelimbs	Retrograde—IO (MAO)
[9]	Rat	A2, C1, C2, C3	PML, CP	Bilateral forelimb, ipsilateral hindlimb, contralateral face, tail	Retrograde –IO (MAO and DAO)
[11]	Rat	C1	CP	Ipsilateral hindlimb, tail	Retrograde—IO (DAO), BPN, Anterograde—NIA
[60]	Cat	C2, C3, D1	Crus 1	Optic chiasm, ipsilateral periorbital region, bilateral forelimbs	Retrograde—IO (MAO, DAO, PO) Anterograde—NIA, NIP, NL
[61]	Ferret	A, X, B, C1, CX, C2, C3, D1, D2	Anterior Lobe	Ispilateral forelimb, contralateral hindlimb	n/a
[12]	Cat	C3	IV/V	Ipsilateral forelimb	Retrograde—IO (DAO) Anterograde—NIA

(continued)

Table 1
(continued)

Reference	Species	Zone/Module	Cerebellar location/ lobule	Physiological pathways	Anatomical projections
[43]	Rat	C1	PML, LS	Ipsilateral forelimb	Retrograde—IO (MAO, DAO), ECN, RF, STN
[46]	Cat	C2, D1, D2	Crus I	n/a	Retrograde—IO (MAO, PO) Anterograde—NI, NL
[50]	Cat	C1, C3	VI and VII, PML	Bilateral forelimbs, periorbital area	n/a
[62]	Rat	A, B, C1, C2, D1	Anterior lobe	Ipsilateral face, bilateral forelimb and hindlimb	n/a
[44]	Cat	C1, C3	V	Ipsilateral forelimb and hindlimb	Retrograde—IO (MAO, DAO), NRTP, BPN, LRN
[63]	Rabbit	C2	LS	Bilateral forelimbs, and periocular area	n/a
[45]	Rat	A2, C1	PML, CP	Ipsilateral forelimb and hindlimb, contralateral face	Retrograde—IO (MAO, DAO), BPN
[64]	Cat	A, B	Anterior lobe	Bilateral forelimb and hindlimb	n/a
[65, 66]	Cat	A, B, C1, C3	Anterior and posterior lobe	Bilateral hindlimb and contralateral VF and DF	n/a
[24]	Rat	A2, C1, C2, C3	LS and PML	Bilateral forelimbs, contralateral face	Retrograde—IO (MAO, DAO) Anterograde—IN
[67]	Rat	A2, C1, C2, D0, D1	PML, CP, crus 2	Bilateral forelimbs and hindlimbs	Retrograde—IO (MAO, DAO, PO); LRN, BPN
[68]	Cat	C1, C2, C3	V	Bilateral forelimb	Retrograde—IO (DAO MAO)
[42]	Cat	C1, C2, C3	PML	Bilateral forelimbs and hindlimbs	Retrograde—IO (MAO, DAO)
[35]	Cat	C1, C2, C3	V	Bilateral forelimb	Retrograde—NIP, NIA Anterograde— NIP, NIA
[69, 70]	Cat	C1, C2	PML		

(continued)

Table 1
(continued)

Reference	Species	Zone/Module	Cerebellar location/ lobule	Physiological pathways	Anatomical projections
				Bilateral forelimb and hindlimb	Retrograde—IO (MAO, DAO), NI Anterograde—NIA, NIP
[71, 72]	Cat	B, X, C1, C2, C3, D1	V	Bilateral forelimb, bilateral hindlimb	Anterograde—LVN, NIA, NIP LN

2.4 Tracer Delivery

1. Polyethylene tubing PE30 and PE50.
2. Syringe pump (e.g. Aladdin Syringe Pump, WPI) or Pneumatic pump system (e.g. Picopump with nitrogen PV800 series WPI).
3. Hamilton Syringes 10 μL.
4. Mineral Oil.

2.5 Surgical Procedures

1. Stereotaxic frame with atraumatic ear bars.
2. Heated blanket.
3. Sodium pentobarbitone or Propofol.
4. Local anesthetic—Lidocaine.
5. Standard surgical tools, including watchmaker forceps and scissors, Rongeurs, drill, retractor, scalpel, clamps, absorbable and non-absorbable suture (sizes 3-0), gel foam, cotton tips, dental acrylic.
6. 0.9% sterile saline.
7. Analgesic.
8. 3-axis Micromanipulator.
9. Stereomicroscope with an eyepiece graticule.
10. Mobile phone camera adaptor for stereomicroscope (e.g. GT Vision).

2.6 Electro-physiological Mapping

1. Hypodermic needles—23G or 25G.
2. Single core wire.
3. Isolated voltage or current stimulator (e.g. Digitimer DS2A, DS3).
4. Platinum-Tungsten or Tungsten in glass microelectrode, impendence of ~0.5 MΩ (e.g. Thomas Recording GmbH, Germany).

5. Amplifier and filter recording and acquisition system.

6. Humbug (Quest Scientific).

2.7 Chronic Implants

1. Single electrodes—insulated stainless-steel (diameter: 50.8 μm. California Fine Wire Company, CA) or 2–4 multitrodes made from insulated tungsten wires (diameter: 12.5 μm California Fine Wire Company). Held in place by cannula 18 mm in length, 34 or 36G, and attached to moveable miniature microdrives.

2. Skull screws.

3. 23G hypodermic needles.

4. Teflon insulated braided stainless-steel wire diameter: 330 μm (Cooner Wire Co, Chatsworth, USA).

5. Trocar.

2.8 Tissue Processing

1. Perfusion set up.

2. Heparinized saline.

3. 4% Paraformaldehyde.

4. 10% Sucrose in phosphate buffer.

5. 30% Sucrose in phosphate buffer.

6. Gelatin.

7. Freezing microtome.

8. Microscope slides and coverslips.

2.9 Microscopy and Imaging

1. Inverted microscope equipped with a mercury UV light source or LED illumination system (e.g. CooLED pE-100 excitation system) and appropriate filter blocks, i.e. rhodamine and wide-band fluorescein.

2. Digital monochrome camera and software.

3. Image analysis software (e.g. ImageJ).

3 Methods

In this section we describe the procedures for studying the zonal and modular physiological organization within the cerebellum (*see* **Note 1**). Studies can examine the organization using solely electrophysiological means or both anatomical and electrophysiological methods, and both can be done chronically.

3.1 Preparation for Anatomical Tracing

1. Glass injection pipettes are constructed using a Narashige vertical pipette puller. Once pulled, clip off the tips either by hand or by using a molten glass bead under a light microscope for

greater accuracy in conjunction with a calibrated eyepiece to measure the tip size. Recommended tip sizes are ~20 μm.

2. Next, prepare the tracers. We use Retrobeads as our retrograde tracer of choice as they transport exclusively in the retrograde direction and produce a highly restricted injection site. They also require no processing for visualization and are non-toxic to neurons. Although Retrobeads are highly resistant to fading, cell labelling cannot be considered permanent (although in our experience labelling in well stored sections can be viewed for ~10 years). The only preparation required for the Retrobeads is to sonicate the vial for 2–3 min before use; this is recommended to improve tracer uptake. Retrobeads can also be mixed with an anterograde tracer of the same color to simultaneously trace both the inputs and outputs from a single injection site. We mix red and green Retrobeads with Fluoro-ruby and Fluoro-emerald, respectively. If using a combined retrograde and anterograde strategy, combine red beads with Fluoro-ruby and green beads with Fluoro-emerald to make up a 20% solution.

3. We fill the glass pipette with tracer by injecting it into the pipette with a 10 μL Hamilton syringe, ensuring that the tip of the Hamilton reaches as far down into the pipette as possible. Flick the pipette with a finger to shake the rest of the tracer the rest of the way down the pipette tip. This also helps to ensure that there are no bubbles in the pipette. To prevent the pipette tips from blocking while preparing the animal for surgery, we place the filled pipette tips into a jar of saline. We do this by inserting the pipettes in a piece of foam that has slits cut into it and glued around the top of the jar containing a small amount of sterile saline. Ensure that the tips of the pipettes are just in the saline. Over time, the capillary action of the glass will draw up the saline, but this can be pushed out once the pipette is loaded in the pneumatic or syringe pump.

4. If using a syringe pump to inject the tracers, prepare the pump by filling a 25 μL Hamilton syringe with mineral oil. Cut a suitable length of PE30 polyethylene tubing that can reach from the syringe pump to the animal and fill the tubing with mineral oil, ensuring that there are no bubbles. Attach one end of the tubing to the 25 μL Hamilton syringe and secure the Hamilton into the pump system. Attach a short length of PE50 (~2 cm) tubing to the free end of the PE30 tubing. Infuse through some mineral oil so that the small piece of PE50 tubing is filled with oil.

3.2 General Surgical Procedures

1. To make up the sodium pentobarbitone, add 1040 μL of 96% ethanol and 2 mL of propylene glycol to 600 mg of pentobarbital salt. Stir well and make up the volume to 10 mL with

sterile dH$_2$O. To prevent precipitation, store the sodium pentobarbitone out of direct light.

2. The animal is weighed and anesthetized with sodium pentobarbitone (i.p.) according to species and weight (rats 60 mg/kg; cats 40 mg/kg). Propofol which is a barbiturate-like anesthetic can be used as an alternative via surgical placement of cannula for i.v. infusion (rats 0.006–0.05 mL/min tail vein; cats 0.04 mL/min, cephalic vein). Although we have used ketamine/xylazine anaesthesia for studies in cerebellar modules [31], sodium pentobarbitone or propofol is preferable for studies of cerebellar zones/modules as barbiturates are known to reduce mossy fiber and parallel fiber related inputs while leaving transmission through the climbing fiber system intact or even enhanced [14, 32–34]. An alternative strategy that we have used in our laboratory is to use a different anaesthesia (e.g. isoflurane) for the duration of the surgical procedures, and then switch to barbiturate during periods of electrophysiological mapping (*see* Subheading 3.3).

3. When a level of surgical anaesthesia has been reached (determined by regular testing of withdrawal and corneal reflexes), the animal is placed in the stereotaxic frame using atraumatic ear bars. Local anesthetic is applied to the ear bars to minimize noxious inputs. During surgery, core temperature is maintained at 37 °C with the aid of a thermo-controlled heated blanket. Eyes are covered with ointment to prevent them from drying out. The fur over the surgical site is shaved and the skin cleaned with an antimicrobial.

4. A midline incision is made to expose the skull and the underlying periosteum cleared with two cotton tips or a blunt scraper. The muscles attached to the occipital bone are detached and reflected away using blunt dissection. A dental drill or bone rongeurs are used to carefully expose the desired areas of the cerebellum under investigation. The dura is removed using either watchmaker forceps and scissors or a hypodermic needle with the tip bent 45°. Before carrying out electrophysiological mapping of the zonal/modular organization of the cerebellum, we routinely sketch a detailed map of the exposure using blood vessels as landmarks, to aid with determining the somatotopical organization and tracer placement. More recently with the advances in smartphone camera capabilities, we also take a photo down the eyepiece of the stereomicroscope with a camera phone. Universal adaptors are now available to connect the smartphone to the microscope eyepiece.

3.3 Electro-physiology Mapping

1. Depending on the cerebellar zones/modules being investigated, we next insert pairs of percutaneous stimulating needles subcutaneously into the appropriate peripheral body sites to

electrically stimulate skin areas that correspond to the receptive field centers of different zones/modules. This can include the forepaws, hindpaws, whisker pads, ipsilateral and/or contralateral to the cerebellar exposure, as well as the base of the tail. Pairs of percutaneous leads are made by cutting the hub off the shaft of two hypodermic needles. We use 25G needles for rats and 23G needles for cats. Each needle is then soldered to a length of single core wire that is long enough to reach the isolated voltage/current stimulator and animal. A brief square pulse (0.1 ms) delivered every 1.5–2 s at an intensity sufficient to generate a small but visible twitch of the body region in question is used. A stimulus intensity of 2T is used routinely for mapping purposes (*see* **Note 2**).

2. The evoked climbing field potentials (*see* **Note 3**), which provide the framework for studying cerebellar zones and modules, are recorded extracellularly from the surface of the cerebellum with a platinum-tungsten or tungsten in glass microelectrode (~0.5 MΩ). The responses are recorded differentially with the indifferent electrode placed nearby on bone or muscle. The signal is amplified, and band pass filtered (30 Hz to 5 kHz) and any 50–60 Hz electrical interference is removed (without signal degradation) by the Humbug. We systematically map the exposure by placing the electrode typically at 0.1–0.2 mm spaced intervals along each exposed cerebellar lobule, moving medial to lateral (e.g. Fig. 2a, b). The peak-to-peak or baseline-to-peak amplitude of the evoked climbing fiber response at different loci on the cerebellar surface of known can be measured and the response size plotted as a function of distance to determine the crossover point between responses from neighboring zones—this is the electrophysiological boundary between zones (e.g. [9, 24, 35]; *see* Fig. 2c). The total number of sites recorded is limited by the area of the exposure, blood vessels, and time constraints associated with recovery experiments if anatomical tract tracing or chronic implants are carried out. In the case of chronic implants, a much less extensive survey of the area is carried out as the aim of such experiments is to generally identify a particular zone/module (e.g. [23, 36, 37]) to guide the optimal placement of the implant (*see* Subheading 3.5).

3. Each cortical zone of a cerebellar module can be defined in terms of the peripheral stimulation site/s that evokes the largest climbing fiber responses within that zone and with particular onset latencies. Our group along with others have studied the physiological responses of a number of cerebellar zones and modules and in specific parts of the cerebellum in a range of species (Table 1). Similar techniques have also been used for mapping the nucleus interpositus component of the C3

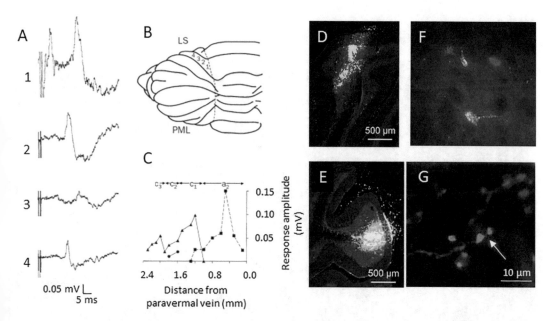

Fig. 2 (**a**) Examples of evoked cerebellar climbing fiber field potentials recorded from lobulus simplex in response to contralateral face stimulation (recording position 1) and ipsilateral forelimb stimulation (recording positions 2–4). Each trace is an average of four consecutive sweeps. The approximate recording loci in lobulus simplex are indicated on the dorsal view of the cerebellum in panel (**b**). The dotted line in (**b**) indicates approximate location of the paravermal vein. (**c**) The spatial separation of the baseline to peak amplitude of the evoked climbing fiber response at different loci on the cerebellar surface plotted against distance from the paravermal vein. The mediolateral width of each electrophysiologically-defined zone is indicated by arrows. Squares = contralateral face stimulation; triangles = ipsilateral forelimb stimulation; circles = contralateral forelimb stimulation. (**d**) Photomicrograph showing an example of red Retrobead injection site. (**e**) Photomicrograph showing an example of a green Retrobead injection site. (**f**) High-power photomicrographs of a neuron double labelled with red and green Retrobeads. (**g**) High-power view of preterminal axon branches anterogradely labelled with Fluoro-Ruby. Arrow indicates synaptic terminals. (Panels (**a–c, g**) reproduced from [24] with permission from Springer Nature; panels (**d** and **e**) reproduced [43] with permission from John Wiley and Sons)

module [38]. Table 2 lists the pros and cons of electrophysiologically approaches to mapping cerebellar connectivity.

3.4 Tracer Delivery (if Carrying out Anatomical Investigations)

1. Guided by the electrophysiological mapping, small injections of tracer material are pressure injected into the cerebellar cortex into the center of the electrophysiologically-defined zone, where the largest cerebellar responses were evoked by peripheral stimulation of a particular body part. To carry out the injections, the pipette is attached to the Picopump pipette holder and the arm of the holder is secured into the micromanipulator. Under microscopic guidance, the pipette is aligned to the target area, just above the surface. If any saline is in the pipette, adjusting the hold pressure of the Picopump will cause the saline to flow out of the pipette.

Table 2
List of pros and cons of electrophysiological mapping

Pros	Cons
Easy to set up and relatively quick to map	Limited mainly to surface mapping and restricted parts of a cerebellar cortical zone. Deeper cerebellar cortical regions such as within fissures are difficult to map.
Ability to physiologically identify key components of anatomical circuitry of individual zones/modules (no reliable landmarks)	Highly sensitive to anesthetics (restricted to climbing fiber inputs not mossy fiber inputs of individual zones)
Consistent physiological features across species (onset latency and pattern of convergence of different zones)	Field mapping resolution restricted at zonal level not microzones (latter requires individual cell recording)
	Access to input and output connections that involve cerebellar cortex; information on projections to cerebellar nuclei or other elements of modules not available by this approach

2. Next, we adjust the timing and ejection pressure on the Pico-pump so that a small, single droplet of tracer is ejected out of the pipette. To determine the volume of the droplet, the radius of the droplet is measured using the microscope eyepiece graticule. The volume of a sphere is used to calculate the volume of the droplet:

$$V = \frac{4}{3}\pi r^3$$

3. For example, a droplet that measures 100 μm has a volume of 0.5 nL; if a total volume of 5 nL volume of tracer is desired, then 20 droplets will need to be delivered in total. The size of the droplet (and therefore its volume) can be increased by altering the pressure and timing of the Picopump. We typically inject 50–250 nL of tracer but more recently, have used volumes of 0.5–30 nL [11].

4. The micromanipulator is then used to advance the micropipette into the cerebellum. Once it is in the brain, we withdraw it slightly to reduce any dimpling. For studies of cerebellar zones/modules, we position the micropipette at a depth approximately 0.2–0.3 mm perpendicular to the pial surface of the cerebellum. The volume of desired tracer is then delivered (*see* **Note 4**). To ensure that the tracer is moving out of the pipette, we carefully monitor the movement of the meniscus in the pipette shank with the microscope. For superficial injections, we are often able to see a small red or green dot appear at the site of the injection. We leave the pipette in situ for 5–10 min to reduce leakage of the tracer back up the pipette

track. Upon the withdrawal of the pipette, we flush the surface of the cerebellum with saline and check that the pipette is still patent by checking the size of the droplet on exit. We repeat this process if injecting a different color tracer at another site. Once the injections have been carried out, we then cover the exposed brain with gel foam and suture the overlying muscle and skin in layers. For larger craniotomies, we recommend dental acrylic to close the exposure. The animal is then closely monitored during recovery and long-acting analgesics are given.

5. If using the syringe pump to infuse the tracers, attach the pipette to the free end of polyethylene tubing and infuse the mineral oil through the end of the pipette tip. Push enough mineral oil out so that the system can accommodate the amount of tracer you will use. Place the tip of the pipette into the vial of tracer taking care not to break the tip. We use a small amount of Blu Tack on the edge of the vial to hold the pipette in place. Withdraw the tracer up into the shank of the pipette. Place the pipette into the pipette holder and insert the holder into the micromanipulator. Proceed with the tracer injection as above for the Picopump, with the only difference being that the syringe pump can be programmed to inject the desired volume of tracer at a rate of 0.25 µL/min.

3.5 Chronic Implants

1. We have carried out studies of cortical and nuclear components of cerebellar modules in awake behaving animals for many years using standard chronic implantation techniques. Prior to the implantation of microdrives [23, 37], microwires (e.g. [22, 39, 40]), or recording chambers [28, 29, 36] into the desired cerebellar module, we routinely use electrophysiological mapping techniques as described above (*see* Subheading 3.3). In cases where we implant microdrives, we use the recording electrodes within the microdrives to record responses to percutaneous electrical stimulation. The recording electrodes are advanced until they are located in a position within the cerebellar cortex in which the largest field potentials are evoked. The microdrive is then fixed to the skull with bone screws and dental acrylic.

2. We also implant chronic stimulating leads to electrically stimulate the forelimb [36] or hindlimb [23]. This allows us to activate the cerebellar zone/module in question. To this end, prior to implantation, pairs of flexible stainless-steel wires at an appropriate length are soldered to connectors within the chronic headpiece or a connector. The other end of the wire is crimped to a pair of 23G hypodermic needles with the plastic hub cut off.

3. At the time of surgery, small incisions are made to the shaved forelimb or hindlimb near the wrist and ankle joint respectively and the superficial fascia reflected to expose the superficial radial and tibial nerves. In the case of rodents, it can be difficult to identify the hindlimb nerves; in this case the wires are implanted either side of the fascia in ankle joint. A trocar is wetted with saline and pushed subcutaneously from the head incision to the incision in the limb. The implant wires are then passed down each limb through the trocar. A small double knot is then tied in each wire at an appropriate length close to where the wire will be implanted to provide enough slack to the final length so as to not impede natural movement and to prevent the wires from breaking. The Teflon insulation distal to the knot is then removed with a scalpel and fine forceps. Using the hypodermic needle, the wire is inserted and positioned around the perineural tissue surrounding the nerve/joint. A second double knot is then tied at the point where the wire emerges from the perineural tissue. This secures the wire in place and the hypodermic needle and any excess wire is cut away. The second wire is secured in the same way but on the opposing side of the nerve or joint. To ensure the wires remain in place over time, the wire is then sutured in place by non-absorbable suture on the inside of each wire knot. The skin margin is then sutured together.

3.6 Survival Time, Perfusion, and Histology

1. For tracer studies, animals have a survival period of 7–10 days to allow sufficient transport of the tracers to occur. A lethal dose of sodium pentobarbitone is then given and the animals are perfused transcardially with heparinized 0.9% saline, followed by 4% paraformaldehyde and then 10% sucrose in phosphate buffer. The brainstem and cerebellum are then removed and stored in 30% sucrose in 4 °C.

2. Once the brains have sunk, we separate the brainstem and midbrain from the cerebellum and section the tissue on a freezing microtome. We routinely cut the cerebellum at 100 μm for visualization of injection sites, whereas the brainstem and/or midbrain is cut at 40–50 μm. We typically section the cerebellum in the sagittal plane and the brainstem/midbrain transversely. If the cerebellum is sectioned coronally or horizontally, the cerebellum can be embedded in gelatin prior to sectioning. To embed the cerebellum, we first embed the cerebellar block in 5% gelatin, followed by embedding in 10% gelatin, and then harden the block in 4% PFA for 2–3 h before transferring it into 30% sucrose to sink overnight. The gelatin-embedded block can then be sectioned as per normal.

3. Sections are collected and mounted serially on microscope slides with gelatin-chrome alum medium (gelatin, chromic

potassium sulfate, dH_2O). The brainstem and midbrain are mounted as two separate series; one series is used for data analysis, the other kept as a spare or used for counterstaining or immunohistochemistry (Retrobeads have a limited compatibility with other histochemical processing). Once sections are mounted, they are left to air dry before viewing.

3.7 Microscopy and Anatomical Mapping

1. For microscopy, slides are temporarily coverslipped with phosphate buffer solution prior to viewing, as more permanent methods can cause fading of cell labelling. Red and green labelling are viewed using the appropriate filter sets. As modules are defined anatomically by their climbing fiber origin within the inferior olive (*see* Fig. 1), retrograde labelling of olivary cells with Retrobeads are mapped onto standard transverse outlines (0.2 mm apart) of the inferior olive which can be created using a camera lucida (e.g. [9, 12, 24, 41, 42]). An advantage of Retrobeads over other tracers is that they produce a well-defined injection site (Fig. 2d, e) and an easily distinguished granular appearance in labelled neurons (Fig. 2f). Other brain areas can also be charted (e.g. pontine nuclei, trigeminal nucleus) and cell labelling mapped onto standard maps [11, 43–45]. The distribution of Purkinje cell terminal fields in regions of the cerebellar nuclei, defined as those territories within the cerebellar nuclei in which labelling was present in fine preterminal axons and their varicosities (Fig. 2g) are also mapped onto standard maps (e.g. [24, 46]).

2. Image analysis software (e.g. ImageJ) can be used to determine the size of the injection site. For each sagittal section in which the core of the injection site is present, the length of the injection site at the Purkinje cell layer is measured and the values from each section summed and the total multiplied by the section interval to obtain an estimate of injection site area (cf. [24, 45]). The image analysis software can also be used to calculate the areas occupied by anterograde terminal labelling (e.g. [17, 24]).

3.8 Chronic Stimulation

1. During chronic studies of cerebellar zones/modules, we stimulate the peripherally implanted stimulating leads using low-intensity electrical stimulation to confirm the zonal/modular identity (*see* **Note 5**) and to carry out investigations on how transmission of peripheral information to a cerebellar zone/module can be altered during behavior. Square pulses of 0.1–0.2 ms duration are delivered typically at 1.5 s intervals or triggered in relation to a behavior (e.g. rearing up movements, [23], Fig. 3). At the first recording session, the stimulus response curve is determined by using incremental increases in stimulus current/voltage. We also carry out the paired pulse

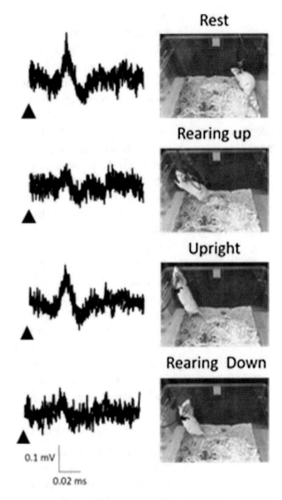

Fig. 3 Evoked climbing fiber fields recorded from the C1 zone in the paramedian lobule of the chronically implanted rat. Climbing fiber fields change in size with rearing behavior. For each phase, shown at the timepoint in the video frame, three overlying traces are plotted following hindlimb stimulation (triangle)

test (*see* **Note 2**). To determine whether movement can affect evoked climbing fiber potentials of the cerebellar module in question, we can use various features of the climbing fiber field. This can include, onset latency, latency to peak, field duration/ width, amplitude of field, and area of the field (Fig. 4).

4   Notes

1. Zones can be subdivided into smaller units called "micro-zones," in which high-resolution electrophysiological mapping within zones have revealed that groups of Purkinje cells have similar climbing fiber receptive fields (e.g. [14, 36, 47–51]).

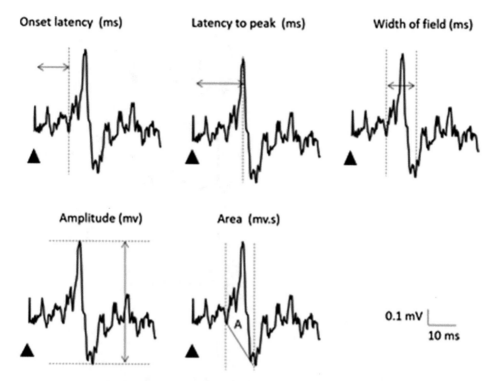

Fig. 4 Examples of measures of evoked climbing fiber fields in chronically implanted rat used to determine whether movement can affect various parameters of the evoked climbing fiber potentials of the cerebellar zone/module in question. Recordings taken from C1 zone in paramedian lobule in response to ipsilateral hindlimb stimulation. Triangle indicates time of stimulus delivery. Dotted lines indicate point/s where measurement is taken, arrow spans time or amplitude of measurement

Field mapping is not of a high enough resolution to reveal the microzonal organization; instead single unit recordings are required.

2. We routinely use single pulse peripheral electrical stimulation to map the cerebellar responses, however, in some cases when it can be difficult to observe a response, using two stimuli delivered 1 ms apart (i.e. 1 kHz) can help facilitate a response. It is also important to ensure that there is no excess cerebrospinal fluid around the recording electrode as this can short the recording. To keep excess fluid away from the recording site, we tear a length of sterile microscope lens tissue and roll it into a thin roll to make a "wick" and place this at the side of the exposure. This absorbs and wicks away the excess cerebrospinal fluid.

3. A number of characteristic features of the evoked field potentials can determine whether they are climbing fiber in origin. This includes an initial sharply rising positive deflection, a reversal in polarity when the electrode is advanced into the molecular layer, an onset latency that is greater than 9 ms,

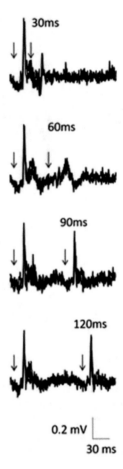

Fig. 5 Example of the paired pulse test in chronically implanted rat. Three overlaying raw traces are given for each with intervals of 30, 60, 90, and 120 ms. Arrows indicate time of stimulus delivery at an intensity of 1.5T. The stimuli were able to elicit a response without resulting in a noticeable twitch or noxious response from the animal

and trial by trial variability in the size of the response with low stimulus intensities. Additionally, when pairs of responses are evoked with supramaximal intensities delivered at intervals between 20 and 120 ms (the paired pulse test), as the interval size decreases, the second response shows a reduction in amplitude (Fig. 5). This pattern is consistent with responses mediated by spino-olivo-cerebellar pathways and the depression is thought to be due to intrinsic properties of inferior olive neurons [25, 52, 53].

4. The micropipettes can become blocked at various stages of the tracer delivery process. To prevent the tip from blocking during alignment with the target region, use a cotton tip to gently moisten the tip with from time to time. If the meniscus does not move when trying to inject the tracer into the brain, we

have often successfully unblocked the tip using the following techniques: moving the pipette up and down a few micrometers with the micromanipulator; pulling the pipette back up into air and dabbing it with a saline-soaked cotton tip; and increasing the pressure and timing of the Picopump (this requires recalibrating the size of the droplet on exit). A final resort is to break the tip back with a pair of watchmaker forceps.

5. Onset latencies of cerebellar modules related to hindlimb stimulation have been found to differ significantly in the anesthetized rat versus awake preparations [23, 37], although a similar phenomenon has not been found for forelimb stimulation. This may be attributed to barbiturate increasing onset latencies [54]. At the end of the chronic recording period it should also be possible to compare the field potential responses recorded in the awake animal with those recorded at the same site but during barbiturate anaesthesia just prior to perfusion.

References

1. Apps R, Hawkes R (2009) Cerebellar cortical organization: a one-map hypothesis. Nat Rev Neurosci 10(9):670–681. https://doi.org/10.1038/nrn2698

2. Cerminara NL, Lang EJ, Sillitoe RV, Apps R (2015) Redefining the cerebellar cortex as an assembly of non-uniform Purkinje cell microcircuits. Nat Rev Neurosci 16(2):79–93. https://doi.org/10.1038/nrn3886

3. Ramnani N (2006) The primate corticocerebellar system: anatomy and function. Nat Rev Neurosci 7(7):511–522. https://doi.org/10.1038/nrn1953

4. Voogd J, Glickstein M (1998) The anatomy of the cerebellum. Trends Neurosci 21(9):370–375. https://doi.org/10.1016/s0166-2236(98)01318-6

5. Apps R, Garwicz M (2005) Anatomical and physiological foundations of cerebellar information processing. Nat Rev Neurosci 6(4):297–311. https://doi.org/10.1038/nrn1646

6. Ruigrok TJ (2011) Ins and outs of cerebellar modules. Cerebellum 10(3):464–474. https://doi.org/10.1007/s12311-010-0164-y

7. Voogd J, Bigare F (1980) Topographical distribution of olivary and cortico nuclear fibers in the cerebellum: a review. In: Courville J, Montigny C, Lamarre Y (eds) The inferior olivary nucleus anatomy and physiology. Raven Press, New York, pp 207–234

8. Voogd J, Ruigrok TJ (1997) Transverse and longitudinal patterns in the mammalian cerebellum. Prog Brain Res 114:21–37. https://doi.org/10.1016/s0079-6123(08)63356-7

9. Atkins MJ, Apps R (1997) Somatotopical organisation within the climbing fibre projection to the paramedian lobule and copula pyramidis of the rat cerebellum. J Comp Neurol 389(2):249–263

10. Buisseret-Delmas C, Angaut P (1993) The cerebellar olivo-corticonuclear connections in the rat. Prog Neurobiol 40(1):63–87

11. Cerminara NL, Aoki H, Loft M, Sugihara I, Apps R (2013) Structural basis of cerebellar microcircuits in the rat. J Neurosci 33(42):16427–16442. https://doi.org/10.1523/JNEUROSCI.0861-13.2013

12. Garwicz M, Apps R, Trott JR (1996) Microorganization of olivocerebellar and corticonuclear connections of the paravermal cerebellum in the cat. Eur J Neurosci 8(12):2726–2738. https://doi.org/10.1111/j.1460-9568.1996.tb01567.x

13. Pijpers A, Apps R, Pardoe J, Voogd J, Ruigrok TJ (2006) Precise spatial relationships between mossy fibers and climbing fibers in rat cerebellar cortical zones. J Neurosci 26(46):12067–12080. https://doi.org/10.1523/JNEUROSCI.2905-06.2006

14. Oscarsson O (1973) Functional organization of spinocerebellar paths. In: Iggo A (ed) Handbook of sensory physiology, vol II, somatosensory system. Springer, New York, pp 339–380

15. Oscarsson O (1979) Functional units of the cerebellum—sagittal zones and microzones. Trends Neurosci 2:143–145. https://doi.org/10.1016/0166-2236(79)90057-2

16. Armstrong DM (1974) Functional significance of connections of the inferior olive. Physiol Rev 54(2):358–417. https://doi.org/10.1152/physrev.1974.54.2.358

17. Apps R, Garwicz M (2000) Precise matching of olivo-cortical divergence and cortico-nuclear convergence between somatotopically corresponding areas in the medial C1 and medial C3 zones of the paravermal cerebellum. Eur J Neurosci 12(1):205–214

18. Dietrichs E (1983) The cerebellar corticonuclear and nucleocortical projections in the cat as studied with anterograde and retrograde transport of horseradish peroxidase. V. The posterior lobe vermis and the flocculo-nodular lobe. Anat Embryol (Berl) 167(3):449–462. https://doi.org/10.1007/bf00315681

19. Sugihara I, Fujita H, Na J, Quy PN, Li BY, Ikeda D (2009) Projection of reconstructed single Purkinje cell axons in relation to the cortical and nuclear aldolase C compartments of the rat cerebellum. J Comp Neurol 512(2):282–304. https://doi.org/10.1002/cne.21889

20. Voogd J, Ruigrok TJ (2004) Cerebellum and precerebellar nuclei. In: Paxinos G, Mai JK (eds) The human nervous system. Elsevier Academic Press, Amsterdam, pp 321–392

21. Ito M (1984) The cerebellum and neural control. Raven, New York

22. Apps R, Lee S (1999) Gating of transmission in climbing fibre paths to cerebellar cortical C1 and C3 zones in the rostral paramedian lobule during locomotion in the cat. J Physiol 516 (Pt 3):875–883. https://doi.org/10.1111/j.1469-7793.1999.0875u.x

23. Lawrenson CL, Watson TC, Apps R (2016) Transmission of predictable sensory signals to the cerebellum via climbing fiber pathways is gated during exploratory behavior. J Neurosci 36(30):7841–7851. https://doi.org/10.1523/JNEUROSCI.0439-16.2016

24. Pardoe J, Apps R (2002) Structure-function relations of two somatotopically corresponding regions of the rat cerebellar cortex: olivo-cortico-nuclear connections. Cerebellum 1(3):165–184. https://doi.org/10.1080/14734220260418402

25. Apps R, Lidierth M, Armstrong DM (1990) Locomotion-related variations in excitability of spino-olivocerebellar paths to cat cerebellar cortical c2 zone. J Physiol 424:487–512.

https://doi.org/10.1113/jphysiol.1990.sp018079

26. Lidierth M, Apps R (1990) Gating in the spino-olivocerebellar pathways to the c1 zone of the cerebellar cortex during locomotion in the cat. J Physiol 430:453–469. https://doi.org/10.1113/jphysiol.1990.sp018301

27. Koutsikou S, Crook JJ, Earl EV, Leith JL, Watson TC, Lumb BM et al (2014) Neural substrates underlying fear-evoked freezing: the periaqueductal grey-cerebellar link. J Physiol 592(10):2197–2213. https://doi.org/10.1113/jphysiol.2013.268714

28. Cerminara NL, Apps R, Marple-Horvat DE (2009) An internal model of a moving visual target in the lateral cerebellum. J Physiol 587(2):429–442. https://doi.org/10.1113/jphysiol.2008.163337

29. Miles OB, Cerminara NL, Marple-Horvat DE (2006) Purkinje cells in the lateral cerebellum of the cat encode visual events and target motion during visually guided reaching. J Physiol 571(Pt 3):619–637. https://doi.org/10.1113/jphysiol.2005.099382

30. Seoane A, Apps R, Balbuena E, Herrero L, Llorens J (2005) Differential effects of transcrotononitrile and 3-acetylpyridine on inferior olive integrity and behavioural performance in the rat. Eur J Neurosci 22(4):880–894. https://doi.org/10.1111/j.1460-9568.2005.04230.x

31. Wise AK, Cerminara NL, Marple-Horvat DE, Apps R (2010) Mechanisms of synchronous activity in cerebellar Purkinje cells. J Physiol 588(Pt 13):2373–2390. https://doi.org/10.1113/jphysiol.2010.189704

32. Gordon M, Rubia FJ, Strata P (1973) The effect of pentothal on the activity evoked in the cerebellar cortex. Exp Brain Res 17(1):50–62. https://doi.org/10.1007/BF00234563

33. Körlin D, Larson B (1970) Differences in cerebellar potentials evoked by the group I and cutaneous components of the cuneocerebellar tract. In: Andersen P, Jansen JKS (eds) Excitatory synaptic mechanisms. Univ. Press, Oslo, pp 237–241

34. Latham A, Paul DH (1971) Effects of sodium thiopentone on cerebellar neurone activity. Brain Res 25(1):212–215. https://doi.org/10.1016/0006-8993(71)90585-3

35. Trott JR, Apps R, Armstrong DM (1990) Topographical organisation within the cerebellar nucleocortical projection to the paravermal cortex of lobule Vb/c in the cat. Exp Brain Res 80(2):415–428. https://doi.org/10.1007/BF00228169

36. Cerminara NL, Garwicz M, Darch H, Houghton C, Marple-Horvat DE, Apps R. Action-based organization and function of cerebellar cortical microcircuits. bioRxiv 2020.04.04.025387. https://doi.org/10.1101/2020.04.04.025387

37. Koutsikou S, Watson TC, Crook JJ, Leith JL, Lawrenson CL, Apps R et al (2015) The periaqueductal gray orchestrates sensory and motor circuits at multiple levels of the Neuraxis. J Neurosci 35(42):14132–14147. https://doi.org/10.1523/jneurosci.0261-15.2015

38. Garwicz M, Ekerot CF (1994) Topographical organization of the cerebellar cortical projection to nucleus interpositus anterior in the cat. J Physiol 474(2):245–260. https://doi.org/10.1113/jphysiol.1994.sp020017

39. Apps R, Lee S (2002) Central regulation of cerebellar climbing fibre input during motor learning. J Physiol 541(Pt 1):301–317. https://doi.org/10.1113/jphysiol.2002.016717

40. Pardoe J, Edgley SA, Drew T, Apps R (2004) Changes in excitability of ascending and descending inputs to cerebellar climbing fibers during locomotion. J Neurosci 24(11):2656–2666. https://doi.org/10.1523/JNEUROSCI.1659-03.2004

41. Apps R (1990) Columnar organisation of the inferior olive projection to the posterior lobe of the rat cerebellum. J Comp Neurol 302(2):236–254. https://doi.org/10.1002/cne.903020205

42. Trott JR, Apps R (1993) Zonal organization within the projection from the inferior olive to the rostral paramedian lobule of the cat cerebellum. Eur J Neurosci 5(2):162–173. https://doi.org/10.1111/j.1460-9568.1993.tb00482.x

43. Herrero L, Pardoe J, Cerminara NL, Apps R (2012) Spatial localization and projection densities of brainstem mossy fibre afferents to the forelimb C1 zone of the rat cerebellum. Eur J Neurosci 35(4):539–549. https://doi.org/10.1111/j.1460-9568.2011.07977.x

44. King VM, Armstrong DM, Apps R, Trott JR (1998) Numerical aspects of pontine, lateral reticular, and inferior olivary projections to two paravermal cortical zones of the cat cerebellum. J Comp Neurol 390(4):537–551

45. Odeh F, Ackerley R, Bjaalie JG, Apps R (2005) Pontine maps linking somatosensory and cerebellar cortices are in register with climbing fiber somatotopy. J Neurosci 25(24):5680–5690. https://doi.org/10.1523/jneurosci.0558-05.2005

46. Herrero L, Yu M, Walker F, Armstrong DM, Apps R (2006) Olivo-cortico-nuclear localizations within crus I of the cerebellum. J Comp Neurol 497(2):287–308. https://doi.org/10.1002/cne.20976

47. Andersson G, Eriksson L (1981) Spinal, trigeminal, and cortical climbing fibre paths to the lateral vermis of the cerebellar anterior lobe in the cat. Exp Brain Res 44(1):71–81. https://doi.org/10.1007/bf00238750

48. Andersson G, Oscarsson O (1978) Climbing fiber microzones in cerebellar vermis and their projection to different groups of cells in the lateral vestibular nucleus. Exp Brain Res 32(4):565–579. https://doi.org/10.1007/bf00239553

49. Ekerot CF, Garwicz M, Schouenborg J (1991) Topography and nociceptive receptive fields of climbing fibres projecting to the cerebellar anterior lobe in the cat. J Physiol 441:257–274. https://doi.org/10.1113/jphysiol.1991.sp018750

50. Hesslow G (1994) Correspondence between climbing fibre input and motor output in eyeblink-related areas in cat cerebellar cortex. J Physiol 476(2):229–244. https://doi.org/10.1113/jphysiol.1994.sp020126

51. Jorntell H, Garwicz M, Ekerot CF (1996) Relation between cutaneous receptive fields and muscle afferent input to climbing fibres projecting to the cerebellar C3 zone in the cat. Eur J Neurosci 8(8):1769–1779. https://doi.org/10.1111/j.1460-9568.1996.tb01320.x

52. Armstrong DM, Eccles JC, Harvey RJ, Matthews PB (1968) Responses in the dorsal accessory olive of the cat to stimulation of hind limb afferents. J Physiol 194(1):125–145. https://doi.org/10.1113/jphysiol.1968.sp008398

53. Eccles JC, Provini L, Strata P, Taborikova H (1968) Topographical investigations on the climbing fiber inputs from forelimb and hindlimb afferents to the cerebellar anterior lobe. Exp Brain Res 6(3):195–215. https://doi.org/10.1007/BF00235124

54. Morissette J, Bower JM (1996) Contribution of somatosensory cortex to responses in the rat cerebellar granule cell layer following peripheral tactile stimulation. Exp Brain Res 109(2):240–250. https://doi.org/10.1007/BF00231784

55. Cerminara NL, Apps R (2011) Behavioural significance of cerebellar modules. Cerebellum 10(3):484–494. https://doi.org/10.1007/s12311-010-0209-2

56. Ackerley R, Pardoe J, Apps R (2006) A novel site of synaptic relay for climbing fibre

pathways relaying signals from the motor cortex to the cerebellar cortical C1 zone. J Physiol 576(Pt 2):503–518. https://doi.org/10.1113/jphysiol.2006.114215

57. Apps R, Hartell NA, Armstrong DM (1995) Step phase-related excitability changes in spino-olivocerebellar paths to the c1 and c3 zones in cat cerebellum. J Physiol 483 (Pt 3):687–702. https://doi.org/10.1113/jphysiol.1995.sp020614

58. Apps R, Trott JR (1997) Topographical organisation within the lateral reticular nucleus mossy fibre projection to the c1 and c2 zones in the rostral paramedian lobule of the cat cerebellum. J Comp Neurol 381(2):175–187

59. Apps R, Trott JR, Dietrichs E (1991) A study of branching in the projection from the inferior olive to the x and lateral c1 zones of the cat cerebellum using a combined electrophysiological and retrograde fluorescent double-labelling technique. Exp Brain Res 87(1):141–152. https://doi.org/10.1007/BF00228515

60. Edge AL, Marple-Horvat DE, Apps R (2003) Lateral cerebellum: functional localization within crus I and correspondence to cortical zones. Eur J Neurosci 18(6):1468–1485. https://doi.org/10.1046/j.1460-9568.2003.02873.x

61. Garwicz M (1997) Sagittal zonal organization of climbing fibre input to the cerebellar anterior lobe of the ferret. Exp Brain Res 117(3):389–398. https://doi.org/10.1007/s002210050233

62. Jorntell H, Ekerot C, Garwicz M, Luo XL (2000) Functional organization of climbing fibre projection to the cerebellar anterior lobe of the rat. J Physiol 522(Pt 2):297–309. https://doi.org/10.1111/j.1469-7793.2000.00297.x

63. Mostofi A, Holtzman T, Grout AS, Yeo CH, Edgley SA (2010) Electrophysiological localization of eyeblink-related microzones in rabbit cerebellar cortex. J Neurosci 30(26):8920–8934. https://doi.org/10.1523/JNEUROSCI.6117-09.2010

64. Oscarsson O (1969) Termination and functional organization of the dorsal spino-olivocerebellar path. J Physiol 200(1):129–149. https://doi.org/10.1113/jphysiol.1969.sp008685

65. Oscarsson O, Sjolund B (1977) The ventral spine-olivocerebellar system in the cat. II. Termination zones in the cerebellar posterior lobe. Exp Brain Res 28(5):487–503. https://doi.org/10.1007/BF00236472

66. Oscarsson O, Sjolund B (1977) The ventral spino-olivocerebellar system in the cat. I. Identification of five paths and their termination in the cerebellar anterior lobe. Exp Brain Res 28(5):469–486. https://doi.org/10.1007/BF00236471

67. Pijpers A, Ruigrok TJ (2006) Organization of pontocerebellar projections to identified climbing fiber zones in the rat. J Comp Neurol 496(4):513–528. https://doi.org/10.1002/cne.20940

68. Trott JR, Apps R (1991) Lateral and medial sub-divisions within the olivocerebellar zones of the paravermal cortex in lobule Vb/c of the cat anterior lobe. Exp Brain Res 87(1):126–140. https://doi.org/10.1007/BF00228514

69. Trott JR, Apps R, Armstrong DM (1998) Zonal organization of cortico-nuclear and nucleo-cortical projections of the paramedian lobule of the cat cerebellum. 2. The C2 zone. Exp Brain Res 118(3):316–330. https://doi.org/10.1007/s002210050286

70. Trott JR, Apps R, Armstrong DM (1998) Zonal organization of cortico-nuclear and nucleo-cortical projections of the paramedian lobule of the cat cerebellum. 1. The C1 zone. Exp Brain Res 118(3):298–315. https://doi.org/10.1007/s002210050285

71. Trott JR, Armstrong DM (1987) The cerebellar corticonuclear projection from lobule Vb/c of the cat anterior lobe: a combined electrophysiological and autoradiographic study. II. Projections from the vermis. Exp Brain Res 68(2):339–354. https://doi.org/10.1007/BF00248800

72. Trott JR, Armstrong DM (1987) The cerebellar corticonuclear projection from lobule Vb/c of the cat anterior lobe: a combined electrophysiological and autoradiographic study. I. Projections from the intermediate region. Exp Brain Res 66(2):318–338. https://doi.org/10.1007/BF00243308

Chapter 5

Designing Behavioral Tasks for Cerebellar Functional Analysis in the Mouse

Court Hull

Abstract

Cerebellar research is rapidly expanding beyond the study of traditional cerebellar-dependent motor behaviors. As it has become widely recognized that the cerebellum contributes to a range of sensorimotor behaviors that also include the social and cognitive domains, it is necessary to design new and flexible behaviors in animal models to study cerebellar function. In particular, the mouse can provide an outstanding model system for studying cerebellar function during behavior due to the wide access for both genetic and experimental manipulations, and the richness of behavioral tasks that mice can be engaged in. This chapter will highlight methods for the design and evaluation of new cerebellar tasks in mice that are also amenable for a variety of in vivo experimental measurements and manipulations such as electrophysiology, optogenetics, and calcium imaging.

Key words Cerebellum, Behavior, Head fixation, Locomotion, Operant tasks, Motor control

1 Introduction

Why is it important to understand cerebellar function in the context of new behaviors? It has long been clear that the cerebellum plays a key role in motor control and motor coordination. As such, well-established cerebellar-dependent sensorimotor association tasks such as eyelid conditioning [1], along with adaptation paradigms such as oculo-motor gain change learning tasks [2] and others [3, 4], have facilitated much of what we know about cerebellar circuit function and its relationship to behavior. However, a growing body of behavioral and anatomical evidence from both human and animal studies has also revealed a much wider role for cerebellar function that also includes non-motor processes [5]. For example, cerebellar processing has been implicated in cognition [6], social processing [7], aggression [8], emotion [9], temporal processing [10], and language [11]. Moreover, impairment of cerebellar function has been linked to several disease states, including autism spectrum disorders [12] and schizophrenia [13]. Even

Roy V. Sillitoe (ed.), *Measuring Cerebellar Function*, Neuromethods, vol. 177, https://doi.org/10.1007/978-1-0716-2026-7_5,
© Springer Science+Business Media, LLC, part of Springer Nature 2022

in the domain of motor control, emerging evidence suggests a more complex role for cerebellar processing that can involve aspects of motor planning and predictions about reward [14]. Hence, if we are to understand the common principles that underlie cerebellar processing and cerebellar learning, it will be necessary to reveal how this structure, despite its strikingly homogenous crystalline architecture, can contribute to such a diverse range of behaviors.

The diversity of behavioral roles that involve cerebellar processing is supported by anatomical evidence showing cerebellar connectivity to many distinct brain regions. In addition to pathways that are clearly involved in motor control, such as descending rubrospinal pathways, the cerebellum is also widely connected to the neocortex disynaptically via the thalamus, including pathways to sensory, motor, and premotor cortical areas [15, 16]. In addition, the cerebellum is also connected with other cortical regions, including prefrontal cortical regions [17]. Moreover, the cerebellum is also indirectly connected via the thalamus to subcortical areas such as the basal ganglia [18, 19]. Recent work has even revealed a direct, monosynaptic connection to the ventral tegmental area (VTA) [20]. Given this diversity of known connections, along with the likelihood that previously unrecognized cerebellar pathways will continue to be discovered, it will be crucial to develop and utilize new tasks in order to study how such circuit pathways are engaged to mediate behavior and learning.

1.1 Established Behaviors for Cerebellar Functional Analysis

Developing new rodent behaviors for the study of cerebellar function requires a starting place, which typically involves one of the two primary motivations discussed above: (1) an identified behavioral phenotype linked to cerebellar processing that has been characterized in humans or animals, such as in patients with cerebellar damage or animals with cerebellar lesion or (2) An anatomical pathway connecting the cerebellum with another brain region that is already implicated in specific aspects of behavior. A third starting place can also come from electrophysiological or other functional measurements that reveal what stimuli or behaviors drive cerebellar neuron firing or how firing patterns change during specific types of learning.

Once such information has been used to define a question about cerebellar function, establishing an appropriate behavior requires: (1) a suitable readout/detector for quantifying behavior, (2) identifying the appropriate cerebellar region, if a known cerebellar pathway was not defined as part of the initial question, and (3) experimental access for recordings and/or manipulation. To illustrate how the field has used these factors to establish new rodent behaviors, it is instructive to review examples from the literature.

2   Innate Behaviors

Innate behaviors are ideal for studies of learning in animal models such as mice for several reasons. First, they can provide for robust, highly reproducible and stereotyped behavioral output, making quantification and comparison across subjects reliable. Second, because they are innate, such behaviors typically require limited training. Third, innate behaviors are, by definition, ethologically relevant, and usually comparable across species. These features thus enable inference of learning mechanisms that are generalizable beyond the rodent. And finally, because cerebellar function has now been implicated in diverse innate responses including social and fear-related behaviors, many tasks for the mouse have already been established that can often be directly translated for cerebellar research without further modification. Examples of such approaches have been adopted to study newly identified connections from the cerebellum involved in innate behaviors, such as the ventral tegmental area [20], periaqueductal gray [21].

2.1   Modification of Innate Behaviors

The cerebellum is required for the calibration or modification of many innate behaviors. Familiar examples include several oculomotor reflexes such as the vestibulo-ocular reflex (VOR), optokinetic reflex (OKR), and smooth pursuit eye movements. A key role for cerebellar learning is often to modify the gain of such reflexes. Cerebellar learning can also be harnessed to modify the timing of innate movements, such as during eyelid conditioning. In this training regimen, an unconditioned stimulus (a corneal airpuff) and its associated unconditioned response (eyelid closure) can be harnessed to generate an eyelid closure in response to a neutral conditioned stimulus through learning that requires the cerebellum [22]. Such tasks fall under the category of traditional cerebellar-dependent behaviors, and have been studied in several animal models including the mouse. Examples of how to setup such behaviors are described in Chapter 9.

2.2   Licking Tasks

Licking is an innate behavior that can be used as a readout for sensory-motor associations. For example, mice can be trained to respond to a learned association by licking in response to a specific sensory stimulus, and by withholding licks to a variant of the same stimulus [23]. Such so-called go/no-go tasks have been used to study the role of multiple brain regions in sensorimotor learning, and now have been adopted to evaluate the cerebellar contribution to sensorimotor learning [24, 25]. In particular, licking tasks have adopted for the study of cerebellar learning because both electrophysiological and imaging studies have revealed representations of licking in the lateral cerebellum [26, 27], and pharmacological experiments have suggested a role for cerebellar circuits in licking tasks such as go/no-go paradigms [25]. Together with the robust, reproducible learning combined such tasks afford, and the with relatively simple methods for quantifying licking as inexpensive optical sensors, licking tasks can offer a valuable platform for studying cerebellar sensorimotor learning.

2.3 Locomotion

Because the cerebellum plays a key role in motor control, and there is ample evidence from human and animal studies showing specific motor deficits associated with cerebellar damage, it has been useful to adapt motor learning paradigms already established in humans and other animal models to the mouse. In particular, because the cerebellum plays a role in coordinating limb movements during locomotion, efforts to adapt locomotion paradigms to the mouse have allowed for circuit-based analysis of how the cerebellum acts to enable such coordination. To study locomotion in the mouse, several approaches have been adopted that are tailored to the both the specific question under investigation, as well the neural recording requirements necessary to address the question.

3 Freely Moving Locomotion

In humans with cerebellar damage, the so-called gait ataxia is a common phenotype [28]. This deficit of locomotion is characterized by uncoordinated walking with reduced coordination between limbs. To study such limb coordination, it is important to measure the fine details of gait, including the complete phases of swing and stance across all limbs, as well as other aspects posture that contribute to coordinated locomotion. Hence, for such studies, freely moving locomotion is necessary, and specialized methods are required to quantify behavior under these conditions.

High-speed video provides the most feature-rich means of monitoring whole animal behavior. However, until recently, the analysis bottleneck of quantifying movement from such videos has been prohibitively challenging, requiring either manual curation or labeling body parts with infrared trackers for automated analysis. However, with the rapid progress in the development of machine learning algorithms capable of trackerless motion capture, and advances in graphical processing units (GPUs) capable of handling high bandwidth data, these barriers have been diminished. As a result, video-based analysis of movement can provide a key method for sophisticated, whole animal tracking during behavior.

Such methods have been used to study the cerebellar contribution to locomotion, using high-speed video to simultaneously capture body movement in profile and from below while mice are running across an open corridor [29]. This setup enables accurate and detailed tracking of all limbs, head, body, and tail, and has allowed for accurate quantification of both normal gait in wild type mice and gait deficits and in a mouse model of ataxia that involves cerebellar Purkinje cell degeneration.

While one of the main barriers to performing such experiments in the past has been the requirement of individual labs to generate custom machine learning algorithms necessary for quantifying movement, there are now several open source options capable of robust behavioral quantification from video. One of the most widespread, DeepLabCut, can be easily implemented with commonly available, relatively inexpensive imaging and computing resources [30].

Beyond measuring naturalistic locomotion, it can also be useful to introduce experimental perturbations to study how cerebellar circuits enable motor learning. In particular, the cerebellum is thought to establish coordinated movement by correcting for external perturbations that occur predictably. This form of learning is termed adaption, and has been studied with a variety of experimental paradigms in humans and animals. One such paradigm is split-belt locomotor adaptation. This is a task in which subjects use a treadmill that can produce either the same or different speeds for the right and left side of the body. When speed differences are introduced, subjects must "adapt" their locomotion to achieve gait symmetry and appropriate interlimb coordination. This paradigm has been extremely valuable for studying the cerebellar contribution to learning in human patients by comparing adaptation in healthy subjects with patients that have cerebellar damage or ataxias [31].

By modifying this paradigm for the mouse, it has become possible to use new tools to study the role of cerebellar circuits in locomotor adaptation [32]. In particular, using similar high-speed video methods as described above in concert with a split-belt treadmill, Darmohray et al. employed cell-type specific chemogenetic manipulation of cerebellar circuits to reveal dissociable features of learning in locomotor adaptation. Notably, such chemogenetic tools are ideal for experiments in freely moving animals, as chemogenetic manipulation can be achieved without invasive procedures that can be challenging during behavior.

3.1 Head-Fixed Locomotion

While some recording and stimulation approaches such as chronically implanted electrodes for electrophysiology, chronic miniscope implantation for calcium imaging, or fiber coupled optical implants for optogenetic experiments can be achieved in freely moving mice, such approaches are significantly more challenging during free movement. These approaches can also impose limitations on optical accessibility/resolution for imaging or single unit isolation for electrophysiology. However, most approaches that can overcome such limitations, such as resonant scanning multiphoton microscopy or single unit electrophysiology, require animal stability. Thus, head-fixed behaviors are also extremely valuable for the study of cerebellar circuit function during behavior.

Locomotion can be studied in head-fixed mice, and in particular when the question at hand does not require analysis of naturalistic gait, or when a recording method such as multiphoton calcium imaging requires head stability. Specific approaches for performing head fixation are discussed below (*see* Section 3.4.1) and this form of restraint can be combined with many different setups that allow for both locomotion and the recording and manipulation of neural activity.

Freely rotating cylindrical platforms have been widely adopted to enable self-generated locomotion in combination with extracellular electrophysiological recordings from the cerebellum. The parts and construction of such cylindrical platforms have been described in detail [33], and such setups have been used for multiple purposes, including to facilitate task engagement during classical conditioning [34], and to study how cerebellar Purkinje cell activity influences ongoing locomotion [35]. Apart from cylindrical platforms, head-fixed locomotion can also be achieved with using a Styrofoam sphere (single axis rotation or free rotation) [36], a flat disc [37], or sloped discs of the type typically used in commercially available small animal running wheels [38]. Regardless of shape, it is necessary to take great care in positioning the mouse such that it can achieve a comfortable stance on the wheel. In addition, moderate food or water restriction can facilitate locomotion, as laboratory mice with ad libitum food and water can often be more sedentary and lethargic.

Finally, running wheels can also be modified by adding a motor that forces specific locomotion speeds. Such motorized wheels can be useful to normalize locomotion speeds across sessions and animals, to ensure that sufficient time is spent in locomotion for data acquisition, and to test for differences between voluntary and involuntary locomotion. Locomotion experiments using a motorized platform have been used to reveal a powerful context dependence of cerebellar learning, demonstrating that the rates of conditioned eyelid responses are dramatically enhanced when conditioning is performed during locomotion [38].

3.2 Operant Behaviors

Operant behaviors also have several advantages for the study of cerebellar learning. Operant tasks are those where a voluntary behavior is guided by either reinforcement or punishment. Hence, a key advantage of such behaviors is that they can involve the formation of arbitrary sensorimotor associations, and therefore expand the range of possibilities for what signals can be harnessed to drive learning. By allowing for such diversity, operant behaviors can be used to more broadly probe the diversity of mechanisms that contribute to cerebellar learning. Moreover, because learning in operant tasks does not involve modification of an unconditioned response, neural activity related to sensory input and motor output can be more readily dissociated. Finally, because the cerebellum is connected to cognitive brain regions such as the neocortex, and has been linked to cognitive disorders such as autism spectrum disorders (ASDs), operant tasks can be used to study the role of the cerebellum in behaviors that require higher order processing. Together, such advantages of operant tasks have led to their more widespread adoption for the study of cerebellar function, and therefore several such tasks have been designed for cerebellar studies in mice. This section will provide examples of such tasks related to forelimb movements, as well as a detailed protocol for establishing a specific operant forelimb task compatible with measurements and manipulation of cerebellar circuits.

3.3 Forelimb Tasks

3.3.1 Skilled Reaching

The role of cerebellar processing in motor control has also been extensively studied using forelimb-based tasks in both humans and non-human primates [39]. Modifying such tasks for the mouse is thus valuable for investigating the underlying circuit mechanisms that contribute to forelimb motor control. However, doing so can also pose significant training challenges, as many forelimb tasks involve non-naturalistic movements that must be shaped by careful training. There are also, however, forelimb tasks that take advantage of naturalistic behaviors such as feeding, and these can typically be established with less extensive training. In addition, as with locomotion tasks, forelimb tasks can be designed for both freely moving and head-fixed mice. This section will provide examples of both freely moving and head-fixed tasks involving forelimb movements that have been designed for the mouse to address questions of cerebellar function.

The cerebellum is required for well-calibrated, fluid movements, and patients with cerebellar damage can exhibit characteristic impairments that include limb oscillations and mis-targeted endpoints of limb movement. Thus, forelimb reaching behaviors can provide a useful tool for investigating the cerebellar contribution to coordinated limb movements. Skilled reaching is a particular class of forelimb movement that typically also involves prehension, such as when mice use their forelimbs to retrieve a food item. Because such goal-directed skilled reaching is a naturalistic and motivated behavior, it can provide a useful means for studying the neural control of forelimb movement.

Skilled forelimb reaching tasks have been modified for mice, both in the head-fixed and freely moving configuration, to study motor control. In particular, such a task has been used in conjunction with optogenetics and electrophysiology to test how cerebellar output neurons in the cerebellar nuclei (CbN) control the ongoing dynamics of movement [40]. The setup for this task involves placing a food pellet on a raised platform that can be accessed by a forelimb reach through a narrow gap between two pieces of clear acrylic. Thus, freely moving mice are compelled to perform a stereotyped forelimb movement in order to retrieve food. In combination with the same type of high-speed videography and analysis described above, such a setup is ideal for investigating the relationship between cerebellar processing and forelimb movement kinematics.

3.3.2 Lever-Based Tasks   Forelimb tasks can also be used to investigate other features of cerebellar processing. For example, the cerebellum plays a well-known role in movement timing, and patients with cerebellar damage also frequently exhibit timing deficits such as delayed movement onset, inability to generate rhythmic movements, and other temporal errors. Because mice can be trained to report timing predictions with forelimb movements, such tasks can provide a means to assess how cerebellar circuits enable proper movement timing.

To establish timing tasks for the study of cerebellar function, one successful approach has been to use setups built around a lever that can be manipulated by the mouse forelimb [41] (Fig. 1). In such setups, mice can be trained to depress the lever, and release it to report their learned timing prediction. Training such behaviors can require several weeks, as they do not harness naturalistic behaviors but instead require arbitrary sensorimotor associations. To achieve such behaviors, flexible, multi-stage training is necessary, and a detailed protocol that illustrates the key principles is described below (*see* Section 3.4.1).

Because such behaviors can be performed in head-fixed mice, allowing for experimental approaches such as video rate multiphoton calcium imaging, single or multiunit electrophysiological recordings, optogenetics, cannulated drug delivery, and other recording and manipulation approaches, operant tasks can have great value in revealing novel features of cerebellar processing. Indeed, such tasks have been key for testing new ideas about how cerebellar circuits operate in the context of timing tasks governed by predictions about reward [41]. Notably, however, interpreting the neural data from such tasks can be challenging, as head-fixed mice often make movements other than the intended forelimb action generated by training. Thus, care must be taken to include experimental and analytical methods that account for movements that may occur outside of the immediate task design (Section 3.4.1).

Beyond simple lever tasks, mouse forelimb movements can be studied with more sophisticated manipulandum, including multi-axis joysticks, moveable wheels, and other such devices [42–44]. Wagner et al. [43] provide an excellent description of how to construct such a robotic manipulandum for mice that can be used for a multitude of operant tasks. By modifying setups based on these types of manipulandum to include variable force transducers,

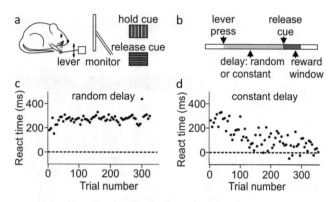

Fig. 1 Reward-guided operant task for head-fixed mice. (**a**) Schematic of experimental setup and visual cues used to instruct behavior. (**b**) Schematic of trial structure. (**c**) Example session from a well-trained mouse when the delay interval is randomized. Reaction time for lever release is measured relative to the onset of the release cue (time = 0). (**d**) Example session from the same mouse when the delay interval is constant (500 ms)

visually guided feedback from a video monitor, reward delivery or punishment to reinforce or discourage specific movements, such forelimb tasks can be used to study a variety of processes. For example, variable force transducers can be used to adjust the difficulty of moving a lever or joystick, thus enabling forelimb-based adaptation tasks. Alternatively, by pairing specific movements with reward or aversive stimuli, it is possible to study the effect of predicted outcome on the cerebellar control of movement. Thus, together such forelimb tasks enable a rich repertoire of behavioral control for mouse studies of cerebellar function.

3.4 Specific Considerations for Establishing Head-Fixed, Forelimb-Based Operant Behaviors

3.4.1 Methods for Head Fixation

We have performed a series of forelimb motor timing experiments under head-fixed conditions in order to use resonant scanning two-photon imaging of climbing fiber input to Purkinje cells [41]. In addition to enabling such imaging approaches, head fixation has the added advantage of limiting other unwanted body movements (though see below for additional control for other movements). To head-fix mice, we designed a titanium headpost that is lightweight and maximizes the surface in contact with the skull while enabling easy access to the cerebellum (Fig. 2). The surgical procedure involves resecting the scalp to expose the skull, and then carefully scraping tissue from the surface of the skull to allow direct contact with the headpost [45]. Complete removal of this tissue is important for a long-lasting implant. We also find that the implants are more stable if performed after P45, when skull growth has slowed. Once the skull is dry, dental acrylic (Metabond) is used to adhere the headpost to the skull.

This headpost implant is compatible with multiple different recording approaches. For instance, as part of the same surgery, it

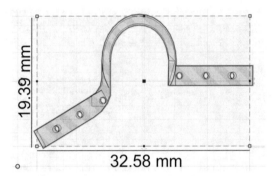

Fig. 2 Headpost design for head-fixed cerebellar behaviors. Open region should face posterior to allow maximum access to the cerebellum. CAD file freely available upon request

is possible to implant electrodes for chronic recording. Alternatively, a cranial window can be implanted over the cerebellum for optical access for either imaging or optogenetic stimulation [45] (see Goldey for detailed cranial windowing procedures). It is often efficient to perform these implants as part of the same surgery to avoid multiple sessions under deep anesthesia. However, especially as students are learning and the procedure goes more slowly, it can be helpful to perform these surgeries serially. An expert surgeon should be able to implant a headpost and cranial window in less than 3 h.

Expression of genetically encoded sensors and opsins can be achieved in multiple ways: using transgenic animals, systemic injection, or targeted viral injections via the cranial window. Targeted injections can be performed at the time of initial surgery to avoid multiple surgeries. However, viral expression is typically best from 2 to 6 weeks following the injection, after which the calcium indicator becomes overexpressed and unusable. Thus, the timing of the viral injection will also depend on the time course of the training and experimental paradigm. In our experiments, because training typically requires a month or more prior to experiments, removal and reimplantation of the cranial window was necessary for viral injections of AAVs encoding genetically encoded calcium indicators (GCaMPs) after initial training of several weeks. This approach also has the advantage that, in the event that an animal does not learn the task, valuable resources such as custom AAVs or electrode implants are not used in advance for animals that will not be part of the final research project.

3.5 Optimizing Experimental Ergonomics

We designed our experimental rig to primarily utilize Thorlabs parts (Fig. 3 and Table 1) such that all pieces are attached to a movable platform with legs. This platform enables positioning of the mouse facing the center of the LCD monitor, and flexible adjustment of rig position under the microscope. The major

Fig. 3 Operant forelimb task behavioral rig. This rig, used for experiments in Heffley et al. [41], is constructed largely from ThorLabs parts as described in Table 1. Side (left) and top (middle) views of the rig with sled removed. Mice are positioned in the sled (right), which is then inserted to the rig for each training session

components of the rig are the sled and tube to support the mouse, headpost clamps, lever assembly, and tube for reward delivery. The tube is made from a 50 mL conical tube (e.g. Falcon tube), and a Velcro strap is wrapped around the tube and sled to loosely secure the mouse in place. The sled then slides into a clamp on the rig, situating the mouse close to the headpost clamps, and can be locked in place.

The headpost clamps should be approximately the height of the tube such that when clamped the mouse's head is above its torso, allowing it to sit comfortably in the tube. The left headpost clamp should always remain fixed in place relative to the platform, and it serves as a landmark to enable reliable positioning of the mouse across sessions. Markings or holes in the fastening arms of the headpost can also be used as landmarks for positioning.

When placing the mouse in the rig, it is necessary to adjust the mouse's head such that the nose is perpendicular to the monitor before tightening the clamp holding the left fastening arm of the headpost. Next, the clamp for the right fastening arm of the head-post can be swiveled into position and tightened. It is extremely important that the two headpost clamps are at the same angle and parallel to each other. If they are not, the headpost will be twisted, this force can break it away from the skull. To ensure that the two headpost clamps are parallel, a standard histology glass slide can be secured in between the headpost clamps and used as a level. Rotate the right headpost clamp until that the glass slide is flat with respect to the table, and then repeat this with the glass slide in the left headpost clamp. Finally, adjust the height of the right headpost clamp until both clamps can be fitted to slide, and then tighten. If the two headposts clamps are parallel, the glass will not crack.

The final positioning should also allow the mouse to easily reach the lever with its right paw. If the lever is too far away, the mouse will be less likely to use it. However, if it is too close, it has the possibility of becoming stuck under the holding tube. It is

Table 1
Parts for constructing operant forelimb task behavioral rig

Part Number (Thorlabs unless noted otherwise)	Description	Quantity Needed	Use
MB810	Aluminum breadboard, 8″ × 10″	1	Base
DTS25	Dovetail translation stage	1	Lever assembly
RS4P	4″ pedestal pillar post	4	Base
PH2E	2″ pedestal post holder	1	Headpost clamps
BE1	Studded Pedestal Base adapter	4	Headpost clamps, reward tube, sled holder
TRK05	Anti-rotation alignment key	1	Headpost clamps
CF125	Short clamping fork	4	Headpost clamps, reward tube, sled holder
CL2	Heavy-duty variable height clamp	2	Sled holder, sled
CL6	Wedged table clamp	1	Electronics assembly
TR3	3″ optical post	8	Headpost clamps, reward tube, sled holder, sled, lever assembly
TR2	2″ optical post	2	Reward tube, sled
TR1.5	1.5″ optical post	1	Lever assembly
TR1	1″ optical post	2	Lever assembly
RA180	Right-angle post end clamp	2	Headpost clamps
RA90	Right-angle clamp	3	Lever assembly
R2	Slip-on post collar	5	Headpost clamps, lever assembly, sled holder, sled
SWC	Rotating clamp	3	Lever assembly, reward tube, sled holder
PJ301	Centered mounting post joint	1	Lever assembly
HW-KIT2	1/4″-20 cap screw and hardware kit	1	All
Altos photonics 4PC69	Plate clamp	2	Headpost clamps
Digikey 626-1027-ND	DB15 male connector	1	Electronics assembly
Digikey 626-1226-ND	DB15 boot	1	Electronics assembly
McMaster 69905K999	Linear push solenoid w/spring (S-07819)	1	Lever assembly
US digital H5-1250-NE-S	1250 CPR ball bearing optical shaft encoder	1	Lever assembly

(continued)

Table 1
(continued)

Part Number (Thorlabs unless noted otherwise)	Description	Quantity Needed	Use
US digital CON-FC5-22	5-pin latching connector (gauge 22)	1	Lever assembly
South Bend 150-FP	7.6 g/0.25 oz. size 8 dipsey sinker (FDS8)	1	Lever assembly
Component supply co. HTX-19T	Stainless steel 304 hypodermic tubing- cut into 12 cm and 1 cm using Dremel tool	2	Reward tube
Custom order sheet metal	0.02″ 301 stainless steel sheet cut into 3 mm × 600 mm	1	Lever assembly
Digikey ATUP-P305T-ND	24 AWG 8 strand wire	1	Electronics assembly
Genesee scientific 28–108	50 mL conical tube	1	Sled
McMaster 3029T51	Velcro strap	1	Sled
Parts not visible in Fig. 3			
Cole Parmer 06422-02	Medium tubing	1	Reward tube
Cole Parmer 06407-70	Small tubing	1	Reward tube
Labjack U6	Digital analog acquisition device	1	Electronics
Labjack CB37 terminal board	Screw terminal connector for U6	1	Electronics
Lee company LHDA1233215H	Reward solenoid	1	Reward
Murata 41 mm piezo	Piezo sensor	1	Movement sensor

helpful to place the lever on its own movable stage in order adjust its distance to the preference of each mouse. This lever is also connected to a digital encoder that monitors its exact position, and can thus be used as a proxy for paw position during manipulations.

Finally, the reward tube should be placed under the mouse's nose, in front of the mouth such that the end of the tube is within licking distance. This distance must be adjusted for each mouse, as snout lengths and headpost positioning is mouse-specific. To measure licking we have used a contact based circuit that is closed when the mouse licks the reward tube. Reward is provided from a reservoir that delivers saccharine solution (10 mM) via gravity, and the reward volume is controlled by a solenoid that opens transiently. The open duration of the reward solenoid thus controls the amount of reward delivered, and should be titrated according to how much reward the mouse consumes based on the number of trials desired (see fluid restriction in the following).

While large animal movements are restricted by the conical tube, Velcro strap and head fixation, animals can still make small body adjustments. To record such movements, we have installed a piezo-electric sensor beneath the mouse, inside the conical holding tube. This provides a sensitive measure of any body movement, and can reliably to register the animal's breathing [46].

3.6 Behavioral Design

3.6.1 Water Restriction

Maintaining robust behavior requires that mice are both motivated and healthy [23]. Water restriction provides a valuable tool to achieve both goals, as laboratory mice under water restriction typically exhibit greater activity, maintain health body weight, and are motivated to perform behaviors that are guided by fluid rewards. To achieve healthy water restriction, water intake can be initially reduced to approximately 2 mL per day (~67 mL per kg for a 30 g mouse) [47] and then further reduced over the course of a week such the mouse achieves 85% initial body weight (or Jackson Labs target: https://www.jax.org/jax-mice-and-services/strain-data-sheet-pages/body-weight-chart-000664). At this stage, mice will actively seek water when it is provided (e.g. from a syringe) prior to training. During training, mice receive saccharine water (10 mM) in proportion to the number of trials performed, and are supplemented after training with the amount necessary to fulfill their daily ration (at least 40 mL/kg).

3.6.2 Habituation/Training

There are multiple viable approaches to habituating mice to a head-fixed experimental setup. Some experimenters have found it useful to extensively handle mice before head fixation or habituate them to head restraint for several sessions prior to training by placing them in head fixation. However, we often find that the mice learn more quickly when they are unfamiliar with the experimental setup, as exploration is necessary to find and manipulate the lever. Regardless of initial strategy, in our motor timing task, mice are required to hold down the lever for at least 400 ms ("required hold time") for a visual target stimulus to change orientations. If the mouse releases the lever at least 100 ms (to account for visual response latency), but no more than 10 s, after the visual target, reward is delivered. This training method achieves the first goal, which is to teach the mouse that pressing and releasing the lever is associated with reward. As the mice learn this motor association, they become more likely to press the lever, and therefore also produce shorter inter-trial intervals. They also become more likely to release the lever in response to the visual stimulus, and therefore exhibit fewer misses (failures to release the lever within 10 s) and more false alarms (failures to hold until the target appears). This process typically occurs over the first few days of training. Key factors for success during this period are animal positioning for easy access to the lever and juice tube, and motivation to earn reward.

3.6.3 Shaping Behavior

Once the animals associate lever manipulation with reward, it is necessary to shape the behavior toward the desired outcome. Hence, the second major training goal is to teach the animals that reward is associated with timing lever release to the visual target. As the mice learn this association, they become more likely to release the lever immediately following the visual stimulus, and this learning is apparent when the distribution of lever release times becomes skewed toward the time immediately following the visual stimulus (Fig. 4). Such behavior is apparent by visualizing the cumulative distribution of lever release times relative to the visual stimulus. This change in release times typically occurs within 4–10 days of initial training. Once this tendency has developed, mice will frequently adopt a timing strategy to release the lever, abandoning stimulus-driven responses if further shaping is not performed. To maintain stimulus-driven responses, a variable delay is added to the required hold time, and this delay is changed by the computer on a trial-by-trial basis. Initially, this variable time range should be relatively small (e.g. 0–300 ms) so as not to require that animals hold the lever for extended periods—the goal is simply to add uncertainty to the time of stimulus presentation and hence lever release and reward. If the animal has learned the stimulus-association, this added variable time interval should not change its behavior substantially, and it should still be more likely to release the lever in response to the visual target. Across several days of training, this variable delay interval can be incrementally increased until the visual target can appear 400–4000 ms after the lever press. This final stage can take up to 30 days of training, and well-trained

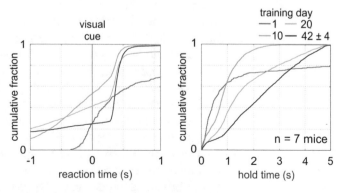

Fig. 4 Cumulative histograms of reaction times (left) and total hold times (right) for mice across training during the variable cue delay paradigm

mice typically exhibit high proficiency at this time point (70–90% correct lever releases). During this final stage, the window for successful (rewarded) lever releases following the visual stimulus can also be narrowed significantly from 10 s to 12 s, though well-trained mice will release immediately following the visual cue regardless of this interval. Once this stage is complete, animals can be considered expert (see below), and engaged in learning sessions where the variable interval is removed to test how mice adapt movement timing to a predictable reward contingency.

3.6.4 Performance Standards/Metrics

The primary metrics to evaluate performance in this behavior are hit rate (the fraction of trials in which the mouse releases the lever in the appropriate time window after the visual target is presented) and false alarm rate (the fraction of trials in which the mouse releases the lever before the visual target). For an expert mouse using a salient visual target, the hit rate should be close to 1 and the false alarm rate should be close to 0. This can be achieved in practice with motivated mice; however, because mice are impatient and inherently exploratory, a false alarm rate of 10–30% is also acceptable performance. To disincentivize misses and false alarms, time outs (2–10 s) that extend the inter-trial interval should be added on error trials. Mice typically will work for 1–1.5 h in such a regime, and perform 400–700 trials. At the end of behavior sessions, however, mice become less motivated. At this point, false alarm rates significantly decrease and misses and reaction times significantly increase. Thus, behavioral data must be cropped from the end of sessions according to thresholds for these criteria.

3.6.5 Detecting Alternate Strategies

The most common alternate strategy that animals adopt is a timing strategy. In other words, rather than a stimulus-reaction strategy, animals will release the lever with a fixed timing that approximates the earliest time when reward is possible. By working rapidly with low hit rates, this strategy can still result in sufficient reward to drive behavior. Such a timing strategy can be readily identified by examining the distribution of hold times. If animals are largely using a stimulus-reaction strategy, the distribution of hold times should be flat; a peak in the distribution suggests a timing strategy. Timing strategies can be effectively discouraged by increasing the required hold time, increasing duration of the variable time interval for cue delivery across trials, and increasing the duration and variability of the inter-trial interval. Each of these approaches negatively impacts timing strategies, but must be balanced against possible decreased motivation that can occur when reward is too infrequent. Other options include increasing the required latency between the visual stimulus and lever release, and decreasing the total reaction window, both of which decrease the probability of rewarding guesses. Together, these approaches allow for robust, well-trained behavior in this task.

References

1. Ohyama T et al (2003) What the cerebellum computes. Trends Neurosci 26(4):222–227

2. Lisberger SG (1988) The neural basis for motor learning in the vestibulo-ocular reflex in monkeys. Trends Neurosci 11(4):147–152

3. Gilbert PF, Thach WT (1977) Purkinje cell activity during motor learning. Brain Res 128 (2):309–328

4. Thach WT, Bastian AJ (2004) Role of the cerebellum in the control and adaptation of gait in health and disease. Prog Brain Res 143:353–366

5. Schmahmann JD (1991) An emerging concept. The cerebellar contribution to higher function. Arch Neurol 48(11):1178–1187

6. Kim SG, Ugurbil K, Strick PL (1994) Activation of a cerebellar output nucleus during cognitive processing. Science 265(5174):949–951

7. Van Overwalle F et al (2014) Social cognition and the cerebellum: a meta-analysis of over 350 fMRI studies. NeuroImage 86:554–572

8. Reis DJ, Doba N, Nathan MA (1973) Predatory attack, grooming, and consummatory behaviors evoked by electrical stimulation of cat cerebellar nuclei. Science 182 (4114):845–847

9. Schmahmann JD, Caplan D (2006) Cognition, emotion and the cerebellum. Brain 129 (Pt 2):290–292

10. Ivry RB, Spencer RM (2004) The neural representation of time. Curr Opin Neurobiol 14 (2):225–232

11. Ackermann H (2008) Cerebellar contributions to speech production and speech perception: psycholinguistic and neurobiological perspectives. Trends Neurosci 31(6):265–272

12. Wang SS, Kloth AD, Badura A (2014) The cerebellum, sensitive periods, and autism. Neuron 83(3):518–532

13. Mothersill O, Knee-Zaska C, Donohoe G (2016) Emotion and theory of mind in schizophrenia-investigating the role of the cerebellum. Cerebellum 15(3):357–368

14. Hull C (2020) Prediction signals in the cerebellum: beyond supervised motor learning. Elife 9:e54073

15. Kelly RM, Strick PL (2003) Cerebellar loops with motor cortex and prefrontal cortex of a nonhuman primate. J Neurosci 23 (23):8432–8444

16. Proville RD et al (2014) Cerebellum involvement in cortical sensorimotor circuits for the control of voluntary movements. Nat Neurosci 17(9):1233–1239

17. Ramnani N (2006) The primate cortico-cerebellar system: anatomy and function. Nat Rev Neurosci 7(7):511–522

18. Bostan AC, Strick PL (2018) The basal ganglia and the cerebellum: nodes in an integrated network. Nat Rev Neurosci 19(6):338–350

19. Hoshi E et al (2005) The cerebellum communicates with the basal ganglia. Nat Neurosci 8 (11):1491–1493

20. Carta I et al (2019) Cerebellar modulation of the reward circuitry and social behavior. Science 363:eaav0581

21. Vaaga CE, Brown ST, Raman IM (2020) Cerebellar modulation of synaptic input to freezing-related neurons in the periaqueductal gray. Elife 9:e54302

22. Medina JF et al (2000) Mechanisms of cerebellar learning suggested by eyelid conditioning. Curr Opin Neurobiol 10(6):717–724

23. Guo ZV et al (2014) Procedures for behavioral experiments in head-fixed mice. PLoS One 9 (2):e88678

24. Tsutsumi S et al (2019) Modular organization of cerebellar climbing fiber inputs during goal-directed behavior. Elife 8:e47021

25. Gaffield MA et al (2016) Chronic imaging of movement-related Purkinje cell calcium activity in awake behaving mice. J Neurophysiol 115 (1):413–422

26. Welsh JP et al (1995) Dynamic Organization of Motor Control within the Olivocerebellar system. Nature 374(6521):453–457

27. Bryant JL et al (2010) Cerebellar cortical output encodes temporal aspects of rhythmic licking movements and is necessary for normal licking frequency. Eur J Neurosci 32(1):41–52

28. Morton SM, Bastian AJ (2004) Cerebellar control of balance and locomotion. Neuroscientist 10(3):247–259

29. Machado AS et al (2015) A quantitative framework for whole-body coordination reveals specific deficits in freely walking ataxic mice. Elife 4:e07892

30. Mathis A et al (2018) DeepLabCut: markerless pose estimation of user-defined body parts with deep learning. Nat Neurosci 21(9):1281–1289

31. Morton SM, Bastian AJ (2006) Cerebellar contributions to locomotor adaptations during splitbelt treadmill walking. J Neurosci 26 (36):9107–9116

32. Darmohray DM et al (2019) Spatial and temporal locomotor learning in mouse cerebellum. Neuron 102(1):217–231. e4

33. Heiney SA et al (2014) Cerebellar-dependent expression of motor learning during eyeblink conditioning in head-fixed mice. J Neurosci 34 (45):14845–14853

34. Ohmae S, Medina JF (2015) Climbing fibers encode a temporal-difference prediction error during cerebellar learning in mice. Nat Neurosci 18(12):1798–1803

35. Sarnaik R, Raman IM (2018) Control of voluntary and optogenetically perturbed locomotion by spike rate and timing of neurons of the mouse cerebellar nuclei. Elife 7:e29546

36. Dombeck DA et al (2007) Imaging large-scale neural activity with cellular resolution in awake, mobile mice. Neuron 56(1):43–57

37. Schneider DM, Nelson A, Mooney R (2014) A synaptic and circuit basis for corollary discharge in the auditory cortex. Nature 513 (7517):189–194

38. Albergaria C et al (2018) Locomotor activity modulates associative learning in mouse cerebellum. Nat Neurosci 21(5):725–735

39. Ebner TJ, Hewitt AL, Popa LS (2011) What features of limb movements are encoded in the discharge of cerebellar neurons? Cerebellum 10(4):683–693

40. Becker MI, Person AL (2019) Cerebellar control of reach kinematics for endpoint precision. Neuron 103(2):335–348. e5

41. Heffley W et al (2018) Coordinated cerebellar climbing fiber activity signals learned sensorimotor predictions. Nat Neurosci 21 (10):1431–1441

42. Wagner MJ et al (2017) Cerebellar granule cells encode the expectation of reward. Nat Publ Group 544(7648):1–18

43. Wagner MJ et al (2020) Skilled reaching tasks for head-fixed mice using a robotic manipulandum. Nat Protoc 15(3):1237–1254

44. Kostadinov D et al (2019) Predictive and reactive reward signals conveyed by climbing fiber inputs to cerebellar Purkinje cells. Nat Neurosci 22(6):950–962

45. Goldey GJ et al (2014) Removable cranial windows for long-term imaging in awake mice. Nat Protoc 9(11):2515–2538

46. Heffley W, Hull C (2019) Classical conditioning drives learned reward prediction signals in climbing fibers across the lateral cerebellum. Elife 8:e46764

47. Glickfeld LL, Histed MH, Maunsell JHR (2013) Mouse primary visual cortex is used to detect both orientation and contrast changes. J Neurosci 33(50):19416–19422

Chapter 6

Learning Paradigms and Genetic Tools for the Study of Cerebellum-Dependent Learning and Memory

Akira Katoh

Abstract

Motor learning, a process known to adapt motor outputs by training with repeated movements, is essential for the enormous capacity of modifications of both body and environment over time, and capable not only of improving ordinary movements, such as infant walking, but also achieving Olympic skill feats or rehabilitation. A variety of recording, stimulation, and perturbation approaches have yielded considerable and convergent evidence for a key role of the cerebellum in motor learning. In the last half century, research on some forms of motor learning has made significant progress and has generated elegant empirical and mathematical models for cerebellum-dependent motor learning. Simultaneously, experimental paradigms and technological advances have made possible the elucidation of motor learning mechanisms at distinct levels. In this chapter, recent progress in the research on cerebellum-dependent learning and memory at the molecular, circuit, and behavior levels, using simple forms of motor learning, are summarized; the corresponding new behavioral paradigms and genetic tools in mice are also described.

Key words Motor learning, Eye movements, Optogenetics, Gait, Long-term depression, Long-term potentiation, Purkinje cells

1 Introduction

Motor learning is the process in charge of minimizing the errors between the target trajectory and the current motor output to achieve a smooth and accurate motor performance, which is essential for physical performance of humans and other animals. We continuously learn to modify our movements, both at the conscious (playing piano, running in a pattern to avoid predators) and unconscious (classical conditioning, reflex modification) levels. Athletes and musicians demonstrate some of the most drastic results of motor learning, but it is required even for daily routine tasks. Not only skilled movements, but also ordinary ones, such as infant's walking, are adapted more smoothly and accurately by motor learning. Motor learning issues can have negative consequences for neurological patients. People with lesions in a

Roy V. Sillitoe (ed.), *Measuring Cerebellar Function*, Neuromethods, vol. 177, https://doi.org/10.1007/978-1-0716-2026-7_6,
© Springer Science+Business Media, LLC, part of Springer Nature 2022

particular brain region, the cerebellum, are severely incapacitated in the ability to modify their motions, having uncoordinated and mistimed motions, compensated by decomposition of complex coordinated behaviors into simple serial movements.

The cerebellum is a widely conserved structure found in all vertebrate species, from fish to mammals, and its relevance is consistent with the severe movement phenotype associated with cerebellar dysfunction. Laboratory interventions in animals through mutations and lesions, in fact, disrupt motor control and motor learning, indicating that the basic architecture of the cerebellum is apt for plasticity and signal processing. Understanding cerebellar functions requires careful analysis of the plasticity mechanisms and signals available during motor learning at sites throughout the neural circuit around the cerebellum.

Among the simple behavioral experimental systems employed in mouse research, eye movements and walking will be outlined, both involving cerebellum-dependent motor learning.

1.1 Reflexive Eye Movements

Eye movements are not only the entrance point of visual information processing that detects the light stimuli surrounding them; vestibular sensation, somatosensory system, autonomic nervous system, and even psychological state are reflected in eye movements. Eye movements can be voluntary or involuntary, and are classified into saccades, smooth pursuit, optokinetic response (nystagmus), vestibulo-ocular reflex, vergence, and visual fixation. In this chapter, two reflexive eye movements used for studies in rodents are described.

The vestibulo-ocular reflex (VOR) is a reflexive eye movement generated by inputs from the vestibular organ that moves the eyes in the direction opposite to the head movement. VOR can be broadly divided into two types depending on the head movement that causes it: angular VOR (rotational VOR) is caused by rotational movement around each axis of front-back, left-right, and up-down, and linear VOR (translational VOR) by translational movement of the entire head in each axial direction. While the former is triggered by the semicircular canal that detects angular acceleration, the latter is triggered by vestibular input from the otolith that detects linear acceleration. Both contribute to stabilization of the visual field when the head moves. In this chapter, we will only refer to the former unless otherwise specified.

VOR has the function of suppressing visual field blurring by moving the eyes in the direction opposite to the head movement. In addition, a relatively simple controlling neural circuit, compared to other motor systems, has been identified. Because of its excellent input and output quantification, VOR can properly decode and read brain responses. If any of the nerves involved in VOR control are impaired, abnormalities are observed. For example, patients with peripheral vestibular dysfunction often show abnormal VOR when the eyes are closed (dark place) [1]. Patients with spinocerebellar degeneration type 3, a polyglutamine disease, have cerebellar and vestibular nerve disorders, and also show abnormalities in VOR [2]. Not only the disease but also brain state changes may be noticeable in VOR. A recent study measured the VOR before and after a driver driving a car felt drowsy, and found that the VOR gain decreased and the VOR gain value varied more before the driver became aware of his drowsiness [3]. Based on these results, detecting VOR as a drowsiness sign signal could lead to the development and application of a feedback system to prevent accidents. Due to the high inter-species VOR conservation, research using experimental animals has been actively conducted. Research combining genetic techniques and highly quantitative VOR measurements in mice can provide important basic data for the future practical application of gene therapy for neurological disorders.

The neural circuit that controls the VOR is well known. Acceleration signals generated by head movement are transmitted from the semicircular canals or otoliths in the inner ear to the vestibular nuclei in the brainstem through primary vestibular nerve fibers. Neurons in the vestibular nuclei output to multiple motor nuclei and ultimately control eye muscle contraction. Some neurons in the vestibular nuclei project to the floccular complex in the cerebellum via mossy fibers, and Purkinje cells that receive vestibular input form inhibitory synapses at vestibular nuclei neurons. Therefore, the main pathway of VOR control is inhibited by the cerebellum. Previous studies have shown that many animal species, including monkeys, rabbits, mice, and goldfish, have similar VOR-regulating neural circuits.

The optokinetic response (OKR) is another reflexive eye movement induced in the same direction with a large movement of the entire visual field. When the visual field moves in one direction, it is accompanied by a rapid phase that resets the line of sight in the opposite direction to a slow phase in the same direction (optokinetic nystagmus: OKN). OKR/OKN is considered the major visual input-dependent eye movement in animals with underdeveloped fovea. In primates with a well-developed fovea, smooth pursuit, a voluntary movement, is mainly activated when there is an optokinetic motion in the visual field. Discriminating OKR/OKN and the smooth pursuit is not easy, but OKR can be observed in particular situations; for example, staring out the window of a moving train. In addition, like VOR, it contributes to the stabilization of the visual field during head movements.

The neural circuit controlling OKR is slightly more complicated than VOR. The visual field motion signals detected in the retina are transmitted to the supraoptic nucleus and optic tract system in the midbrain. The nucleus reticularis tegmenti pontis receives these output signals and transmits them to the vestibular nuclei that control the motor nuclei in turn responsible for the extraocular muscles, similar to VOR output pathway. Retinal signals are also sent to the visual cortex via the thalamus, and the output is transmitted to the optic tract/accessory optic tract system and the dorsolateral pontine nuclei. The inhibitory collateral tract from the pontine reticular nucleus and dorsolateral pontine nucleus to the mossy fiber-cerebellum is also common with VOR, supporting the cooperation between OKR and VOR. In addition, there are inputs from the supraoptic nucleus, and the supraoptic nucleus to the cerebellum, via the inferior olivary nucleus in the brain stem, involved in cerebellar synaptic plasticity for VOR and OKR motor learning as described later.

1.2 Gait Activity

Walking is one of the basic exercises wherein each foot achieves articulated movements, including several joints, and is adaptively generated in a diverse and fluctuating environment. As a result of cooperatively controlling the time and spatial patterns of many muscle activities, walking is performed smoothly. To perform smooth and stable locomotion, coordination not only between parts of a limb but also different limbs is required. In other words, walking consists of reproducible and repetitive movements of all limbs.

In rodents, rotarod and stationary beam tests have been used to assess walking ability [4]. In the rotarod test, a mouse is placed on a rotating beam and must move in synchrony. In the stationary beam test, a mouse is placed on a stationary beam and should either walk along. In these tasks, the latency before falling and the number of falls are measured. In the stationary beam test, movement time is measured as well, defined as the time needed to walk across the beam since some animals just remain on the stationary beam without walking to compensate the loss in balance. In the rotarod test, careful observation is also required since the deficit may be masked due to its hanging on tightly to the rotating rod.

Gait is reported to be controlled with information within a neural circuit formed between the spinocerebellum and the spinal cord, called the spinocerebellar loop [5]. The basic rhythm patterns during gait are intrinsically generated in spinal neuronal networks that form the central pattern generator [6, 7]. However, such rhythmic behaviors sometimes require changes to adapt to a new environment, with the cerebellum playing an important role in limb coordination and phase control mechanisms during walking. Information on the activity of the central pattern generator during gait is transferred to the cerebellum through the ventral spinocerebellar tract (VSCT) as an efferent copy. In addition, information derived from various somatosensory systems' receptors is fed back to the cerebellum via the dorsal spinocerebellar tract (DSCT). In fact, the DSCT mediates global parameters of hindlimb kinematics and is considered important for appropriate limb movement control during locomotion. These feedback signals into the cerebellum are supposed to play roles in discriminating the sensory information caused by environmental disturbance from reafferent information resulting from actual walking [8].

Output signals from VSCT and DSCT enter the cerebellar cortex via mossy fibers. These signals are transferred through parallel fibers, axons of cerebellar granule cells, into Purkinje cells, which in turn output to neurons in the deep cerebellar nuclei and descending tract neurons in the brainstem. The activities of interneurons and motor neurons in the spinal cord are regulated by signals circulating in the spinocerebellar loop.

When such information flow into the cerebellum via mossy fibers is partially lost, motor deficits during gait occur. In adult *Lurcher* mice where Purkinje cells were totally absent, walking was characterized by short steps with exaggerated hindlimb flexion in the swing phase [9]. In addition, both the interlimb step ratio, defined as the step length of the reference limb divided by the step length of the opposite limb, and interlimb coupling, defined as the temporal relation of one footfall with respect to another limb footfall, varied more than in normal mice.

1.3 Adaptive Motor Learning and Relevant Region in the Brain

If you have renewed your glasses and the magnification of your field of view has changed, or if the environment around you sways with you while riding a vehicle, the field of view will be continuously blurred for a long time when the normal VOR works during head movements. In these situations, where the visual field becomes unstable due to the VOR caused by the vestibular input, in other words, when there is a mismatch between the vestibular and the visual stimuli that are repeatedly given, the VOR has a mechanism that flexibly responds to the situation by adaptively changing its circuit characteristics for a long period of time with using visual stimulus.

The process of reducing the error signal and bringing the motor output closer to the target trajectory by repeatedly applying the same stimulus is called motor learning. Although VOR and OKR are relatively simple motor systems, they have a motor learning function that adaptively changes their dynamic characteristics by training, like other complex motor systems. Many studies of VOR and OKR using experimental animals have been conducted to approach the basic mechanism of motor learning due to their relatively simple control mechanisms, high conservation in many animal species, and high quantity of input and output [10]. OKR is a closed-loop system wherein the eye movement output itself affects the stimulus input. In other words, it is a feedback system in which the visual image on the retina causes OKR, and the output reduces the visual image. On the other hand, the VOR is an open-loop system in which the output does not affect the vestibular input causing it, and the VOR does not change only with the vestibular input that causes the VOR. When combined with OKR driven by the visual stimulus, it changes the characteristics of the VOR circuit.

Adaptive VOR motor learning can be reproduced at the laboratory level with more quantitative stimuli. The adaptive change of the horizontal VOR in the laboratory is induced by combining the left-right sinusoidal head rotation with the rotation of random dots or striped visual stimuli placed around the animal. VOR gain increases when out-of-phase visual stimuli with vestibular stimuli are repeatedly administered for tens of minutes to several hours, and VOR gain decreases when in-phase visual stimuli with vestibular stimuli are given. An adaptive increase in OKR gain is also evoked by the continuous rotation of random dots or striped visual stimuli placed around the animal in a left-right sinusoidal shape with a stationary head.

Which brain areas are required for induction of the changes by motor learning for reflexive eye movements? As both VOR and OKR require visual stimuli to induce adaptive motor learning, the region responsible for motor learning must receive visual information. Based on the neural circuits that control VOR and OKR, the motor learning field of these reflexive eye movements is expected to lie in the cerebellum or vestibular nuclei. There is a model that feedback due to the visual blur as an error signal to the cerebellar Purkinje cells changes the transmission efficiency of the parallel fiber-Purkinje cell synapse resulting in adaptive changes of the VOR/OKR output [11]. Ablation of the flocculus in the cerebellum abolished adaptive gain increases in the VOR, without significantly changing basal VOR gain [12–16]. Similarly, functional flocculus inactivation with high-frequency stimulation of climbing fibers also blocked adaptative changes in the VOR [17]. Mice deficient in protein kinase C γ, having developmental defects in the elimination of supernumerary climbing fiber–Purkinje cell synapses so that multiple climbing Purkinje cells' fibers persists into adulthood [18], showed impaired motor learning in the VOR [19], without any other abnormalities observed in the cerebellum. On the other hand, VOR motor learning was established even if the vestibular vermis, which receives input related to head movement, was removed [20]. These results suggest the importance of the cerebellar floccular complex for adaptive changes in the VOR, but do not prescribe whether changes occur in the flocculus, or signals in the flocculus simply guide changes elsewhere in the brain.

Manipulation of the inferior olive (IO), the source of climbing fibers, provides some evidence for the necessity of climbing fiber activity in motor learning in the VOR. Site-specific degeneration of the IO abolished adaptive gain increases in the VOR [21, 22] and OKR [13]. Lesions of a nucleus upstream of the IO, the nucleus of the optic tract (NOT), also abolished adaptive gain decreases in the VOR [23]. However, in all these studies, basal properties such as visual stimulus tracking were altered, and thus these lesions may have resulted in a lack of learning simply through a performance deficit.

As mentioned above, gait movement is smoothly and stably performed by cooperatively controlling the temporal and spatial patterns of activity of several limb muscles. A wild-type mouse on a rotating rod will try to walk on it but fall immediately at first. After about 10 trials, however, it will be able to walk stably on the rod due to motor learning.

Mechanical ablation of the cerebellum or inactivating neuronal activity/synaptic transmission in the cerebellum demonstrated its importance for acquiring the ability to stay on the rotating rod. For example, the loss of Purkinje cells in a spontaneously occurring mutant mouse, *Lurcher*, caused those mice difficulties for walking on the rotarod [24, 25]. However, they also revealed abnormal online gait control, providing no information about the role of the cerebellum in motor learning.

Once motor learning is acquired, it is considered to be maintained for a certain period of time and expressed when needed. This process is called "motor memory." For example, someone may not be able to ride a bicycle well at first but never fall down after repeated practice. Once learnt, the skill will be retained even if you do not do it for a long time. Walking and reflexive eye movements also show such a motor memory process.

Previous studies reported that pharmacological suppression of the activity of the cerebellum after short-term training abolished all acquired changes [26, 27], but other studies reported partial memory loss with similar treatments [28]. Recently, lidocaine bilateral injections, which can block sodium channels, were used to block floccular activity in monkeys after acute training with complete abolition of expressed VOR gain increases after infusion [27]. Another study in chronically adapted squirrel monkeys found that ipsilateral inactivation of the flocculus with muscimol resulted in an adapted response decrease in brainstem neurons, although the data between animals was quite variable [29]. Blocking cerebellar activity in goldfish with lidocaine infusion after acute training caused expression of learned increases and decreases in VOR gain to return to naïve values [26]. However, this suppression of a learned decrease in VOR gain was temporary, lasting only a few minutes, while the suppression of increased VOR gain lasted up to 3 h after infusion. Thus, memory storage encoding VOR gain increases and decreases may depend on different extents of the flocculus.

Storage of differently timed information may also depend on the extent of the flocculus. One study used rapid vacuum aspiration to remove the cerebellum from goldfish after either gain-up or gain-down training, finding a reduction in the long-latency portion of the adapted VOR gain, but no change in the short-latency part

[28]. As the cerebellar side loop presumably mediates the longer-latency portion of learned responses, and brainstem changes would mediate the shorter-latency portion, this suggests distributed plasticity mechanisms correlate with different motor learning latencies. On the other hand, a study employing 7 Hz climbing fiber stimulation to temporarily deactivate the cats' flocculus, found that motor learning induced over 3 days in the VOR was completely insensitive to flocculus deactivation [17]. Thus, lesion data suggest a distributed plasticity in the system, both inside and outside the cerebellum.

There are reports on the memory retention characteristics of OKR adaptive learning. Wild-type mice training with a continuous sinusoidal visual stimulus for 1 h increases OKR gain. When this training was performed daily for 5 days, the OKR gain before the start of training on the final day was significantly increased compared to the first, and when the same 1-h training was performed thereafter, short-term learning occurred and OKR gain further increased [30]. To identify storage sites for short-term and long-term memory, lidocaine, which temporarily blocks neurotransmission by inhibiting sodium channels, was bilaterally injected into flocculi after 1-h training on the final day. Lidocaine administration abolished the OKR gain increase caused by short-term learning of the day but did not affect the increased OKR gain caused by long-term learning resulting from the 4-days training. This result suggests that the memory of short-term learning is stored in the flocculus, but the motor memory after long-term training is stored somewhere other than the flocculus. In addition, the gain increase in OKR induced by 800 continuous cycles of sinusoidal visual stimuli without rest disappeared with lidocaine administration to the flocculi after learning, whereas the OKR gain increase induced by 800 cycles of visual stimuli divided into four sessions at 1-h intervals was maintained after lidocaine administration. These results suggest that the memories "learned" in the cerebellum by distributed learning rapidly shift to another brain region where memory fixation occurs.

1.4 Potential Synaptic Plasticity Mechanisms for Cerebellum-Dependent Motor Learning

When sensory signals transmitted via mossy fibers (somatosensory signals for walking, vestibular sensory signals for VOR and visual signals for OKR) are input to Purkinje cells via parallel fibers, simultaneous inputs to the same Purkinje cell from climbing fibers reduce synaptic transmission efficiency between parallel fibers and Purkinje cells. This synaptic plasticity, called long-term depression (LTD), was predicted by Marr and Albus' theoretical studies [31, 32] and experimentally proven by Ito et al. [11, 33, 34]. In recent years, various molecules involved in LTD have been identified using genetically mutated mice.

As mentioned above, motor learning in VOR/OKR and gait is cerebellar dependent, and LTD of parallel fiber-Purkinje cell synapses has been proposed as a basic mechanism supporting cerebellar-dependent motor learning. As climbing fibers transmit information of various error signals, and LTD is a synaptic plasticity triggered by climbing fiber input, LTD could be a motor learning mechanism triggered by an error signal when the motor output deviates from the target trajectory, and that improves motor output. In fact, several studies have reported that animals with LTD deficiency or those who lost climbing fibers show motor learning disabilities. Furthermore, this LTD model can be applied to many motor learning responses other than VOR, OKR or gait.

Metabotropic glutamate receptor subtype 1 (mGluR1) is strongly expressed in Purkinje cells and is essential for inducing LTD at parallel fiber-Purkinje cell synapses [35]. mGluR1 gene knock-out (KO) mice have abnormal innervation of climbing fiber in Purkinje cells and severe impairment in motor learning on rotarod [36]. In addition, mGluR1 KO mice rescued with an mGluR1-transgene under the L7 promoter, which expressed mGluR1 only in Purkinje cells, showed recovered motor learning on the rotarod depending on the amount of mGluR1 expression as well as normal innervation of climbing fibers and LTD at parallel fiber-Purkinje cell synapses [37].

Involvement of Ca^{2+}/calmodulin-dependent protein kinase II (CaMKII) in LTD induction has been reported. In αCaMKII-KO juvenile mice, LTD as well as motor learning in the VOR were impaired [38]. Nitric oxide released from parallel fibers acts through protein phosphatase inhibition in Purkinje cells' dendritic spines enhancing AMPAR phosphorylation [39]. LTD at parallel fiber-Purkinje cell synapses was abolished in neuronal nitric oxide synthase (nNOS) gene KO mice [40], and their motor learning in the OKR was impaired [41]. Similarly, transgenic mice named L7-PKCi mice, with Purkinje cell-selective protein kinase C inactivation, showed a lack of LTD at parallel fiber-Purkinje cell synapses. Training these mice with horizontal visual-vestibular antiphase stimuli for 1 h did not increase the VOR gain [42]. Interestingly, when the same stimuli were repeated for 1.5 h/day for 8 days (mice reared in the dark except during training), VOR gain significantly changed compared to that before training, although not as much as in wild-type mice [43].

The architecture of the cerebellum is often called "crystalline" because it is composed of regular repeated modules, organized uniformly in a number of functional units. Each unit is connected with distinct brain areas in an orderly manner, and local units consist of a relatively small number of cell types. For one Purkinje cell, there are hundreds to thousands of parallel fiber inputs and one climbing fiber input, and this basic unit does not change in any region of the cerebellum. Studies on the cerebellum suggest that these modules behave similarly, processing different information but in similar ways, which raises the hope that studying the function of the cerebellum during one behavior may lead to the comprehension of a general algorithm that may mediate many other forms of motor learning. In fact, the cerebellum has been typically considered uniform in terms of functions, such as synaptic plasticity. Recently, however, even in the seemingly homogeneous cerebellum, the same forms of synaptic plasticity follow different rules in different regions of the cerebellum enabling responses to various situations, such as error correction with different feedback time windows from tens to hundreds of milliseconds.

A local increase in calcium levels is required for LTD induction at parallel fiber-Purkinje cell synapses. A climbing fiber input causes a strong depolarization in a Purkinje cell, resulting in a large increase in calcium concentration in its dendrites via voltage-gated calcium channels, which continued for tens to 200 ms after the climbing fiber input to induce LTD. This means that, in the case of VOR training, if the time lag between the sensory input via parallel fibers corresponding to the vestibular input and the error signal via the climbing fiber is within 200 ms, LTD induction can be established. Suvrathan et al. [44] reported that LTD was induced with a wide timing difference of 0–150 ms between the parallel fiber input and the climbing fiber input in cerebellar vermis slices, conventionally used in previous studies, whereas LTD induction required the climbing fiber input to be almost exactly 120 ms after the parallel fiber input in the flocculus slice. Time shifts of climbing fiber inputs >20 ms could not induce LTD in the flocculus. Similar results were obtained in in vivo behavioral experiments using non-anesthetized monkeys. Interestingly, individual cells were differentially tuned in the vermis and flocculus with the timing of climbing fiber input required for LTD induction. In the cerebellar vermis, Purkinje cells showing tuning characteristics in different time windows were mixed, whereas in the flocculus, most Purkinje cells had the limited tuning characteristics of 120 ms time windows. Although it is not clear why such diversity exists even within the cerebellum, different rules among cerebellar-dependent motor learning seem to exist due to the different strictness of signal inputs, supposed to be quite simple overall, and each possibly selected in an experience-dependent manner.

In parallel fiber-Purkinje cell synapses, not only LTD but also long-term potentiation (LTP), enhancement of synaptic transmission efficiency, occurs. In the absence of climbing fiber input, LTP occurs at parallel fiber-Purkinje cell synapses [45, 46]. During training to increase VOR gain by a combination of out-of-phase head and visual stimuli, there is climbing fiber input during head movement in the ipsilateral direction. Thus, LTD is supposed to occur at parallel fiber-Purkinje cell synapses activated when the head moves in the ipsilateral direction, while LTP occurs at the parallel fiber-Purkinje cell synapses activated when the head moves in the contralateral direction. Conversely, the opposite pattern occurs during training to reduce the VOR gain [47]. Since both presynaptic and postsynaptic LTP of parallel fiber-Purkinje cell synapses have been reported, these LTD and LTP are not exactly the reverse process, but one may mask the other. Schonewille et al. [48] posed that the main aspect of VOR motor learning is LTP rather than LTD, as some genetically manipulated mice, such as PICK1 KO mice, have shown an adaptive learning of VOR gain comparable to normal mice despite their impaired LTD. In addition, another type of synaptic plasticity, rebound potentiation, occurs at synapses between Purkinje cells and basket/stellate cells, inhibitory inter-neurons in the cerebellar cortex, and Purkinje cell depolarization continuously enhances the efficiency of inhibitory synaptic trans-mission. The impairment of VOR adaptive learning has also been associated with a lack of rebound potentiation [49]. Transgenic mice of the GABARAP gene, a GABAA receptor-binding protein, showed normal LTD, but impaired rebound potentiation at inhib-itory interneuron-Purkinje cell synapses and impaired adaptive changes in VOR gain. These results show that LTD is not the only mechanism that triggers cerebellum-dependent motor learning, but multiple mechanisms may work in parallel to establish motor learning, as backup systems compensating for each other.

The cerebellum is essential for the acquisition of adaptive motor learning for gait, VOR, and OKR as mentioned above. Glutamate receptors subtype-δ2 (GluD2) are selectively expressed in Purkinje cells and known to be essential for inducing LTD at parallel fiber-Purkinje cell synapses. Mutant KO mice for GluD2 have a number of deficits related to the cerebellum: reduction of synapses between parallel fiber-Purkinje cells, abnormal innervation of climbing fibers, impaired LTD at parallel fiber-Purkinje cell synapses, cerebellar ataxia, etc. [50]. Furthermore, anti-GluD2 antibody administration into the subarachnoid space over the ver-mis in the cerebellum of wild-type mice induced minor gait ataxia and impaired motor learning on rotarod persisting for 2–4 h after antibody administration [51]. Although such cerebellar deficits disappeared and those mice could stably walk on the rotating rod as well as control mice 6 h after antibody administration, mice receiving anti-GluD2 antibodies could not stay on the rotating rod 24 h after antibody administration without showing gait ataxia and needed to learn again how to walk on the rotarod, whereas control mice remembered. These results suggest that a GluD2-mediated signaling pathway contributes to the walking motor memory.

Boyden et al. [52] elucidated that learned changes acquired after 30 min training for increasing VOR gain were lost by $\geq$50% after 24 h in Ca^{2+}/calmodulin-dependent protein kinase type 4 (CaMKIV) gene-deficient mice, whereas it was retained by $\geq$80% in wild-type mice. Since LTD at parallel fiber-Purkinje cell synapses was normally induced, the efficacy of synaptic transmission returned to normal 90 min later, while it usually lasts >3 h in wild-type mice. These results suggest that the memory of the learning effect acquired by motor learning is retained in the cerebellum for 24 h, through an LTD mechanism as well as learning acquisition. Interestingly, VOR memory retention characteristics also differed by stimulation frequency during training, but there was no difference in memory retention characteristics between CaMKIV gene-mutated mice and wild-type mice for gain-down motor learning.

2   Methods

As mentioned above, synaptic plasticity mechanisms, LTD/LTP at parallel fiber-Purkinje cell synapses, and rebound potentiation at inhibitory interneuron-Purkinje cell synapses have been thought to be candidates for neuronal bases of cerebellum-dependent motor learning based on a number of studies using mutant mice. In cases where a specific gene is deleted during development, however, not only a particular form of synaptic plasticity, such as LTD, but also other potential changes derived from neural circuit formation or their compensatory changes may influence cerebellum-dependent motor learning. To selectively focus on the role of synaptic plasticity in motor learning, a new genetic tool, optogenetics, has recently been developed and applied for in vivo experiments.

In animal experiments in the laboratory, it is possible to create a visual-vestibular interaction environment that cannot occur in nature by designing various visual-vestibular combined stimuli. Although there are criticisms that such artificial stimuli cannot evaluate biologically meaningful functions, it is also an effective method that can reveal the characteristics of the system that cannot be investigated by natural stimuli. Recently, researchers have developed a number of behavioral paradigms to more deeply understand cerebellar functions for motor learning.

2.1 How to Measure Reflexive Eye Movements in Mice

In the case of animal research, the animal's head is fixed on a table, and VOR or OKR is induced by applying a sinusoidal or step-like rotational vestibular or visual stimulus, respectively, and the infrared video camera method or embedded search coil method is employed to measure the eye movements (Fig. 1a, b). In addition, Payne and Raymond recently described [53] a new method for measuring eye movements in head-fixed and freely moving mice, using a small magnet surgically implanted on the eye; changes in the magnet angle as the eye rotated were detected by a magnetic field sensor.

In the case of sinusoidal horizontal or vertical vestibular stimulation, often preferred because of the high linearity of input and output, VOR/OKR can be described by two parameters: gain is the speed ratio of the eyeball and the vestibular/visual stimuli, and phase is the timing difference between the eyeball and these stimuli (Fig. 1c).

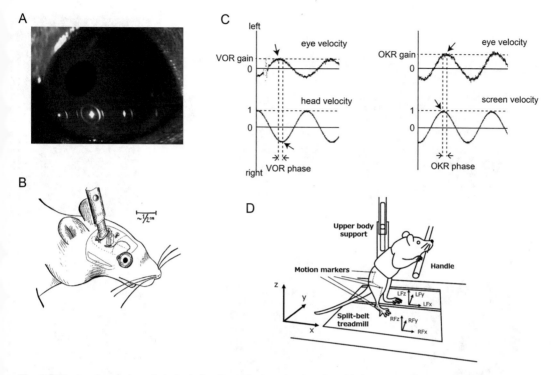

Fig. 1 Behavioral experiments to test simple motor learning in mice. (**a**) An image of mouse eye recorded with infrared CCD camera. (**b**) Illustration of implanted coil on mouse eye for search coil method (provided from Dr. Edward Boyden). (**c**) Gain and Phase in the VOR and OKR. (**d**) Experimental setup for rodent's walking on split-belt treadmill by the hindlimbs. It rested their forepaws on the bar in front of them and wore a harness connected with a flexible beam fixed on the treadmill (provided from Dr. Dai Yanagihara)

2.2 New Approach to Synaptic Plasticity Mechanisms for Cerebellum-Dependent Motor Learning

Optogenetics is a technique in which light-gated ion channels or pumps are expressed in a particular type of cells by genetic engineering and are manipulated by light with a specific wavelength. The development of optogenetic technique has made it possible to manipulate the activity of specific neurons reversibly with precise spatiotemporal control. Using optogenetics, neural activity can be manipulated at the millisecond time scale by using opsins, e.g. light-activated channelrhodopsin-2 (ChR2) or halorhodopsin, and we now can directly connect neural activity and behavioral properties.

There are a couple of ways to control opsin expression. Generally, the most robust and stable expression is achieved in a transgenic mouse line. Using viral vectors for opsin delivery can implement more localized and time-limited expression of opsins around the injection site. In either way, under the control of a specific promotor, expression of the opsin used is only in a genetically defined subpopulation of cells rather than in all cells.

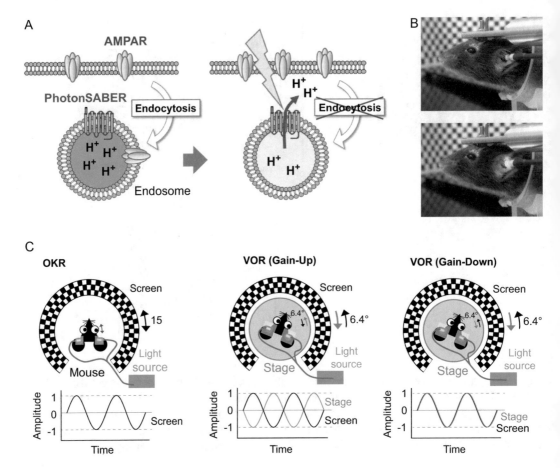

Fig. 2 Photon-SABER Regulates Endosomal pH and motor learning by Light Stimulation. (**a**) Schematic drawing of Photon-SABER. Photon-SABER de-acidifies the endosomal lumen through light-stimulated H+ pump activities and inhibits the endocytosis of AMPA receptor (AMPAR). (**b**) Images of fiberoptic illumination to the bilateral flocculi in the mouse during training. (**c**) Schematic drawing of the experimental setup. The Purkinje cell-specific Photon-SABER mouse, which had optic fibers placed above the bilateral flocculi, was surrounded by a checked-pattern screen. The eye movements made during sinusoidal oscillation of the screen (15°) were monitored for motor learning in the OKR (left). The head was moved 6.4° and a frequency of 0.5 Hz and the screen was moved in the opposite direction from the head for VOR gain increase (middle), and the screen was moved in the same direction as the head for VOR gain decrease (right) (cited from Ref. 56 and modified)

Nguyen-Vu et al. [54] achieved ChR2 expression selectively in Purkinje cells using adeno-associated virus (AAV) carrying loxP-flanked ChR2-EYFP under the Ef1α promoter injected into the cerebellum of L7-Cre mice. Kimpo et al. [55] could induced ChR2 expression selectively in climbing fibers using AAV with ChR2-EYFP under the CamKIIα promoter injected into the IO. Both examined to understand whether Purkinje cells or climbing fibers or both would encode instructive signals to induce motor learning in the VOR by using selective stimulation during training without visual stimuli.

We recently have developed another type of optogenetically-activated system, Photon Sensitive ASR-Based Endocytosis Regulator (Photon-SABER) [56] (Fig. 2a). Internalization of postsynaptic AMPA-type glutamate receptors (AMPA receptors) by activity-dependent and clathrin-mediated endocytosis requires to induce LTD at parallel fiber-Purkinje cell synapses and acidification of the endosomal lumen is essential for the endocytosis. We generated Purkinje cell-specific Photon-SABER transgenic mice to control the induction of LTD during training for motor learning.

2.3 New Training Paradigms for Adaptive Motor Learning in VOR and OKR

In general, cerebellum-dependent motor learning is supposed to be simple compared to other learning systems. There are many variations in VOR motor learning other than increasing or decreasing gain. The ideal eye movement is equal to the movement of the visual stimulus relative to the head. Normal viewing conditions, where the visual stimulus is stationary, require an eye movement equal in velocity to the head movement (gain of 1) and exactly in phase with oppositely directed head movement (defined as a phase of $0°$), and is hence referred to as $\times 1/0°$, or simply $\times 1$. A visual-vestibular stimulus called $\times 2$, with a velocity of the visual stimulus twice in amplitude and perfectly out-of-phase to the head movement, is conventionally used to increase the VOR gain by motor learning. To take another example, the $\times 1/90°$ lead paradigm would require a VOR with a gain of 1 and a phase such that eye velocity leads oppositely directed peak head velocity by $90°$. When visual-vestibular stimuli with a series of timing differences between them are presented to mice, after a short training period, the timing when the eye should turn during head movements can be learned by changing the VOR phase [57, 58] (Fig. 3).

Cross-axis training stimuli provide visual-vestibular stimuli in a combination of horizontal head rotation and vertical visual stimuli or vice versa [59]. In the former, for example, horizontal head rotation induces vertical VOR after several hours of training in monkeys, which normally does not occur [60, 61].

Learning effects observed even under conditions different from those during training is called generalization. We quantitatively compared the stimulus generalization of learned increases and decreases in the VOR gain in mice. Before and after training, VOR was measured by rotating the head and body of the mouse in total darkness at 0.5, 1, 2, and 5 Hz. Changes in VOR performance were induced by pairing head rotations with a moving visual stimulus at a training frequency of 1 Hz to see if different sets of synapses or different kind of plasticity mechanisms are used during increases and decreases in movement amplitude [62].

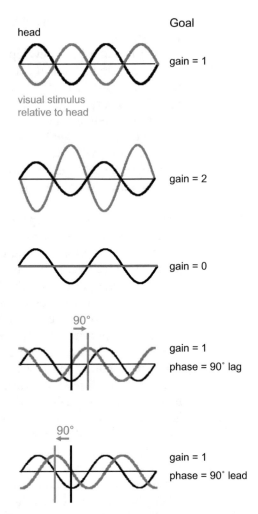

Fig. 3 Protocol for motor learning of VOR gain and phase. The eye movement required to stabilize image motion is equal to the movement of the visual stimulus relative to the head. For example, during ×1/90° lead training, the visual stimulus movement had the same amplitude as head movement but was phase shifted to lead oppositely directed head movement by 90°. The eye movement required to stabilize an image under normal viewing conditions with an earth-stationary visual stimulus is also shown (top)

2.4 New Methods to Approach Cerebellum-Dependent Control in Gait Activity

Although the rotarod test has been known as a relatively well-validated measurement, a detailed description of the kinematics of locomotion, meaning the dynamics of interlimb coordination, can be performed by another measurement, a treadmill test. Recent studies have found that rodents could walk without difficulty on a treadmill [63] (Fig. 1d). To analyze the kinematics of the locomotion pattern, pieces of reflective markers are usually placed on the shaved skin of the hindlimb, femoral head, knee joint, ankle joint, metatarsophalangeal joint, and toe and recorded with a high-speed digital image camera [64].

A split-belt treadmill can also be used to observe the dynamics of interlimb coordination. Moreover, if the disturbance is applied to a certain part with a certain strength at every step (such as altering the speed of one out of four belts), it is possible to investigate the process of predictive, adaptive motor learning with disturbances during gait [65].

3   Results and Discussion

3.1   Synaptic Plasticity Mechanisms for Adaptive Motor Leaning

In cerebellar slices of Photon-SABER mouse [56], LTD was inhibited only when illuminated with yellow light as the synaptic AMPA receptor uptake on endosomes of the Purkinje cell was reversibly inhibited with yellow light illumination, while other characteristics including neural circuit formation were not affected. LTD inhibition by yellow light local illumination of the floccular complex (Fig. 2b), only during training, impaired gain increase in the OKR and VOR (Fig. 4a, b). Interestingly, these mice did not differ from wild-type mice in adaptive learning of VOR gain-down even under yellow light illumination (Fig. 4c). The results observed under directly and reversibly inhibited LTD with light strongly support the model that the neural basis of motor learning for OKR and VOR gain up would be LTD at cerebellar parallel fiber-Purkinje cell synapses. In addition, depending on the direction of changes of VOR gain, increase or decrease, different mechanisms seem to be involved in adaptive motor learning. Furthermore, VOR gain-down may be related to synaptic plasticity triggered by input from the cerebellar lobe to the vestibular nuclei.

3.2   New Aspects of Adaptive Motor Learning in the VOR and OKR

As mentioned above, when visual-vestibular stimuli with a series of timing differences between them are presented to mice, after a short training period, the timing when the eye should turn during head movements can be learned by changing the VOR phase [57]. Interestingly, a training stimulus requiring a VOR phase such that the eye velocity leads to an opposite peak head velocity tended to decrease VOR gain as well as VOR phase lead, and the one the eye is required to be delayed to the head tended to decrease VOR gain and VOR phase lag (Fig. 5a). We reported that mutant mice heterozygous for P/Q-type voltage-gated Ca^{2+} channels exhibited lower ability to change the VOR phase but normal changes in VOR gain [57, 58]. The learning of movements' timing should be as important as learning its amplitude, and, based on clinical studies, the cerebellum seems to be involved in motor learning of movement timing. Since P/Q-type voltage-gated Ca^{2+} channels are highly expressed in the cerebellum and known to control Ca^{2+} levels in Purkinje cells, plasticity mechanisms in the cerebellum related to P/Q signaling may contribute to adaptive motor learning of the VOR phase. Moreover, these results suggest

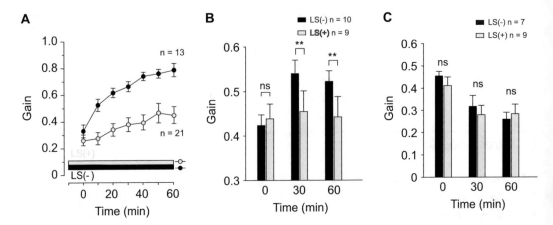

Fig. 4 Fiberoptic bilateral illumination on flocculi inhibits OKR adaptation and VOR gain increase but not VOR gain decrease in Photon-SABER mice. (**a**) The panel shows the time course of OKR adaptation (gain) under light-stimulated LS(+) ($n = 21$ mice) and control LS($-$) ($n = 13$ mice) conditions. (**b**) The time course of VOR gain adaptation in the gain increase protocol. The gain was measured before (0) and 30 and 60 min after training without (LS[$-$], $n = 10$ mice) or with (LS[+], $n = 9$ mice) light stimulation. (**c**) The time course of VOR gain adaptation in the gain decrease protocol. The gain was measured before (0) and 30 and 60 min after training without (LS[$-$], $n = 7$ mice) or with (LS[$-$], $n = 9$ mice) light stimulation. **$p < 0.001$; ns, not significant; two-way ANOVA followed by Bonferroni post hoc test. Data are presented as mean $\pm$ SEM (cited from Ref. 56 and modified)

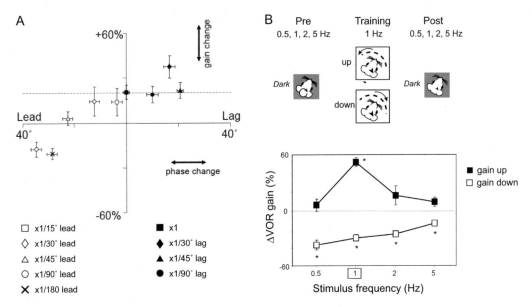

Fig. 5 Changes in VOR gain and phase by VOR phase training and generalization pattern of VOR gain changes in wild-type mice. (**a**) Learned changes in the phase (abscissa) and gain (ordinate) of the VOR in wild-type mice induced by ten different visual-vestibular training paradigms. Error bars indicate standard error. (**b**) Decreases in VOR gain generalize across test frequency more than increases in gain. Changes in VOR gain were induced by a training stimulus frequency of 1 Hz

that learned changes in the temporal dimension and amplitude of movements are differentially regulated through distinct signaling pathways.

Cross-axis training was examined in mice by rotating the visual projection around the roll axis of the head-fixed coordinate frame while maintaining the vestibular stimulus around the horizontal axis, and induced increases in torsional VOR gain as well as vertical, but decreases in horizontal VOR gain [59]. This learned change is considered a manifestation of the spatial coordinate conversion function of the VOR control system. Since the functional directions for sensing angular acceleration formed by the semicircular canals as detectors of vestibular inputs do not completely coincide with the directions of contraction/extension of the six extraocular muscles as effectors, a coordinate transformation is performed to generate an appropriate eye movement to match the visual stimulus during the head movement detected by the semicircular canals in the three-dimensional space.

In mice, after training to decrease VOR gain, we found that learned changes generalized to vestibular stimulus frequencies other than the frequency of the visual-vestibular stimulus used during training [62]. In contrast, after training to increase VOR gain, significant learned changes occurred only at the frequency of the visual-vestibular stimulus used in training, and the learning effect became smaller as the distance from the training frequency increased (Fig. 5b). Our results suggest that increases and decreases in VOR gain are not exact inverses at the circuit level. At one or more sites, the plasticity mechanisms supporting decreases in VOR gain must be less synapse-specific, or affect neurons more broadly tuned for head rotation frequency, than the mechanisms supporting increases in gain.

3.3 Cerebellum-Dependent Control on Gait

Darmohray et al. [65] demonstrated *pcd* mutant mice, a recessive mutant with moderate ataxia resulting from complete postnatal loss of Purkinje cells and subsequent partial loss of cerebellar granule cells, were impaired at locomotor adaptation trained by asymmetries in both space and time of step length during walking on the split-belt treadmill.

The *ho15J* mouse is a mutant mouse with an autosomal recessive genetic lineage, lacking exon 2 of the gene encoding GluD2. In those mice, the number of parallel fiber-Purkinje cell synapses is reduced to approximately 40% of that in wild-type mice and they show gait ataxia [66]. Although there were no differences in general temporal parameters of locomotion between *ho15J* and wild-type mice, abnormal excessive toe elevation and severe hyperflexion of the ankle during the swing phase characterize gait ataxia in *ho15J* mice by kinematic analysis of hindlimb movements during treadmill locomotion.

Takeuchi et al. [67] also found the mice with a null mutation of the gene for the cerebellin 1 precursor protein (Cbln1), predominately produced and secreted from cerebellar granule cells, showed 80% reduction in the number of parallel fiber-Purkinje cell synapses compared with wild-type mice and revealed abnormal hindlimb movements that were characterized by excessive toe elevation during the swing phase, and by severe hyperflexion of the ankles and knees. When recombinant Cbln1 protein was injected into the cerebellum of cbln1-null mice, the step cycle, stance phase durations and the angular excursions of the knee during a cycle period were partially rescued. These results suggest that signal transduction from the spinal cord into the cerebellum through the spinocerebellar loop, including neural circuits of mossy fiber-parallel fiber-Purkinje cells, is critical for online gait control as well as moto learning on gait.

4   Conclusions

In this chapter, the characteristics of simple motor systems, two reflexive eye movements, VOR and OKR, and gait activity are described. The abilities of eye movements and gait to adapt to new environments by motor learning as well as online control are necessary for the regulation of physical movements. The cerebellum plays a key role in motor learning, a process that improves motor output by repeated training, and in motor memory that maintains those learned changes for a certain period of time. Developing approaches using simple motor systems in mice collate the power of genetics, molecular biology, neural circuits, and behavior to analyze how the cerebellum contributes to more complex behaviors and, in the future, are expected to help develop training methods for athletic performance improvement and rehabilitation of motor dysfunction.

Acknowledgements

I would like to thank Dr. Jennifer Raymond for providing me an opportunity to contribute a chapter on this volume. I also thank Drs. Michisuke Yuzaki, Shinji Matsuda, Wataru Kakegawa, Dai Yanagihara and Edward Boyden for kindly providing me illustrations presented in this chapter; Editage (www.editage.com) for English language editing; and Mika, Jukana, Rinan Kato and Botan for encouragement. I am supported by JSPS KAKENHI Grant Number JP18KK0286.

References

1. Cousins S, Kaski D, Cutfield N, Seemungal B, Golding JF, Gresty M, Glasauer S, Bronstein AM (2013) Vestibular perception following acute unilateral vestibular lesions. PLoS One 8(5):e61862. https://doi.org/10.1371/journal.pone.0061862

2. Buttner N, Geschwind D, Jen JC, Perlman S, Pulst SM, Baloh RW (1998) Oculomotor phenotypes in autosomal dominant ataxias. Arch Neurol 55(10):1353–1357

3. Hirata Y, Nishiyama J, Kinoshita S (2009) Detection and prediction of drowsiness by reflexive eye movements. Conf Proc IEEE Eng Med Biol Soc 2009:4015–4018. https://doi.org/10.1109/iembs.2009.5333504

4. Lalonde R, Strazielle C (2019) Motor performances of spontaneous and genetically modified mutants with cerebellar atrophy. Cerebellum 18(3):615–634. https://doi.org/10.1007/s12311-019-01017-5

5. Arshavsky YI, Gelfand IM, Orlovsky GN (1983) The cerebellum and control of rhythmical movements. Trends Neurosci 6:417–422. https://doi.org/10.1016/0166-2236(83)90191-1

6. Delcomyn F (1980) Neural basis of rhythmic behavior in animals. Science 210(4469):492–498. https://doi.org/10.1126/science.7423199

7. Grillner S (1975) Locomotion in vertebrates: central mechanisms and reflex interaction. Physiol Rev 55(2):247–304. https://doi.org/10.1152/physrev.1975.55.2.247

8. Stecina K, Fedirchuk B, Hultborn H (2013) Information to cerebellum on spinal motor networks mediated by the dorsal spinocerebellar tract. J Physiol 591(22):5433–5443. https://doi.org/10.1113/jphysiol.2012.249110

9. Fortier PA, Smith AM, Rossignol S (1987) Locomotor deficits in the mutant mouse, Lurcher. Exp Brain Res 66(2):271–286. https://doi.org/10.1007/BF00243304

10. Boyden ES, Katoh A, Raymond JL (2004) Cerebellum-dependent learning: the role of multiple plasticity mechanisms. Annu Rev Neurosci 27:581–609. https://doi.org/10.1146/annurev.neuro.27.070203.144238

11. Ito M (1982) Cerebellar control of the vestibulo-ocular reflex--around the flocculus hypothesis. Annu Rev Neurosci 5:275–296

12. Barmack NH, Pettorossi VE (1985) Effects of unilateral lesions of the flocculus on optokinetic and vestibuloocular reflexes of the rabbit.

J Neurophysiol 53(2):481–496. https://doi.org/10.1152/jn.1985.53.2.481

13. Katoh A, Kitazawa H, Itohara S, Nagao S (1998) Dynamic characteristics and adaptability of mouse vestibulo-ocular and optokinetic response eye movements and the role of the flocculo-olivary system revealed by chemical lesions. Proc Natl Acad Sci U S A 95(13):7705–7710. https://doi.org/10.1073/pnas.95.13.7705

14. Lisberger SG, Miles FA, Zee DS (1984) Signals used to compute errors in monkey vestibuloocular reflex: possible role of flocculus. J Neurophysiol 52(6):1140–1153

15. Nagao S (1983) Effects of vestibulocerebellar lesions upon dynamic characteristics and adaptation of vestibulo-ocular and optokinetic responses in pigmented rabbits. Exp Brain Res 53(1):36–46

16. Rambold H, Churchland A, Selig Y, Jasmin L, Lisberger SG (2002) Partial ablations of the flocculus and ventral paraflocculus in monkeys cause linked deficits in smooth pursuit eye movements and adaptive modification of the VOR. J Neurophysiol 87(2):912–924. https://doi.org/10.1152/jn.00768.2000

17. Luebke AE, Robinson DA (1994) Gain changes of the cat's vestibulo-ocular reflex after flocculus deactivation. Exp Brain Res 98(3):379–390

18. Kano M, Hashimoto K, Chen C, Abeliovich A, Aiba A, Kurihara H, Watanabe M, Inoue Y, Tonegawa S (1995) Impaired synapse elimination during cerebellar development in PKC gamma mutant mice. Cell 83(7):1223–1231. https://doi.org/10.1016/0092-8674(95)90147-7

19. Kimpo RR, Raymond JL (2007) Impaired motor learning in the vestibulo-ocular reflex in mice with multiple climbing fiber input to cerebellar Purkinje cells. J Neurosci 27(21):5672–5682. https://doi.org/10.1523/JNEUROSCI.0801-07.2007

20. Cohen H, Cohen B, Raphan T, Waespe W (1992) Habituation and adaptation of the vestibuloocular reflex: a model of differential control by the vestibulocerebellum. Exp Brain Res 90(3):526–538

21. Haddad GM, Demer JL, Robinson DA (1980) The effect of lesions of the dorsal cap of the inferior olive on the vestibulo-ocular and optokinetic systems of the cat. Brain Res 185(2):265–275. https://doi.org/10.1016/0006-8993(80)91067-7

22. Tempia F, Dieringer N, Strata P (1991) Adaptation and habituation of the vestibulo-ocular reflex in intact and inferior olive-lesioned rats. Exp Brain Res 86(3):568–578. https://doi.org/10.1007/BF00230530

23. Yakushin SB, Reisine H, Buttner-Ennever J, Raphan T, Cohen B (2000) Functions of the nucleus of the optic tract (NOT). I. Adaptation of the gain of the horizontal vestibulo-ocular reflex. Exp Brain Res 131(4):416–432. https://doi.org/10.1007/s002219900303

24. Cendelin J, Korelusova I, Vozeh F (2008) The effect of repeated rotarod training on motor skills and spatial learning ability in Lurcher mutant mice. Behav Brain Res 189(1):65–74. https://doi.org/10.1016/j.bbr.2007.12.013

25. Yoshida T, Katoh A, Ohtsuki G, Mishina M, Hirano T (2004) Oscillating Purkinje neuron activity causing involuntary eye movement in a mutant mouse deficient in the glutamate receptor delta2 subunit. J Neurosci 24(10):2440–2448. https://doi.org/10.1523/JNEUROSCI.0783-03.2004

26. McElligott JG, Beeton P, Polk J (1998) Effect of cerebellar inactivation by lidocaine microdialysis on the vestibuloocular reflex in goldfish. J Neurophysiol 79(3):1286–1294

27. Nagao S, Kitazawa H (2003) Effects of reversible shutdown of the monkey flocculus on the retention of adaptation of the horizontal vestibulo-ocular reflex. Neuroscience 118(2):563–570

28. Pastor AM, de la Cruz RR, Baker R (1994) Cerebellar role in adaptation of the goldfish vestibuloocular reflex. J Neurophysiol 72(3):1383–1394

29. Partsalis AM, Zhang Y, Highstein SM (1995) Dorsal Y group in the squirrel monkey. II. Contribution of the cerebellar flocculus to neuronal responses in normal and adapted animals. J Neurophysiol 73(2):632–650. https://doi.org/10.1152/jn.1995.73.2.632

30. Shutoh F, Ohki M, Kitazawa H, Itohara S, Nagao S (2006) Memory trace of motor learning shifts transsynaptically from cerebellar cortex to nuclei for consolidation. Neuroscience 139(2):767–777. https://doi.org/10.1016/j.neuroscience.2005.12.035

31. Albus J (1971) A theory of cerebellar function. Math Biosci 10:25–61

32. Marr D (1969) A theory of cerebellar cortex. J Physiol 202(2):437–470

33. Ekerot CF, Kano M (1985) Long-term depression of parallel fibre synapses following stimulation of climbing fibres. Brain Res 342(2):357–360

34. Hirano T (1990) Effects of postsynaptic depolarization in the induction of synaptic depression between a granule cell and a Purkinje cell in rat cerebellar culture. Neurosci Lett 119(2):145–147

35. Aiba A, Kano M, Chen C, Stanton ME, Fox GD, Herrup K, Zwingman TA, Tonegawa S (1994) Deficient cerebellar long-term depression and impaired motor learning in mGluR1 mutant mice. Cell 79(2):377–388

36. Kano M, Hashimoto K, Kurihara H, Watanabe M, Inoue Y, Aiba A, Tonegawa S (1997) Persistent multiple climbing fiber innervation of cerebellar Purkinje cells in mice lacking mGluR1. Neuron 18(1):71–79. https://doi.org/10.1016/s0896-6273(01)80047-7

37. Ichise T, Kano M, Hashimoto K, Yanagihara D, Nakao K, Shigemoto R, Katsuki M, Aiba A (2000) mGluR1 in cerebellar Purkinje cells essential for long-term depression, synapse elimination, and motor coordination. Science 288(5472):1832–1835. https://doi.org/10.1126/science.288.5472.1832

38. Hansel C, de Jeu M, Belmeguenai A, Houtman SH, Buitendijk GH, Andreev D, De Zeeuw CI, Elgersma Y (2006) alphaCaMKII is essential for cerebellar LTD and motor learning. Neuron 51(6):835–843. https://doi.org/10.1016/j.neuron.2006.08.013

39. Lev-Ram V, Makings LR, Keitz PF, Kao JP, Tsien RY (1995) Long-term depression in cerebellar Purkinje neurons results from coincidence of nitric oxide and depolarization-induced Ca^{2+} transients. Neuron 15(2):407–415. https://doi.org/10.1016/0896-6273(95)90044-6

40. Lev-Ram V, Nebyelul Z, Ellisman MH, Huang PL, Tsien RY (1997) Absence of cerebellar long-term depression in mice lacking neuronal nitric oxide synthase. Learn Mem 4(1):169–177. https://doi.org/10.1101/lm.4.1.169

41. Katoh A, Kitazawa H, Itohara S, Nagao S (2000) Inhibition of nitric oxide synthesis and gene knockout of neuronal nitric oxide synthase impaired adaptation of mouse optokinetic response eye movements. Learn Mem 7(4):220–226

42. De Zeeuw CI, Hansel C, Bian F, Koekkoek SK, van Alphen AM, Linden DJ, Oberdick J (1998) Expression of a protein kinase C inhibitor in Purkinje cells blocks cerebellar LTD and adaptation of the vestibulo-ocular reflex. Neuron 20(3):495–508

43. Van Alphen AM, De Zeeuw CI (2002) Cerebellar LTD facilitates but is not essential for

long-term adaptation of the vestibulo-ocular reflex. Eur J Neurosci 16(3):486–490

44. Suvrathan A, Payne HL, Raymond JL (2016) Timing rules for synaptic plasticity matched to behavioral function. Neuron 92(5):959–967. https://doi.org/10.1016/j.neuron.2016. 10.022

45. Coesmans M, Weber JT, De Zeeuw CI, Hansel C (2004) Bidirectional parallel fiber plasticity in the cerebellum under climbing fiber control. Neuron 44(4):691–700. https://doi.org/10. 1016/j.neuron.2004.10.031

46. Lev-Ram V, Mehta SB, Kleinfeld D, Tsien RY (2003) Reversing cerebellar long-term depression. Proc Natl Acad Sci U S A 100(26):15989–15993. https://doi.org/10. 1073/pnas.2636935100

47. Ke MC, Guo CC, Raymond JL (2009) Elimination of climbing fiber instructive signals during motor learning. Nat Neurosci 12(9):1171–1179. https://doi.org/10.1038/ nn.2366

48. Schonewille M, Gao Z, Boele HJ, Veloz MF, Amerika WE, Simek AA, De Jeu MT, Steinberg JP, Takamiya K, Hoebeek FE, Linden DJ, Huganir RL, De Zeeuw CI (2011) Reevaluating the role of LTD in cerebellar motor learning. Neuron 70(1):43–50. https://doi. org/10.1016/j.neuron.2011.02.044

49. Tanaka S, Kawaguchi SY, Shioi G, Hirano T (2013) Long-term potentiation of inhibitory synaptic transmission onto cerebellar Purkinje neurons contributes to adaptation of vestibulo-ocular reflex. J Neurosci 33(43):17209–17220. https://doi.org/10. 1523/jneurosci.0793-13.2013

50. Kashiwabuchi N, Ikeda K, Araki K, Hirano T, Shibuki K, Takayama C, Inoue Y, Kutsuwada T, Yagi T, Kang Y et al (1995) Impairment of motor coordination, Purkinje cell synapse formation, and cerebellar long-term depression in GluR delta 2 mutant mice. Cell 81(2):245–252. https://doi.org/10. 1016/0092-8674(95)90334-8

51. Hirai H, Launey T, Mikawa S, Torashima T, Yanagihara D, Kasaura T, Miyamoto A, Yuzaki M (2003) New role of delta2-glutamate receptors in AMPA receptor trafficking and cerebellar function. Nat Neurosci 6(8):869–876. https://doi.org/10.1038/nn1086

52. Boyden ES, Katoh A, Pyle JL, Chatila TA, Tsien RW, Raymond JL (2006) Selective engagement of plasticity mechanisms for motor memory storage. Neuron 51(6):823–834. https://doi.org/10.1016/j. neuron.2006.08.026

53. Payne HL, Raymond JL (2017) Magnetic eye tracking in mice. elife 6:e29222. https://doi. org/10.7554/eLife.29222

54. Nguyen-Vu TD, Kimpo RR, Rinaldi JM, Kohli A, Zeng H, Deisseroth K, Raymond JL (2013) Cerebellar Purkinje cell activity drives motor learning. Nat Neurosci 16(12):1734–1736. https://doi.org/10. 1038/nn.3576

55. Kimpo RR, Rinaldi JM, Kim CK, Payne HL, Raymond JL (2014) Gating of neural error signals during motor learning. Elife 3:e02076. https://doi.org/10.7554/eLife.02076

56. Kakegawa W, Katoh A, Narumi S, Miura E, Motohashi J, Takahashi A, Kohda K, Fukazawa Y, Yuzaki M, Matsuda S (2018) Optogenetic control of synaptic AMPA receptor endocytosis reveals roles of LTD in motor learning. Neuron 99(5):985–998. e986. https://doi.org/10.1016/j.neuron.2018. 07.034

57. Katoh A, Chapman PJ, Raymond JL (2008) Disruption of learned timing in P/Q calcium channel mutants. PLoS One 3(11):e3635. https://doi.org/10.1371/journal.pone. 0003635

58. Katoh A, Jindal JA, Raymond JL (2007) Motor deficits in homozygous and heterozygous p/q-type calcium channel mutants. J Neurophysiol 97(2):1280–1287. https://doi.org/10.1152/ jn.00322.2006

59. Hubner PP, Khan SI, Migliaccio AA (2014) Velocity-selective adaptation of the horizontal and cross-axis vestibulo-ocular reflex in the mouse. Exp Brain Res 232(10):3035–3046. https://doi.org/10.1007/s00221-014-3988-8

60. Fukushima K, Sato T, Fukushima J, Kurkin S (2000) Cross axis VOR induced by pursuit training in monkeys: further properties of adaptive responses. Arch Ital Biol 138(1):49–62

61. Peterson BW, Baker JF, Houk JC (1991) A model of adaptive control of vestibuloocular reflex based on properties of cross-axis adaptation. Ann N Y Acad Sci 627:319–337

62. Kimpo RR, Boyden ES, Katoh A, Ke MC, Raymond JL (2005) Distinct patterns of stimulus generalization of increases and decreases in VOR gain. J Neurophysiol 94(5):3092–3100. https://doi.org/10.1152/jn.00048.2005

63. Leblond H, L'Esperance M, Orsal D, Rossignol S (2003) Treadmill locomotion in the intact and spinal mouse. J Neurosci 23(36):11411–11419

64. Fujiki S, Aoi S, Funato T, Sato Y, Tsuchiya K, Yanagihara D (2018) Adaptive hindlimb split-

belt treadmill walking in rats by controlling basic muscle activation patterns via phase resetting. Sci Rep 8(1):17341. https://doi.org/10.1038/s41598-018-35714-8

65. Darmohray DM, Jacobs JR, Marques HG, Carey MR (2019) Spatial and temporal locomotor learning in mouse cerebellum. Neuron 102(1):217–231. e214. https://doi.org/10.1016/j.neuron.2019.01.038

66. Takeuchi E, Sato Y, Miura E, Yamaura H, Yuzaki M, Yanagihara D (2012) Characteristics

of gait ataxia in delta2 glutamate receptor mutant mice, ho15J. PLoS One 7(10): e47553. https://doi.org/10.1371/journal.pone.0047553

67. Takeuchi E, Ito-Ishida A, Yuzaki M, Yanagihara D (2018) Improvement of cerebellar ataxic gait by injecting Cbln1 into the cerebellum of cbln1-null mice. Sci Rep 8(1):6184. https://doi.org/10.1038/s41598-018-24490-0

Chapter 7

Activity-Dependent Chromatin Mechanisms in Cerebellar Motor Learning

Pamela Valnegri, Tomoko Yamada, and Yue Yang

Abstract

Neural circuit activity representing sensorimotor experiences trigger molecular mechanisms that drive long-lasting changes in brain circuits underlying learning and memory. Recent advancements in molecular genetics have led to development of rich toolboxes, e.g. optogenetics and CRISPR, that enable precise temporal and cell type-specific control of neural circuit activity and downstream activity-dependent mechanisms. One such molecular mechanism, the organization of three-dimensional (3D) genome architecture, has emerged as a powerful regulator of rapid and coordinated gene expression in response to neural circuit activity. Here, we describe how to perform optogenetic stimulation of granule neurons at the input layer of the cerebellar cortex in mice and how to profile activity-dependent changes in neuronal genome architecture. In addition, we will discuss how to genetically knock out chromatin regulators specifically in granule neurons in adult mice to study the functions of genome organization in activity-dependent gene expression and cerebellar-dependent motor learning.

Key words Optogenetics, CRISPR-Cas9 genetics, AAV injection, Granule neurons, Genome architecture, Gene expression, Next-generation sequencing, Anterior dorsal cerebellar vermis, Delay tactile startle conditioning

---

1 Introduction

The ability to learn in response to sensorimotor experience is critical for organisms to navigate and adapt to their environment. Activity-dependent programs of gene expression play critical roles in the brain to stably modify neural circuits and generate adaptive behavior [1, 2]. These genetic programs alter the cell state of neurons, including increasing their intrinsic excitability to facilitate co-allocation of neuronal ensembles during the formation of memories [3, 4]. Additional genetic programs regulate late-phase synaptic plasticity to drive memory consolidation [5–8].

The nuclear mechanisms regulating activity-dependent gene expression have been the subject of intensive study [2, 9–11]. An emerging principle in chromatin biology is that genes and their

Roy V. Sillitoe (ed.), *Measuring Cerebellar Function*, Neuromethods, vol. 177, https://doi.org/10.1007/978-1-0716-2026-7_7,

regulatory elements in the genome are organized in a 3D, hierarchical manner. The folding of the genome at multiple spatial scales plays important roles in controlling genetic programs that are necessary for organismal development [12, 13]. However, we are just beginning to understand the genome architecture configurations of diverse neurons in the brain, including how these configurations are modified by neural circuit activity.

As a critical site for nonconscious motor learning, such as riding a bicycle or playing the piano, the cerebellum offers opportunities to bring novel insights in our understanding of fundamental processes in the brain that regulate memory encoding and storage. Granule neurons in the cerebellum are an ideal system to study the molecular mechanisms underlying neural circuit plasticity and learning and memory due to their abundance for biochemical interrogation and well-defined neural circuit architecture. These neurons are located at the input layer of the cerebellar cortex and encode sensorimotor representations of the environment as well as the internal state of the animal [14–19]. Beyond local organization within the cerebellar cortex, topographic organization of cerebellar circuits across different lobules is important for a wide-range of sensory, motor, and cognitive tasks [20, 21].

Anterior dorsal cerebellar vermis (ADCV) is the superficially located region in lobule IV/V that has been implicated in body posture and positioning [22, 23] and is amenable to optical approaches to monitor neural circuit physiology and actuate changes in neural circuit activity. We recently identified key functions for ADCV in a novel associative learning paradigm called delay tactile startle conditioning [18]. We subsequently performed biochemical profiling of microdissected ADCV to elucidate the activity-regulated nuclear mechanisms including changes in genome architecture in granule neurons in mice. These experimental approaches provide an integrated view on how sensorimotor experience activates granule neurons in the adult cerebellum to reshape genome organization and thereby orchestrates transcriptional programs important for memory (Fig. 1). In this chapter, we will describe methods to probe the relationships between patterned neural circuit activity, alterations in chromatin structure and gene expression, and mouse behavior. We will cover mouse surgery, optogenetic stimulation, and viral vector delivery techniques, as well as approaches to set up our cerebellar-dependent motor learning paradigm and biochemically profile the genome-wide chromatin and transcriptome landscapes from the relevant cerebellar areas.

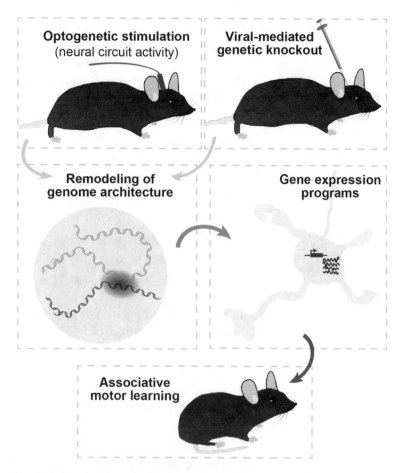

Fig. 1 Summary of methods presented in this chapter. Techniques include surgical approaches to implant a fiber-optic cannula for optogenetic stimulation and to inject adeno-associated viruses for CRISPR-mediated gene silencing. Following neural circuit activation or genetic knockout, additional techniques described include biochemical profiling of genome architecture and the transcriptome. Furthermore, mice subjected to molecular genetics approaches can be tested in a new cerebellar-dependent associative motor learning paradigm

---

2 Materials

2.1 Equipment and Tools

2.1.1 Surgery

1. 150 W fiber-optic halogen illuminator (Amscope, HL-150-B).

2. Binocular articulating arm pillar clamp stereo zoom microscope (Amscope, SM-8BZ-144S).

3. Digital microtorque (Nail Drill, DM45 SETB).

4. Stereotaxic frame and mouse adaptor (Stoelting, 51730 and 51625).

5. Cannula holder (Kopf, 1766-AP).

6. Glass beads sterilizer (Fisher Scientific, 14-955-341).

7. Heating pad (Wuhostam).

8. Standard surgical tools, including forceps, micro scalpel, extra fine scissor, mini-screwdriver, cotton-tipped applicators.

9. C&B Metabond Quick Cement System (Parkell Inc., S398, S399, S371, S379, and S387).

10. Depilatory cream.

11. Eye ointment.

12. Headplates (Emachineshop.com, custom machined from stainless steel, $1'' \times 5/16'' \times 1/32''$).

13. Screws (Amazon, AMS120-1B, 000-120 $\times$ 1/16 SL BIND MS SS).

14. Ball mill, carbide (Circuitmedic, 115-6050).

15. Nylon surgical sutures (5-0 sterile nylon).

16. Vetbond tissue adhesive (3M Corp).

2.1.2 Optogenetic Stimulation

1. Mono fiber-optic cannula (Doric Lenses, MFC_200/260-0.22_0.5 mm_ZF2.5_FLT).

2. Diode blue 450 nm laser (Opto Engine, 200 mW, 3% power stability).

3. Non-contact receptacle style laser to multimode fiber coupler (Oz Optics, HPUC-23AF-450-M-13.9AS-11).

4. Ø200 μm core, 0.22 NA ferrule patch cable (Thorlabs, M80L01).

5. Acousto-optic modulator (Quanta Tech, for control of DPSS lasers but not required for diode lasers).

6. Arduino Due board.

7. Photodiode sensor and optical power console (Thorlabs, S130C and PM100A).

2.1.3 Viral Delivery

1. Nanoject II Auto-Nanoliter Injector (Drummond).

2. Glass Pipette Puller (Sutter Instruments).

2.1.4 Delay Tactile Startle Conditioning Apparatus

1. Equipment previously described in [18, 24, 25] with modifications detailed in Subheading 3.3.

2.1.5 Tissue Collection and Biochemical Procedures

1. 2 mL glass dounce homogenizer (Fisher Scientific, K8853000002).

2. Bioruptor Pico sonication system (Diagenode).

3. Thermomixer (Eppendorf, 5382000023).

4. 2100 Bioanalyzer (Agilent).

5. Qubit 4 Fluorometer (Thermo Fisher Scientific).

2.1.6 Hi-C or PLAC-Seq, Additional Reagents Described in [26, 27]

1. Agilent High Sensitivity DNA Kit (Agilent, 5067-4626).

2. NEBNext Ultra™ II DNA Library Prep Kit (New England Biolab, E7645S).

3. NEBNext® Multiplex Oligos for Illumina, Index Primers Set 1 (New England Biolab, E7335S).

4. QIAquick PCR Purification Kit (Qiagen, 28104).

2.1.7 RNA-Seq

1. Trizol (Thermo Fisher Scientific, 15596026).

2. NEBNext rRNA Depletion Kit (New England Biolabs, E6310S).

3. NEBNext Ultra™ II Directional RNA Library Prep Kit for Illumina (New England Biolabs, E7760S).

4. Illumina NextSeq 550 platform.

2.2 Mouse Genetic Lines

For optogenetic stimulation, LSL-ChR2(H134R)-YFP (Ai32) mice [28] harboring the channelrhodopsin-2 (H134R) transgene downstream of a floxed STOP cassette are bred with Math1-Cre driver mice [29] to generate Ai32-GC mice expressing ChR2 in granule neurons in the cerebellar cortex. For viral delivery of gRNA in genetic knockout experiments, LSL-Cas9-EGFP mice [30] are bred with GABA(A)Rα6-Cre driver mice [31] to generate Cas9-EGFP-GC mice expressing Cas9-EGFP specifically in granule neurons. LSL-ChR2(H134R)-YFP, Math1-Cre, and LSL-Cas9-EGFP mouse lines are purchased from Jackson laboratory. Both male and female mice are standardly used.

2.3 General Surgical Procedures

Mice should be at least 5 weeks old at the time of surgery and at most 6 months old. Mice are anesthetized with ketamine/xylazine (100/10 mg/kg intraperitoneal) and artificial tear ointment is generously applied to the eyes to prevent drying. Carprofen (5 mg/kg subcutaneous) is given to mice undergoing viral injection to reduce brain inflammation. Depilatory cream or an electric shaver is used to remove hair at the surgical site from the frontal bone to the occipital bone. The surgical area is sterilized by wiping the skin with three alternating swipes of 70% ethanol and betadine. The scalp overlaying the skull is removed and lidocaine HCl/epinephrine (1%/1:100,000) solution may be topically applied to provide local analgesia and to reduce bleeding. Under a stereo microscope, the fascia is removed and the skull surface is dried.

Mice first undergo headplate implantation, as this facilitates head restraint during subsequent surgical procedures (*see* Subheadings 2.4 and 2.5), while also permitting mice to freely locomote on a cylindrical treadmill while head-fixed during optogenetic stimulation (*see* Subheading 3.1) or delay tactile startle conditioning (*see* Subheading 3.3). A custom headplate with a center hole is placed over bregma [25]. To strengthen attachment between the headplate and the skull, the skull is thinned 0.4 mm in depth and 0.7 mm in diameter at the center of the headplate and embedded with a small screw. Metabond cement is applied to secure the head-plate and implanted screw to the skull. The anterior location of the headplate leaves more caudal skull areas overlying the cerebellum unobstructed for cannula implantation or viral injection.

Following additional surgical procedures detailed below, saline solution (37 °C, 0.3–0.5 mL subcutaneous) is provided to replenish the animal's fluids and maintain its weight, and buprenorphine (0.5–1.0 mg/kg subcutaneous) is given to minimize inflammation. Mice are kept on a heat pad and monitored until they fully recover from surgery. Mice should recover for at least 5 days following surgery as brain inflammation subsides before behavioral experiments.

2.4 Fiber-Optic Cannula Implantation

For fiber-optic cannula implantation, the animal is placed in a custom stereotaxic device. The headplate is angled as needed to position the mouse's head and fully expose the skull above the cerebellum. Under microscope guidance, the skin above the occipital bone is resected and a 0.7 mm diameter hole through the skull over the anterior cerebellum is drilled, with care taken not to break through the dura. A mono fiber-optic cannula (0.22 NA, 200 μm core) projecting 0.5 mm below the skull surface is stereotaxically implanted and secured to the skull using Metabond cement. Because skull thickness is ~0.6–0.7 mm, the tip of the cannula may be situated above the dura without damaging it (Fig. 2a).

2.5 Viral Injection

Injection micropipettes with a tip diameter of 10–20 μm are pulled from 3.5" Drummond glass micropipettes using a pipette puller. The micropipette is back filled with mineral oil and placed on the wire plunger of a Nanoject II microinjection system, which is mounted to a stereotaxic frame. After emptying the micropipette, it is filled with solution containing virus for multiple animals. During mouse surgery, the skin above the occipital bone is resected and a 0.7 mm diameter hole is made in the skull overlying the cerebellar region of interest under a stereo microscope. Adeno-associated virus (AAV) is injected at different depths below the dura surface. After the injection, the scalp is sutured using surgical nylon sutures in an interrupted pattern for skin closure and glued with Vetbond.

3 Methods

3.1 Optogenetic Stimulation of Granule Neurons in ADCV

Optogenetic approaches can be used to study mechanisms that regulate neural circuit function and behavior. For example, optogenetic activation of granule neurons in ADCV revealed critical roles of genome architecture regulation in vivo for delay tactile

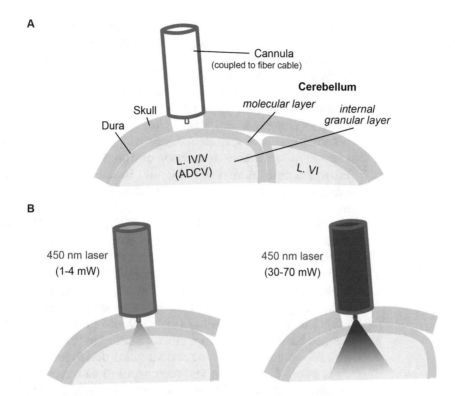

Fig. 2 Fiber-optic cannula implantation. (**a**) The fiber-optic cannula is composed of a ferrule that rests snugly on the skull surface above ADCV and an optical fiber that projects 0.5 mm facing the dorsal surface of ADCV. A fiber cable that relays the laser beam may be mated with the cannula using a mating or interconnect sleeve (not shown). (**b**) Schematic depicting how laser power at the tip of the optical fiber corresponds to activation of granule neuron cellular compartments located at different depths in ADCV. Low laser power (~1–4 mW) triggers selective activation of parallel fibers in the molecular layer. Higher laser power (~30–70 mW) induces depolarization of granule neuron cell somas down to a depth of ~1 mm in ADCV

startle conditioning [18]. Ai32-GC mice, as described above, selectively express channelrhodopsin-2 fused to YFP in granule neurons and a subset of mossy fibers in the cerebellar cortex. To confirm the expression of the channelrhodopsin-2 transgene, a convenient method is to visualize the bright YFP fluorescence that is found throughout the cerebellar cortex, but largely is absent in other brain areas.

A 450 nm diode laser is used to activate channelrhodopsin-2 rapidly and transiently. Alternatively, a DPSS laser (e.g. 473 nm) that is closer to the maximal excitation wavelength of some variants of channelrhodopsin-2 may be used in conjunction with an acousto-optic or electro-optic device to switch on and off the beam. The advantage of diode lasers is that the beam can be modulated with sub-millisecond precision without adding expensive shutter systems. The laser beam is coupled to a multimode fiber cable, with ~50–90% coupling efficiency, depending on the properties of the beam. An Arduino Due board connected to the laser power supply for diode lasers or to the shutter controller for DPSS lasers is used to trigger the pattern of photostimulation.

Optogenetic stimulation protocols should induce neuronal activity that mimic physiological patterns observed in vivo. When the mouse is at rest, granule cells are mostly inactive, producing low numbers of action potentials [32]. During mouse locomotion, excitatory mossy fiber input evokes sparse high-frequency action potential bursts in granule neurons with ~12 spikes/burst and an average instantaneous frequency of 106 Hz. Thus, the average burst duration in vivo is ~100 ms. In whole-cell recordings of granule neurons in acute cerebellar slices from Ai32-GC mice, we found that 100 ms of optogenetic stimulation drives granule neuron burst firing at ~90 Hz during this interval [18]. However, optical stimuli should be pulsed rather than continuously delivered, because the latter may lead to depolarization block of bursting and also heating of brain tissue [33]. The slow closing rate of channelrhodopsin-2 (for H134R, τ is ~17.9 ms) [34] ensures persistent channel opening for tens of milliseconds after activation. Thus, a 50 Hz train of 10 ms pulses over a 100 ms period causes granule neuron burst spiking but minimizes heating of the cerebellum.

The intensity of the optical stimulus determines the populations of granule neurons and their axons or somas that are preferentially activated. Light intensity from the optical fiber is rapidly attenuated in brain tissue [35]. Parallel fiber axons of granule neurons that are superficially located in the top 150 μm of ADCV will be most readily recruited by the optical stimulus. At low light intensities at the tip of the optical fiber (~1–4 mW) measured using an optical meter, optogenetic stimulation of granule neuron parallel fibers in the molecular layer with ~10 mW/mm^2 blue light (https://web.stanford.edu/group/dlab/cgi-bin/graph/chart. php, Fig. 2b) as the conditioned stimulus can produce a rapid form of associative conditioning when paired with an unconditioned stimulus [18]. However, these low light intensities fail to provide sufficient somatic depolarization in the internal granule layer, as assessed using immediate-early gene markers that respond to somatic calcium. At higher light intensities at the tip of the optical fiber (~30–70 mW), we can trigger immediate-early gene expression in granule neuron somas within a ~0.8 mm^3 volume throughout the medial ADCV. Due to light attenuation in the internal granule layer and our fiber-optic cannula parameters, this would equate to ~100 mW/mm^2 near the top of the internal granule layer and ~15 mW/mm^2 near the bottom of the ADCV (Fig. 2b). Note that at all light intensities considered here, no phototoxicity nor activation of gene expression programs related to tissue stress is observed. By contrast, at laser powers >150 mW and markedly with continuous rather than pulsed optical stimulation, we find visible tissue damage and the induction of astrocyte and microglia reactive programs (*see* also Subheading 3.2).

Finally, we note a few additional considerations for the photo-stimulation of Ai32-GC mice. The superficial location of ADCV makes it an accessible site for optical control with minimal surgical damage to the cerebellar cortex. Using surgical procedures described above, a fiber-optic cannula is placed above ADCV with the dura intact. Following recovery from surgery, mice are habituated on a cylindrical treadmill with head restraint for 30 min and at least a day prior to optogenetic experiments. Habituation is necessary to reduce stress during optogenetic stimulation and other behavioral experiments, particularly to mitigate the possibility that the headplate detaches from hyperactive animals that attempt to escape head restraint. During optogenetic stimulation, head-fixed animals may freely locomote on the treadmill. Immediately after optogenetic stimulation of ADCV at the appropriate laser intensity, the ADCV is dissected under microscope guidance and biochemical analyses may be performed. The choice of the laser intensity should reflect the locus of cellular or synaptic plasticity that is being tested.

3.2 Conditional CRISPR-Mediated Gene Knockout Using Viral Infection in ADCV

CRISPR-mediated genome editing represents a powerful tool to elucidate the functions of genetic and epigenetic regulators in the control of adaptive mouse behavior. Unlike a classical germline knockout approach, targeted deletion of proteins in a cell type-specific manner and at a time after brain development would effectively address their functions in the adult brain. This can be achieved using intersectional CRISPR genetics, where vectors encoding small guide RNA (sgRNA) are packaged in AAV and stereotaxically delivered to conditional Cas9 transgenic animals. Granule neuron-specific Cas9 expression in Cas9-EGFP-GC mice, detailed in Subheading 2.2, can be confirmed using EGFP immunohistochemical analyses. AAV encoding sgRNA targeting a gene of interest or a control sgRNA together with the red fluorescence protein mCherry are injected into the ADCV of Cas9-EGFP-GC mice at least 4 weeks old, thereby bypassing development. For robust granule neuron expression, AAV1 and AAV9 serotypes have been successfully used in the literature [15, 18], albeit with limited targeting specificity for cell types in the cerebellar cortex. However, because Cas9-EGFP-GC transgenic mice selectively express Cas9 enzyme in granule neurons in the cerebellar cortex, sgRNA-mediated gene knockout occurs only in adult granule neurons.

An important step during stereotaxic virus injection is to direct the spread of the virus to the target area and minimize tissue damage. This is achieved by controlling three main parameters: the speed of insertion and removal of the needle in the brain, the amount of virus injected, and the speed of injections. The micropipette is inserted into the brain at 1 mm per min. Upon arrival at the target site, the micropipette should settle at least 30 s before beginning injections. To target the entire ADCV, 50 nL of virus is injected every 10 s, with a total of 10 injections at each of three different depths (0.7 mm, 0.4 mm, and 0.2 mm below the dura surface) starting from the deepest location and moving superficially. Per animal, we use viral titers of 0.5–1×10^{13} vg to achieve sufficient sgRNA expression in granule neurons throughout the ADCV. After injections, wait 2 min before slowly raising the needle at 1 mm per min.

After several days, spread of the virus can be visualized using mCherry fluorescence. To verify that the viral injection procedure does not significantly cause tissue damage, the extent of tissue inflammation can be evaluated by assessing astrocyte and microglia reactivity using either RNA-seq or immunohistochemical analyses of markers including Gfap and Iba1. Severe cases of tissue damage can be observed using a stereo microscope, where necrotic tissue is discolored. Following viral delivery, allow 1 to 2 weeks for the expression of sgRNA and mCherry. An additional 2 weeks is needed to create deleterious mutations at Cas9 cleavage sites due to errors in nonhomologous end-joining repair in postmitotic neurons [36]. Thus, typically 4 weeks after injections, knockout of the gene of interest is achieved in all infected granule neurons and mice may then be subjected to behavioral paradigms or biochemical procedures. The efficiency of gene knockout is evaluated by quantifying the levels of targeted RNA or protein at the site of injection.

3.3 Delay Tactile Startle Conditioning

To assess how neural circuit activity and molecular mechanisms in adult ADCV granule neurons regulate cerebellar-dependent learning, mice are subjected to delay tactile startle conditioning. The behavioral apparatus is adapted from the mouse eyeblink conditioning setup [18, 24] with a few modifications. One to three days after habituation on a cylindrical treadmill while head-fixed, mice are placed in a dark enclosure and noise restricted room. The delay tactile startle conditioning paradigm uses an unconditioned tactile stimulus (US) that contacts the mouse's nose. With a short latency of <20 ms, a robust unconditioned startle response (UR) is observed (Fig. 3). The neural circuits underlying the tactile startle response including the trigeminal nerve and ventrocaudal pontine reticular formation have been mapped in rodents [37]. In head-fixed mice, stereotyped features of the tactile startle response are rapid total body movements, including a bilateral hindlimb movement that propels the animal backwards on the treadmill. For the tactile stimulus, a blunted elliptical nose cone padded with soft material at the cone tip is used to touch the mouse's nose. In early designs, we used a plush toy's nose as the tactile stimulus, but we find a 15 mL conical falcon tube taped with kimwipes at the cone tip for padding also works well. A stepper motor is used to control the timing and speed of the tactile stimulus. For the initially neutral conditioned stimulus (CS), a blue LED light may be used, as in eyeblink conditioning [25].

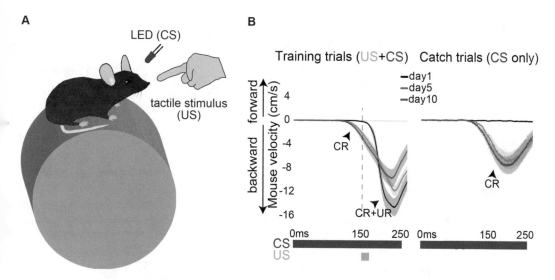

Fig. 3 Delay tactile startle conditioning. (**a**) Schematic of the behavioral apparatus. The unconditioned stimulus (US) comprises a motorized system that touches the nose of a head-fixed mouse resting upon a treadmill coupled to a rotary encoder. A visual LED cue is presented as the conditioned stimulus (CS). (**b**) The timing of the CS, US, conditioned response (CR), and unconditioned response (UR) during training trials with paired CS and US presentation, and interspersed catch trials with CS presentation only, over multiple days of associative conditioning. Robust mouse backwards movement is induced by the US on day 1, while the CS is an initially neutral stimulus. Following several daily training sessions, mice learn to move backwards in response to the CS. (Reproduced from [18])

Daily mouse behavioral sessions include 80 trials of CS and US pairing, with ~10 interspersed catch trials that consist of the CS in the absence of the US. Mouse locomotion is tracked using a rotary encoder coupled to the cylindrical treadmill. A CS-US inter-stimulus interval (ISI) of 150–400 ms is used with shorter ISIs resulting in faster learning. After a few days, mice exhibit a conditioned startle response (CR) to the CS, with a delay of 80–100 ms from CS onset when using a visual CS cue (Fig. 3b). The CR is quantitatively measured using the rotary encoder and is defined as a turning of the treadmill in the direction of the mouse moving backwards. We typically restrict analyses of the CR to after the delay period and during the ISI. The percentage of trials containing CRs in each daily session is used as the index of motor learning. Following the behavior paradigm, the ADCV may be microdissected and subjected to biochemical procedures including gene expression and genome architecture profiling described in the next section.

3.4 Biochemical Procedures to Examine the Chromatin Landscape

To study 3D genome organization across multiple scales (Fig. 4a), several biochemical approaches have been developed [38, 39]. Among them, the most popular procedure is in situ chromosome conformation capture with high-throughput sequencing (Hi-C), which requires a series of biochemical reactions

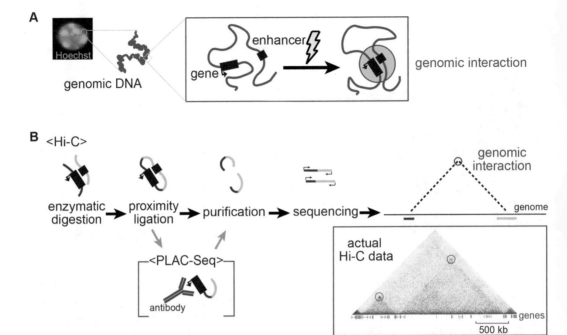

Fig. 4 Approaches to profile 3D genome architecture. (**a**) *Left*, global genome organization of a granule neuron nucleus stained with Hoechst dye. *Middle*, a large stretch of a chromosome. *Right*, a gene and its regulatory enhancer located within a topologically associated domain. In response to neuronal activity, a subset of enhancers dynamically interacts with gene promoters to regulate transcription. (**b**) *Top*, the strategy for in situ chromosome conformation capture with high-throughput sequencing (Hi-C). Brain tissues are first dissected, crosslinked, and dissociated. Permeabilized nuclei are then subjected to enzymatic digestion, ligation of proximal genomic fragments, and purified. Following paired-end sequencing of ligated DNA, bioinformatics analyses are performed to identify genomic interactions. *Bottom left*, the PLAC-Seq approach requires an additional immunoprecipitation step that enriches for specific regulatory regions of the genome, e.g. gene promoters or enhancers, bound by post-translationally modified histone proteins. *Bottom right*, Hi-C contact map at a ~2 megabase genomic locus generated from cerebellar tissue

including cell fixation, enzymatic digestion of the genome, and ligation of genomic fragments that are in close proximity (Fig. 4b). After sequencing the ligated genomic fragments, standardized bioinformatics pipelines including Juicer or HiC-Pro may be used to generate genome architecture maps [40, 41]. In addition to the Hi-C technique, a modified version of Hi-C called proximity ligation-assisted ChIP-Seq (PLAC-Seq) is used to enrich for interactions with specific gene regulatory regions including promoters or enhancers (Fig. 4b). Protocols for Hi-C and PLAC-Seq using cultured dissociated cells have been published [26, 27]. In this section, we highlight several critical steps optimized for the profiling of brain tissues with limited cell numbers, e.g. ADCV, which contains 250,000–500,000 cells.

Tissues are mechanically sheared in fixation buffer containing 100 μL of formaldehyde solution (11.1% formaldehyde, 50 mM

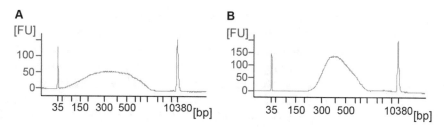

Fig. 5 Assessing DNA quality during library construction for genome architecture profiling. (**a**) Bioanalyzer trace of sonicated DNA fragments after ligation. The desired length for DNA fragments is between 200 and 500 bp. (**b**) Bioanalyzer trace of the final library for sequencing. The desired average size of fragments is ~400 bp

Hepes (pH 7.9), 100 mM NaCl, 1 mM EDTA, 0.5 mM EGTA), and 1 mL of PBS using a dounce homogenizer (pestle A) for 25 times, transferred to a new microtube, and incubated for 15 min at room temperature with rotation. After adding 110 µL of glycine solution (1.25 M glycine, 10 mM Tris–HCl (pH 8)) to stop the reaction, the pellet is washed with PBS twice, flash frozen in liquid nitrogen, and stored at −80 °C.

After collecting multiple crosslinked frozen samples, the tissues are thawed on ice and two to three ADCV are pooled together for each biological replicate. The pooled tissues are lysed in 1 mL of cold lysis buffer (10 mM Tris–HCl pH 8.0, 10 mM NaCl, 0.2% Triton-X with proteinase inhibitor). Single nuclei are dissociated using dounce homogenizers (pestle A for ten times and then pestle B for 20 times) and filtered using a 40 µm nylon mesh. The lysate is then incubated on ice for 15 min and spun down at slow speed (200 × *g* for 5 min) to avoid cell clumping. The pelleted nuclei are then subjected to enzymatic reactions including restriction enzyme digestion, modified nucleotide fill-in, and DNA ligation.

The next step is DNA fragmentation, where nuclei are sonicated until DNA sizes are ~500 bp (Fig. 5a). With a Bioruptor Pico, we use 15–20 cycles of 10 s sonication on and 50 s off per cycle. However, the duty cycles must be optimized based on the sonication system. It is important to test sonication efficiency using a high sensitivity DNA kit on a Bioanalyzer because the DNA concentration from a small dissected region of the brain is low. For testing, ~5% (5–10 µL) of the sonicated lysate is treated with 2 µg of Proteinase K in 90–95 µL of elution buffer (10 mM Tris–HCl pH 8.0, 350 mM NaCl, 0.1 mM EDTA, 1% SDS) and incubated overnight at 65 °C for de-crosslinking. The test sample DNA is purified using a PCR purification kit and the eluates are analyzed with a Qubit and Bioanalyzer to estimate DNA concentration and quality (Fig. 5a). For Hi-C, lysates containing 20 to 100 ng of DNA are used in subsequent steps. For PLAC-Seq, ~80% of the entire lysate is used for each biological replicate. When comparing two samples (e.g. unstimulated versus optogenetically-stimulated samples), equivalent amounts of DNA should be used across conditions. In PLAC-Seq, lysates are mixed with 2 μg of H3K4me3 antibody, which enriches for promoter-associated genomic interactions, overnight at 4 °C. The antibody–DNA complexes are purified using protein G Sepharose beads with extensive washes, and bound DNA fragments are eluted with elution buffer (10 mM Tris–HCl pH 8.0, 350 mM NaCl, 0.1 mM EDTA, 1% SDS) at 65 °C for 15 min with continuous vortex.

Hi-C or PLAC-Seq sequencing libraries are generated by immobilizing DNA fragments on MyOne Streptavidin T1 beads with the NEBNext Ultra™ II DNA Library Prep Kit for Illumina according to the manufacture's instruction with modifications. Briefly, end repair, adaptor ligation, and PCR reactions are performed on the beads-immobilized DNA fragments. After the adaptor ligation reaction step, DNA fragments are washed twice with wash buffer (5 mM Tris–HCl pH 8.0, 1 M NaCl, 0.5 mM EDTA, 0.05% Tween 20) at 55 °C for 2 min, followed by washing with H_2O. DNA fragments are amplified by PCR using indexed primers and PCR products are purified. The concentration and quality of the indexed libraries are then assessed using a Qubit and Bioanalyzer (Fig. 5b) and sequenced on an Illumina sequencer (e.g. NextSeq 550). 36 bp paired-end reads are sufficient to map DNA fragments to genomic regions. Longer read lengths modestly improve accuracy during genome alignment but should be trimmed at sequences overlapping with restriction enzyme cut sites. The depth of sequencing depends on the desired genome resolution and the genome architecture features studied. To visualize genomic loops, we typically sequence at least 450–2000 million reads per condition for Hi-C and 200–300 million reads per condition for PLAC-Seq, with two to three biological replicates per condition.

In addition to genome architecture, the effects of optogenetic stimulation or genetic perturbation on gene expression can be assessed using RNA-Seq. Briefly, total RNA is extracted from a brain region of interest (e.g. the ADCV) using Trizol according to the manufacturer's instructions, and the quality and quantity of RNA can be measured using a Bioanalyzer with a RNA 6000 Pico kit. ~100 ng of RNA is treated with the NEBNext rRNA Depletion Kit to enrich for both nascent and mature mRNA. cDNA libraries prepared with NEBNext Ultra™ II Directional RNA Library Prep Kit for Illumina are sequenced using an Illumina sequencer. 36 bp paired-end reads are sufficient for detecting the expression of most genes. Longer read lengths are recommended for gene families with high sequence similarity or for analyses of alternatively spliced transcripts. At least three to four biological replicates per condition are recommended for downstream statistical analyses.

4 Conclusions

In this chapter, we describe the integration of molecular neuroscience with systems neuroscience approaches to advance our understanding of the molecular and cellular mechanisms that orchestrate long-term memory. We detail the use of optogenetics and acute CRISPR-Cas9 genetic perturbation in a defined region of the adult mouse cerebellum, followed by behavioral analyses and transcriptome and genome architecture profiling of cerebellar tissue. This suite of techniques allows us to examine how patterned activity in granule neurons dynamically controls chromatin and gene expression programs necessary for cerebellar-dependent motor learning. Because mutations of genome architectural regulators and chromatin enzymes are linked to diseases of human cognition [42–44], studies on the regulation of the epigenome will provide new windows on how nuclear processes contribute to the structure and function of a healthy brain.

References

1. Lisman J et al (2018) Memory formation depends on both synapse-specific modifications of synaptic strength and cell-specific increases in excitability. Nat Neurosci 21:309–314
2. Yap EL, Greenberg ME (2018) Activity-regulated transcription: bridging the gap between neural activity and behavior. Neuron 100:330–348
3. Han JH et al (2007) Neuronal competition and selection during memory formation. Science 316:457–460
4. Cai DJ et al (2016) A shared neural ensemble links distinct contextual memories encoded close in time. Nature 534:115–118
5. Kelleher RJ 3rd, Govindarajan A, Tonegawa S (2004) Translational regulatory mechanisms in persistent forms of synaptic plasticity. Neuron 44:59–73
6. Zhai S et al (2013) Long-distance integration of nuclear ERK signaling triggered by activation of a few dendritic spines. Science 342:1107–1111
7. Mardinly AR et al (2016) Sensory experience regulates cortical inhibition by inducing IGF1 in VIP neurons. Nature 531:371–375
8. Sun X et al (2020) Functionally distinct neuronal ensembles within the memory engram. Cell 181(410–423):e417
9. Herre M, Korb E (2019) The chromatin landscape of neuronal plasticity. Curr Opin Neurobiol 59:79–86
10. Tyssowski KM, Gray JM (2019) The neuronal stimulation-transcription coupling map. Curr Opin Neurobiol 59:87–94
11. Heinz DA, Bloodgood BL (2020) Mechanisms that communicate features of neuronal activity to the genome. Curr Opin Neurobiol 63:131–136
12. Zheng H, Xie W (2019) The role of 3D genome organization in development and cell differentiation. Nat Rev Mol Cell Biol 20:535–550
13. Bashkirova E, Lomvardas S (2019) Olfactory receptor genes make the case for interchromosomal interactions. Curr Opin Genet Dev 55:106–113
14. Sawtell NB (2010) Multimodal integration in granule cells as a basis for associative plasticity and sensory prediction in a cerebellum-like circuit. Neuron 66:573–584
15. Giovannucci A et al (2017) Cerebellar granule cells acquire a widespread predictive feedback signal during motor learning. Nat Neurosci 20:727–734
16. Wagner MJ et al (2017) Cerebellar granule cells encode the expectation of reward. Nature 544:96–100
17. Wagner MJ et al (2019) Shared cortex-cerebellum dynamics in the execution and learning of a motor task. Cell 177(669–682):e624

18. Yamada T et al (2019) Sensory experience remodels genome architecture in neural circuit to drive motor learning. Nature 569:708–713

19. Markwalter KH et al (2019) Sensorimotor coding of Vermal granule neurons in the developing mammalian cerebellum. J Neurosci 39:6626–6643

20. Stoodley CJ, Schmahmann JD (2010) Evidence for topographic organization in the cerebellum of motor control versus cognitive and affective processing. Cortex 46:831–844

21. King M et al (2019) Functional boundaries in the human cerebellum revealed by a multidomain task battery. Nat Neurosci 22:1371–1378

22. Chambers WW, Sprague JM (1951) Differential effects of cerebellar anterior lobe cortex and fastigial nuclei on postural tonus in the cat. Science 114:324–325

23. Mauritz KH, Dichgans J, Hufschmidt A (1979) Quantitative analysis of stance in late cortical cerebellar atrophy of the anterior lobe and other forms of cerebellar ataxia. Brain 102:461–482

24. Heiney SA et al (2018) Single-unit extracellular recording from the cerebellum during Eyeblink conditioning in head-fixed mice. In: Sillitoe R (ed) Extracellular recording approaches, Neuromethods, vol 134. Humana Press, New York, NY, pp 39–71

25. Heiney SA et al (2014) Cerebellar-dependent expression of motor learning during eyeblink conditioning in head-fixed mice. J Neurosci 34:14845–14853

26. Rao SS et al (2014) A 3D map of the human genome at kilobase resolution reveals principles of chromatin looping. Cell 159:1665–1680

27. Fang R et al (2016) Mapping of long-range chromatin interactions by proximity ligation-assisted ChIP-seq. Cell Res 26:1345–1348

28. Madisen L et al (2012) A toolbox of Cre-dependent optogenetic transgenic mice for light-induced activation and silencing. Nat Neurosci 15:793–802

29. Pan N et al (2009) Defects in the cerebella of conditional Neurod1 null mice correlate with effective Tg(Atoh1-cre) recombination and granule cell requirements for Neurod1 for differentiation. Cell Tissue Res 337:407–428

30. Platt RJ et al (2014) CRISPR-Cas9 knockin mice for genome editing and cancer modeling. Cell 159:440–455

31. Funfschilling U, Reichardt LF (2002) Cre-mediated recombination in rhombic lip derivatives. Genesis 33:160–169

32. Powell K et al (2015) Synaptic representation of locomotion in single cerebellar granule cells. eLife 4:e07290

33. Herman AM et al (2014) Cell type-specific and time-dependent light exposure contribute to silencing in neurons expressing Channelrhodopsin-2. eLife 3:e01481

34. Lin JY (2011) A user's guide to channelrhodopsin variants: features, limitations and future developments. Exp Physiol 96:19–25

35. Yizhar O et al (2011) Optogenetics in neural systems. Neuron 71:9–34

36. Incontro S et al (2014) Efficient, complete deletion of synaptic proteins using CRISPR. Neuron 83:1051–1057

37. Yeomans JS et al (2002) Tactile, acoustic and vestibular systems sum to elicit the startle reflex. Neurosci Biobehav Rev 26:1–11

38. Bonev B, Cavalli G (2016) Organization and function of the 3D genome. Nat Rev Genet 17:661–678

39. Kempfer R, Pombo A (2020) Methods for mapping 3D chromosome architecture. Nat Rev Genet 21:207–226

40. Durand NC et al (2016) Juicer provides a one-click system for analyzing loop-resolution hi-C experiments. Cell Syst 3:95–98

41. Servant N et al (2015) HiC-Pro: an optimized and flexible pipeline for Hi-C data processing. Genome Biol 16:259

42. Singh VP, Gerton JL (2015) Cohesin and human disease: lessons from mouse models. Curr Opin Cell Biol 37:9–17

43. Gabel HW et al (2015) Disruption of DNA-methylation-dependent long gene repression in Rett syndrome. Nature 522:89–93

44. Ronan JL, Wu W, Crabtree GR (2013) From neural development to cognition: unexpected roles for chromatin. Nat Rev Genet 14:347–359

Chapter 8

In Vitro Voltage Imaging of Subthreshold Activity in Inferior Olive Neurons with ANNINE-6plus

Kevin Dorgans, Bernd Kuhn, and Marylka Yoe Uusisaari

Abstract

Since the 1970s, a multitude of fluorescence-based methods for observing changes in neuronal membrane voltage have been developed, progressively improving in terms of signal sensitivity, speed, and targeting. While significant development has been seen in the past decade in engineering novel variants of genetically encoded voltage indicators (GEVI), synthetic dyes can, in many cases, be more straightforward and flexible tools to be used. Here, we describe in detail an improved dye-labeling method where the synthetic, pure electrochromic voltage-sensitive dye ANNINE-6plus is introduced into the brain region of interest via stereotactical injection in a mouse. Specifically, we have optimized the labeling protocol for effective voltage imaging of subthreshold activity in inferior olive. The homogenous labeling and excellent signal-to-noise ratio allow this method to reveal olivary subthreshold oscillations in high spatiotemporal resolution. In addition to the procedures related to dye introduction and optical imaging, we give a detailed description of the process of acute slice preparation from the mouse inferior olive.

Key words Inferior olive, Voltage imaging, ANNINE-6plus, One-photon, In vitro, Hot slicing, Stereotactic microinjection

---

1 Introduction

Neuronal code is built on transmembrane ion current dynamics that drive fluctuations in the voltage across neuronal membrane. Thus, direct or indirect measurement of the electrical processes lies at the foundation of neuroscience, and several complementary approaches have been developed in the decades since Hodgkin and Huxley [1] presented their groundbreaking neuromethod for intracellular voltage recording.

Overall, the current methods for observing transmembrane voltage changes in excitable cells such as neurons can be classified into either electrical or optical categories, each with distinct features that render them useful for particular purposes. While electrical recordings are unsurpassed in the sensitivity and temporal resolution, they necessitate physical contact with the target neuron, which

Roy V. Sillitoe (ed.), *Measuring Cerebellar Function*, Neuromethods, vol. 177, https://doi.org/10.1007/978-1-0716-2026-7_8,
© Springer Science+Business Media, LLC, part of Springer Nature 2022

often leads to either unsurmountable difficulties in implementing the method or to limited reach in spatial terms. To overcome these limitations various optical methods have been implemented, each of them relying on distinct biosignals (such as intracellular calcium concentration and neurotransmitter release) that reflect electrical activity and that have been made observable by means of fluorescent light emission. Voltage imaging methods [2] in general excel in providing relatively direct measures of transmembrane voltage changes. The electrical events that can be observed with voltage imaging range from fast action potentials and slow state shifts to subtle, localized subthreshold fluctuations such as postsynaptic potentials. Theoretically, voltage imaging could be used to instantaneously reveal the full scale of electrical behavior of entire neurons and networks.

Fortunately, the lure of voltage imaging methodologies' theoretical possibilities has attracted the attention of numerous researchers since the 1970s [3] aiming to increase the efficiency and applicability of the method. Several laboratories developed fast, synthetic voltage-sensitive dyes [4–6] with a voltage-sensing mechanism based on the molecular Stark effect [7]. These dyes report voltage changes linearly and with nanosecond temporal resolution. However, they cannot be easily targeted to a specific cell or cell-type which hampers their applicability. Now, also genetically encoded voltage indicators are available with promising features [8]. The key advantage of genetically encoded indicators is cell-type specific and even subcellular targeting. However, their brightness and temporal resolution remain to be challenging.

Here, we focus on the synthetic voltage-sensitive dye ANNINE-6plus [6]. The details of voltage imaging with ANNINE dyes and protocols for different types of experiments were described before [7, 9]. ANNINE dyes can be filled into single neurons, for example Purkinje neurons [10] or injected into the tissue for bulk loading [11, 12]. The dye is compatible with two-photon microscopy or whole-field imaging methods [7, 9].

The inferior olive (IO), while not anatomically a part of the cerebellum, can be functionally considered to be a cerebellum-related structure as its neurons exclusively target the cerebellum [13]. Its uncontested importance in cerebellar function, mediated by powerful excitatory synapses between IO neurons and cerebellar Purkinje neurons (PNs), is contrasted with sparseness of research into its function as a network. Granted, a number of elegant classical works [14–16] have revealed multiple aspects of IO function that deviate from "standard textbook neuronal networks," such as graded action potential waveforms, dendrite-originating axons, and gap junction-dependent subthreshold oscillations (STOs) of the membrane voltage. However, despite the significance these features carry in investigations of overall cerebellar computation, it is striking that nearly all physiological studies published thus far have been limited to single electrode recordings.

A few exceptions to this general trend have appeared [17–19] where voltage-sensitive dyes were used to reveal network-wide propagation and synchronization of IO STOs. These works have stood as important cornerstones of what is currently known about IO network dynamics, even if lacking in spatiotemporal precision and sensitivity in comparison to more recent works conducted in other structures. However, as the olivo-cerebellar research field is rapidly expanding into fields of study outside the classical regime of motor control, investigations of IO function should not be left aside. For the purpose of re-instigating research interest into the important structure, we describe a protocol employing organic dye-based, whole-field voltage imaging in in vitro brain slice preparation, optimized for IO.

2 Materials

2.1 Preparation of ANNINE-6plus (A6+) Stock Solution

1. ANNINE-6plus powder [6, 7, 9] (B. Kuhn lab. OIST, JP).

2. DMSO (Dimethylsulfoxide), anhydrous (Life Technologies, UK).

3. Ultracentrifugation and sonication-resistant vials (0.5 ml; Eppendorf).

2.2 In-Vivo Stereotactic Injection of A6+ into Inferior Olive of Adult Anesthetized Mouse

2.2.1 Injection Setup

1. Isofluorane (1-chloro-2,2,2-trifluoroethyl difluoromethyl ether) (IsoFlo, Zoetis, US).

2. Low-flow anesthesia system (Somno-suite Low-flow anesthesia system, Kent Scientific, US).

3. Robotic stereotaxic frame (Neurostar, Tübingen, Germany).

4. Injectomate microinjection system (Neurostar, Tübingen, Germany).

5. Computer-assisted targeting software (StereoDrive, Neurostar, Germany).

6. Heating pad complemented and temperature sensor device (Biological Temperature Controller TMP-5b, Supertech instruments, UK).

2.2.2 Pipette Preparation

1. Quartz capillaries (outer diameter: 1.14 mm, inner diameter 0.53 mm non-polished quartz capillaries with filament, model Q14-53-10-NP, Sutter Instrument, USA).

2. Laser puller (P-2000, Sutter Instrument, USA).

3. Pipette beveler (BV-10 Beveler, Sutter Instrument).

4. Mineral oil.

2.2.3 Surgery	1. Myochlorine gel (Sato Pharmaceutical, JP).

2. Xylocaine (Xylogel 0.05%).

3. Dental Drill Surgic XT Plus (Micro High Speed, NSKTech, Australia).

4. Vetbond Tissue Adhesive (3 M).

5. IR heating lamp (Exo Terra Heat-Glo Infrared Spot Lamp, 120-V).

6. 2× Dumont 5/45 Forceps (11251-35, Fine Science Tools, Vancouver, Canada).

2.3 Acute Brainstem Slice Preparation at Physiological Temperature

1. Campden vibratome (Vibrating microtome, model 7000 smZ-2, Campden Instruments, UK).

2. Ceramic blades (38 mm × 7 mm × 0.5 mm ceramic blades, model 7550-1-C, Campden Instruments, UK).

2.3.1 Slicing Device

3. Slicer heater element (Physiological temperature controller, model 7611A, Campden Instruments, UK).

4. Recovery chamber (BSK1, Scientific System Designs INC, CA).

2.3.2 Surgical Tools

1. Extra Fine Bonn Scissors (14084-08, Fine Science Tools, Vancouver, Canada).

2. Hardened Fine Scissors (14091-09, Fine Science Tools, Vancouver, Canada).

3. Dumont 55 Forceps (11295-51, Fine Science Tools, Vancouver, Canada).

4. Iris Forceps, 10 cm WPI (14091-09, World Precision Instruments, Vancouver, US).

2.3.3 Chemicals

1. D-(+)-Glucose powder (G8270, Sigma-Aldrich, GE).

2. Sodium chloride powder (7647-14, Sigma-Aldrich, GE).

3. Sodium bicarbonate powder (S5761, Sigma-Aldrich, GE).

4. Calcium chloride hexahydrate (442909, Sigma-Aldrich, GE).

5. Magnesium sulfate heptahydrate (230391, Sigma-Aldrich, GE).

6. Potassium chloride, anhydrous (P9333, Sigma-Aldrich, GE).

7. Potassium dihydrogen phosphate (P5655, Sigma-Aldrich, GE).

2.4 In Vitro Voltage Imaging	1. 488 nm laser source (3 W custom 488 nm laser; Genesis MX-488, Coherent, US).
2.4.1 Illumination	2. 561 nm emission high-pass filter (RazorEdge ultrasteep long pass edge filter, Semrock).
	3. Bright-field microscope (BXW51, Olympus, JP).
	4. 60× water immersion objective (LumPlanFL N 60 NA 1.00 W, Olympus, JP).
	5. 5× air objective (MPlan N 5× 0.15 NA A, Olympus).
2.4.2 Camera	1. High-speed CMOS camera (MiCAM03, BrainVision, JP).
	2. sCMOS camera (Zyla4.2, Andor, UK).

3 Methods

3.1 Preparation of ANNINE-6plus (A6+) Solutions

For a successful staining of IO neurons, the hydrophobic voltage-sensitive dye A6+ needs to be completely dissolved. To do this, 33 μl of pure DMSO is added in an ultrasonication-resistant vial containing 400 μg of A6+ powder. The solution is agitated and sonicated at 35 °C for 15–30 min to obtain a final concentration of 13 mg/ml A6+ in DMSO. If an orange precipitate is visible in the vial after 30 s centrifugation (6200 rpm, corresponding to 1310–2960 × g (with 2ml tubes in CENTRY 103 Minicentrifuge, Gilson) mini-centrifuge), the process is continued by vortexing the vial for an additional 1 min and sonicated until A6+ is completely dissolved. This stock solution can be stored for a few months at −20 °C.

The final staining solution (400 μg/ml A6+ in 3% DMSO solution) is prepared by adding distilled water (MilliQ Advantage, Merck Millipore, Germany) at 1:33 ratio to the stock solution, sonicated at 35 °C for 15–30 min, and aliquoted to 50 μl tubes (Safe-stock tubes 0.5 ml, Eppendorf). These aliquots can be stored at −20 °C and should be carefully sonicated for at least 10 min before usage.

3.2 In-Vivo Stereotactic Injection of A6+ into Inferior Olive of an Adult Anesthetized Mouse

3.2.1 Pipette Fabrication

For a reproducible and precise targeting of IO, special care is given to the injection pipette fabrication (*see* **Note 1**). IO spans from 6.5 to 6.7 mm dorso-ventral from skull surface and is indeed one of the deepest locations in the mouse brain. This localization makes microinjection challenging for several reasons. First, commonly used reference atlases are often unreliable guides when it comes to extremely deep regions such as the IO, as the distance exacerbates the subtle variability in skull shape between mouse strains. Second, a glass injection pipette can easily bend as it passes through the brain tissue. For reproducible targeting of IO, we only use quartz capillaries instead of borosilicate for our injection pipettes. Quartz pipettes are harder and thus bend less while being moved within the brain parenchyma, leading to a more reliable tip position at depths beyond 5 mm. Furthermore, quartz pipettes can be beveled so that their sharp edge will easily penetrate the membranes covering the brain.

To pull quartz capillaries into injection pipettes, a laser puller (P-2000, Sutter Instrument, USA) is used with a one-step program (to be crafted using guidelines provided by the manufacturer). The resulting pipette should have a shaft of at least 10 mm. For A6+ injections in brainstem, the pipette tip requires special attention to obtain a large opening without broadening the shaft (Fig. 1a). After pulling, the pipette should be similar to patch-clamp electrodes (and have a resistance approaching 10 MΩ before beveling) rather than juxtacellular ultrathin electrodes ($R > 20$ MΩ). Then, around 1 mm of the tip is cut under a magnifying glass using precision scissors. The remaining size to the capillary is longer than 8 mm with a shank diameter not bigger than 0.5 mm (bigger size will impair pipette penetration in the drilled area and cause compression damages in brain tissues). Pipettes are lightly bevelled for a few seconds (BV-10 Beveler, Sutter Instrument, US); care should be taken to avoid pipette bending during the bevelling procedure. The finalized pipette shaft should be at least 7 mm long. The prepared pipette should be stored in a dust-free container, avoiding any contact with the tips, and used on the same day of fabrication.

3.2.2 Anesthesia and Mouse Well-Being

Proper surgical procedures are essential for rapid animal recovery from surgery and thereby reproducibility of experimental outcomes. All animal procedures should be approved by the local institutional animal care and use committee before starting any experiments. Approved surgical procedures and availability of anesthetic compounds varies between countries and institutions. However, we want to highlight the fact that a key to minimal effects on animal health is limiting the surgical operation duration. Mice are prone to dehydration and have problems in maintaining their homeostatic status, and all efforts should be made to develop a fast and efficient workflow for injections.

At the beginning of the surgical operation, the animal (P40, c57bl6/j male mice) is anesthetized. We use 5% isoflurane for induction (delivered into the induction box, Sliding top chamber, Somno-suite Low-flow anesthesia system, Kent Scientific, USA). Usually it does not take longer than 200 to 300 s until the mouse is unresponsive and breathing calmly (postural body muscles relaxed and no reaction to strong hind-paw and tail pinches). At this point the animal is placed in a stereotaxic frame (Neurostar, Tübingen, Germany) with constant delivery of isoflurane through a nose cone with a constant airflow of 0.2 L/min. At the beginning, isoflurane is set to 1.5%, adjusted to 1.2%–1.8% to maintain breathing at around 1 breath per minute). Mouse body

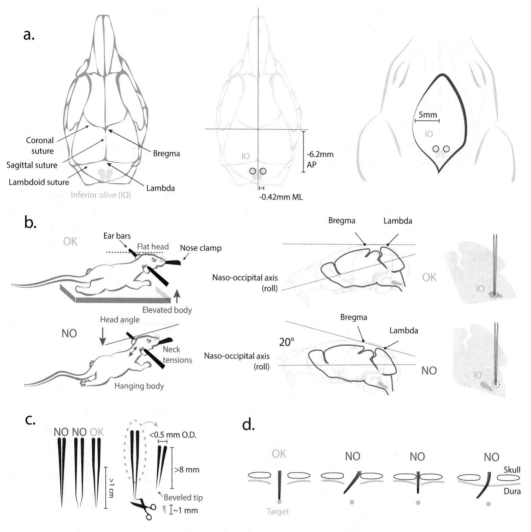

Fig. 1 Tips for performing correct stereotaxic injections in Inferior Olive. (**a**) *Left panels*: Schematics representing the dorsal view of a mouse skull and the positioning of Inferior Olive relative to reference landmarks and skull sutures. *Right panel*: Skin incision has to be long enough to clear the area a few millimeters around the sutures and landmarks. (**b**) Injections are correctly targeted if skull is flat (not tilted) and mouse neck is not stretched (no tensions in upper body). (**c**) Injection pipette schematics: the quartz capillary is pulled over 1 cm distance. Then the edge is cut and beveled. The whole tip measures at least 8 mm. (**d**) Injection pipette is centered within a neat cranial aperture cleared from connective tissues. Correct targeting is performed only if the pipette is centered, doesn't touch the skull and is not deviated by connective tissues

temperature is maintained with a heating pad complemented to a temperature sensor device (Biological Temperature Controller TMP-5b, Supertech instruments, UK). We use heating pad with rectal probe (Digital thermometer, Sato Keiryoki, JP), and find that setting the target temperature to 38 °C is most stable for keeping the animal in healthy condition. To protect the animal's eyes from drying and infection, they are covered with myochlorine gel (Sato

Pharmaceutical, JP). To prevent dehydration, 200 μl of 0.9% NaCl solution is administered sub-cutaneously. Finally, before commencing the surgery, the skin covering the dorsal extent of skull is shaved, disinfected and locally anethetized with gel application of Xylocaine (Xylogel 0.05%).

3.2.3 Mouse Body and Skull Alignment

Targeting brain regions by stereotactic surgery always requires detailed attention to animal head positioning. This is even more critical in extremely deep brain regions such as inferior olive, as even slight deviation from proper alignment leads to missing the target region (Fig. 1b).

After attaching the mouse head stably in the stereotactic frame, its body position needs to be adjusted. This is not a common practise when targeting superficial structures such as the cerebellar cortex or the hippocampus, but tension or torsion in upper vertebrae and neck musculature can pull and distort brainstem orientation (Fig. 1c). Generally speaking, any position of the body can be acceptable as long as the neck is not stretched and the position can be reliably reproduced with every operation; we place the body slightly elevated with respect to animal size (the mouse should not be held in the air by the ear bars. The animal weight has to be distributed between chest and upper limbs to limit the tension in neck).

Only after the body is well-positioned (should not be moved throughout the surgery), skull skin is carefully incised with scalpel blade over 1 cm length from between the ears to occipital bone (Fig. 1d—right panel). The exposed skull surface is cleaned with physiological saline solution in a region reaching at least 5 mm of the occipital bone and at least a 7 mm opening on mediolateral axis to allow clear visualization of cranial sutures in order to identify bregma and lambda landmarks.

Next, the skull alignment is adjusted. We use the Neurostar robotic stereotactic frame with computer-assisted targeting. It allows not only fine alignment of the skull but also to a certain degree compensates for slight deviations from perfectly flat arrangement. This compensation is not necessary for successful IO targeting, but the researcher needs to take extra care on leveling bregma and lambda landmarks and making sure that the skull is straight and not turned around the axial direction.

In brief, the anatomical landmarks Bregma and Lambda are used to level the skull. Next, the position is adjusted so that two points bilaterally from the (2 mm right and left) are vertically aligned. A common rule of thumb is that for leveled alignment the points should not differ more than 0.2 mm in vertical position; however, in our experience this is not sufficient for successful targeting of deep brainstem locations. Thus, the difference should be made as small as possible with the equipment available.

3.2.4 Craniotomy and Dye Delivery into the Inferior Olive

When the skull is completely immobilized and the frame carefully calibrated, a drilling mark is made at the intersection between skull surface and the pipette, while pipette is lowered towards injection target (*see* **Note 2**). The coordinates we use for targeting the dorsal and principal nucleus of the IO are (relative to Bregma) -6.2 mm AP, ±0.42 mm ML, and 6.7 mm DV. A circular, neat, 2-mm hole is drilled with a diamond burr of a dental precision drill with soft circular movements to avoid sudden skull penetration and potentially damaging the dura. When the drilling is complete, the dura mater is pierced with sharp forceps or a needle tip to allow pipette entry into the brain tissue without any bending.

Before starting the injections, the beveled pipette (prepared in Subheading 2.2.3) is filled with mineral oil with a syringe attached to flame-pulled plastic capillary and tightly fixed to the injection device of the stereotaxic frame. Next, the dye is backfilled by dipping the tip into a small drop (around 2 µl) of thawed and re-sonicated A6+ solution on a piece of parafilm (Sigma-Aldrich, GE) and slowly retracting the syringe plunger of the injection device (*see* **Note 3**). To ascertain that the tip is not blocked and the dye solution can be smoothly ejected, a small protraction of the plunger (corresponding to 10 µl) should result in an appearance of a small droplet at the tip of the pipette. The injection should proceed immediately after preparing the pipette, to prevent the tip from clogging. For best results, it is recommended to prepare a new pipette for each insertion into the brain.

The pipette is inserted into the brain tissue, taking care that it will not bend by avoiding any contact with skull, and then slowly (5 mm/min) driven into the target depth (Fig. 1e). It is advisable to briefly (at least 30 s) pause before commencing injection in order to let any possible tissue deformations reside. Then 100 nl of A6+ solution is injected with a flow rate of 100 nl/min at 4 locations 50 µm apart in vertical direction (from 6.7 mm DV to 6.5 mm DV), starting from the ventralmost site. Note that the somewhat higher flow rate than commonly used when injecting viral vectors is recommended in order to maximize the spread of the lipophilic A6+. Furthermore, several smaller injections instead of a single larger one is used to obtain a large, homogenously-labeled brain region.

After the injection is completed, it is recommended to again pause for at least 1 min (recommended 5 min) before commencing the pipette retraction. When the pipette tip is entirely outside of the brain, it is also essential to check if A6+ has been successfully ejected from the pipette (*see* **Note 4**).

3.2.5 Post-operative Care

When bilateral injection is completed, the skull is cleaned and the skin hydrated with physiological saline. After the skin is cleaned and dried of excess liquid, the surgical wound is closed with *N*-butyl cyanoacrylate glue (super glue or VetBond, 3 M) (*see* **Note 5**). To improve post-operative recovery, 300 µl physiological saline for animal hydratation as well as 7.5 mg/kg Carprofen for analgesia are administrated sub-cutaneously. The animal is placed in a recovery cage under an infrared heating lamp for at least 10 min, in its original bedding and observed until abdominal breathing accelerates and the mouse recovers normal locomotor activity. The animal is regularly monitored during the first 72 h after surgery or until sacrifice, whichever occurs first.

3.2.6 Acute Brainstem Slice Preparation at Physiological Temperature

The art of preparing acute brain slices culminates in preserving brain tissue in nearly-unchanged condition through the whole process of tissue extraction, slicing, slice transfer, and incubation over several hours of experimentation. It requires the experimenter to be aware of the extreme fragility of the brain slices, paying exquisite attention to 4 major parameters: (1) osmotic changes, (2) pH changes, (3) mechanical damage (such as tissue stretching), and (4) temperature changes. Common methods for slice preparation procedures make use of several different solutions for slice incubation during brain dissection, slicing, recovery period, and the actual experimentation. Various physiological solutions are used for brain extraction, slicing, slice recovery, and incubation/experimentation. The detailed composition of these solutions are tuned with the intent to either prevent excitotoxicity by including glutamate blockers (like kynurenic acid) or adjusting cation content [20] or to reduce oxidative stress by adding ascorbic acid or glutathion as anti-oxidant agents for fragile tissues [20, 21]. However, we do not recommend using multiple solutions nor the rapid cooling to +4 °C, as subjecting the brain tissue to the inevitable shifts in pH, osmolarity, and temperature when changing the incubation solution can permanently alter network physiology [22]. Our "hot slicing" method (described in detail in [23, 24]) aims to minimize the environmental changes experienced by the brain tissue, as well as minimizing any mechanical insults, and is suitable for any brain region sectioning, providing superior results especially when using brains from adult or old animals. Here, we describe the details as optimized for IO and brainstem slices.

3.2.7 Preparation of Standard Physiological Solution (SPS)

50 ml of 1 M stock solutions of KCL, $MgSO_4$, KH_2PO_4, CaCl are prepared and stored for long period at 4 °C. The salts must be completely dissolved (by mixing and vortexing) before cooling the solutions. On the day of the experiment, final SPS is prepared by adding 7.36 g NaCl, 1.8 g Glucose, and 2.18 g $NaHCO_3$ into 900 ml of distilled water and mixed carefully. Next, 3.4 ml of KCl and 1.2 ml KH_2PO_4 (both 1 M) are added, before gassed with 5% CO_2–95% O_2 for at least 10 min in order to stabilize the pH to around 7.2–7.3. Then, 1.3 ml $MgSO_4$ (1 M) and 2 ml $CaCl_2$ (1 M) are added, observing that there is no cloudiness in the solution, and

distilled water is added to reach 1 L. It is recommended to check the osmolarity of the solution (should be 310 mOsm), and if necessary, add more water. From this point, the solution should be kept constantly gassed until the end of experiment to keep pCO_2/pO_2 constant.

3.2.8 Workspace Preparation

Before commencing acute slice preparation, it is essential to ascertain that all tools, slicing and incubation chambers are thoroughly cleaned and disinfected, as a common cause for rapid slice quality deterioration after slicing is bacterial growth. Furthermore, all tools must be in impeccable condition (no rust, no cracks) and within easy reach during dissection.

A critical piece of equipment for successful slice preparation is the vibratome slicer, and utmost care should be taken to ascertain that it is functioning optimally and that the blade used for cutting is in perfect shape. Regardless of the model of slicer used, every effort should be made to reduce vibrations of the blade in vertical (z) dimension, as well as any unnecessary vibrations from roughness of the blade. We use the Campden vibratome (Vibrating microtome, model 7000 smZ-2, Campden Instruments, UK) that allows adjustment of the slicing blade position so that the blade edge moves nearly perfectly horizontally, with vertical vibration amplitude of less than 0.2 μm. Also, we recommend using ceramic blades (38 mm × 7 mm × 0.5 mm ceramic blades, model 7550-1-C, Campden Instruments, UK), that are superior in smoothness to any stainless steel blades. (Note that ceramic blades are very brittle and even a slight touch with a pipette or a brush can damage the edge.) The blade alignment should be confirmed before each use.

When beginning slice preparation, the standard physiological solution (SPS) that has been thoroughly gassed is warmed to 37 °C (acceptable range: 34–35, but absolutely below 37 °C) and poured into the slicing chamber where gassing continues. The Campden slicer provides a dedicated heater element (Physiological temperature controller, model 7611A, Campden Instruments, UK); if this is not available, it is possible to maintain the temperature relatively constant by using the chamber usually meant to hold ice as a container for warm water. Care must be taken that the slicing chamber does not overheat.

A thoroughly cleaned recovery chamber (model BSK1, Scientific System Designs INC, CA) that allows continuous, smooth gas delivery as well as temperature control is prepared by filling it with warmed, gassed SPS. It is important to prepare the chamber well in advance so that any bubbles that will form on the mesh can be removed before placing the slices in it, as they will ruin the slices if coming into contact with them. It is also critical to adjust gas flow so that the slices will not move on the mesh during incubation.

3.2.9 Brainstem Dissection

Animal is anesthetized with lethal dose of anesthetic agent (as instructed by the institutional protocols; we use a small amount of isofluorane (200–400 µl) placed in an airtight box with the animal). When unconscious and nonresponsive (confirmed with a *strong* toe-pinch), the mouse is swiftly decapitated with sharp surgical scissors with blade length at least 40 mm (Surgical scissors-serrated, 14 cm length, 40 mm cutting edge, Fine Science Tools, Vancouver, Canada). It is imperative that these scissors must be in excellent condition, otherwise the decapitation will invariably pull on the skin, muscles and vertebra that lead to brainstem deformation. After decapitation, the head is briefly rinsed in warmed SPS and moved onto a deep petri dish with clean SPS before opening the skull.

To expose the IO-containing brainstem safely, small scissors (with hardened fine blades of at least 20 mm and sharp straight tips, Fine Science Tools, Vancouver, Canada) are used to laterally incise cervicals to the parietal bone through the occipital condyle right above the ear canal (Fig. 2a). During this step, it is imperative to avoid damaging brain tissue with the scissor blade on the inside of the skull, by aligning it along the bone walls, and using the external blade of the scissors to perform the cutting while the inside part slides dorso-ventrally along the bone with minimal movement. Then, the parietal bone is cut laterally and the whole dorso-posterior part of the skull removed with small forceps (Fig. 2b). If properly done, colliculi and cerebellar cortex are visible on dorsal brain surface and cover the whole brainstem. Next, the caudal part of the brain is separated from the rostral part using a new, clean razor blade or scalpel blade cutting through the colliculi while the skull is held perfectly horizontal and still (Fig. 2c). A sharp and clean vertical cut will separate the brainstem/cerebellar cortex block from the rest of the brain. At this stage, the tissue block is still attached to the skull by cranial nerves and conjunctive tissue (Fig. 2d). Brainstem/cortex block will now be separated with thin forceps from the bone, avoiding any tissue stretching. Through the procedure, the tools must only be in touch with cerebellar cortex or lateral and dorsal parts of the brainstem to avoid damaging the IO, and absolutely avoid touching the ventral part. If the tissue block does not separate easily from the bone, the cranial nerves need to be carefully cut one by one with thin forceps. Note that the dissection of the brainstem from the skull should not take longer than 150–250 s (from decapitation to full extraction).

Brainstem Slicing

Within a few seconds of extracting the brainstem from the skull, it is lifted on a spatula with the rostral face facing down, and its surfaces are lightly dried by dabbing with absorbent paper without squashing the tissue. Next, a small amount of cyanoacrylate superglue (around 5 mm diameter drop) is placed on the completely dry surface a tissue block holder. The amount should be such that it

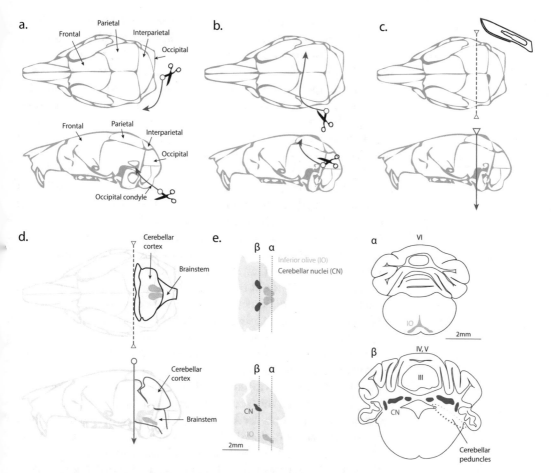

Fig. 2 Procedures of brainstem extraction and slicing. (**a–c**) Dorsal (top) and lateral (bottom) view of mouse skull and the three steps of skull cutting. (**a**) *step 1*: First scissor cut from occipital condyle (posterior) to parietal bone (anterior). (**b**) *step 2*: second scissor cut is along parietal bone, from one side to another. (**c**) After extraction of bone with forceps, scalpel blade separates brainstem. (**d**) The block of tissue contains brainstem and cerebellar cortex. Scalpel cut has to be vertical to keep IO intact. (**e**) The first slice containing IO (slice alpha) is within the first 500 µm of sectioning (from dorsal part to ventral, if brainstem is glued of coronal section). When cerebellar cortex and brainstem are joined together with cerebellar peduncles of the same slice (slice beta), this is one of the last ones containing IO

covers an area larger than the brainstem tissue block. The brainstem block is then delicately dropped straight onto the glued region, then immediately covered with the warm, gassed SPS by applying it gently on top with a pipette. Note that perfect execution of this step is necessary for the "hot slicing" method to be successful, and we recommend investing time in practicing this fine motor skill. Next, the tissue block holder is placed in slicing chamber so that the ventral side of the brainstem is facing the vibratome blade and slicing can begin. The optimal vibratome operation parameters vary between brain regions and specifics of the equipment used. We find that 80 Hz vibration frequency with 1.25 mm horizontal amplitude is optimal when cutting IO slices at 200–300 μm thickness. Advancing speed should be set as low as possible (0.1 mm/s). Note that with such slow cutting speed the whole procedure lasts at least 20–30 min for 4 slices. The slicing chamber must be kept at constant temperature and well-gassed, while it must be made sure that the gas bubbles do not cause liquid movements that could damage the tissue.

Slow slicing is beneficial for slice quality. However, extended duration of slice procedure has a negative effect on neuron survival as oxygen diffusion is limited within the brain tissue block. To reduce slicing time, we suggest to discard as much as possible of the tissue caudal to the brainstem. The IO spans roughly 1 mm rostro-caudally (Fig. 2e) starting from the level where the brainstem joins cerebellum (easily identifiable by the visible cerebellar peduncles). With 300 μm slices, this means that to optimize the slicing time, no more than 4 slices should be produced.

Each cut slice should be collected immediately after separating from the blade using either a plastic or glass Pasteur pipettes, while avoiding collision on incubation chamber and any mechanical stress by pressure or glass contact. Presence of thick myelinated fibers in the brainstem causes that slices often stick to glass when collected; in that case, delicate pressure movements will help detaching the slice. Direct contact with any tools should be avoided. Collected slices are placed in the recovery bath with the warmed, gassed solution SPS and not used for experiments earlier than 60 min after slicing has completed. This "resting" period provides time for the most damaged, superficial cells to disintegrate, and the surviving cells to recover from the slicing shock, resulting in significantly clearer view. Also, at this moment the researcher is advised to take a break to allow uninterrupted experimenting for the rest of the day.

3.2.10 Assessing Slice Quality

Slicing and the ensuing mechanical and biochemical stress invariably leads to some extent of cell death and membrane degradation, and only through patient practice and procedure optimization it is possible to reach reproducible high quality acute slice experiments. When improving on the technique, the researcher should check the slice quality before attempting to conduct any physiological experiment and exclude sub-standard slices (*see* **Note 6**). Here we provide some useful features to look for to ensure that the slices used in experiments are of uniform quality.

1. Slices that accidentally swirl inside bath, float on liquid meniscus, stick on dehydrated surfaces should not be used for any recording related to cell physiology. When transferring and positioning the slices in recording chamber, care must be taken to avoid any mechanical stress such as slice desiccation, excessive movements or slice compression.

2. General characteristics of slices and neurons as visualized under DIC optics. The slice perimeter should be smooth and the overall appearance of the slice should be bright and uniform. While a "bubbly" surface indicates significant damage, the density of neurons is function of slice quality (at least 5–10 neurons visible in a field of view when using 40x objective); if only few are visible, the network is likely more damaged to obtain meaningful experimental results. Neurons with smooth and spherical membranes are dead and a slice regions containing large amount of such cells should not be used. Sometimes a portion of primary dendrite root section (2–5 μm wide) is visible around somatic area of healthy neurons (*see* **Note 7**).

3. The stiffness of neuronal membrane is an excellent metric for estimating slice quality. This can be tested by extruding intracellular solution from the patch pipette while approaching. Fragile membranes fall apart in front of an approaching pipette approach; also, unhealthy neurons will react to the change in local osmolarity due to solution change and disintegrate by swelling. Slices with such neurons should be discarded in entirety, as there is no point in trying to obtain recording in such a poor network condition.

4. "Healthy IO neurons" membrane voltage stabilizes quickly at negative potentials (−45 to −65 mV) after establishing whole-cell patch-clamp configuration in current clamp mode. Further assessment can be done by observing the IO neurons' stereotypical electrophysiological behavior, such as the classic IO spike with significant calcium shoulder lasting up to 100 ms, spikelets and a prominent after-hyperpolarization. The signature behavior, subthreshold oscillations (5 mV to 20 mV) with sinusoidal waveforms.

3.2.11 In Vitro Voltage Imaging

Experimental Setup

For the voltage imaging experiments, slices that are judged to be of acceptable quality are viewed with a bright-field microscope (BXW51, Olympus) magnified with 5× (MPlan N 5× 0.15 NA air, Olympus) or 60× objective (LumPlanFL N 60× 1.00 NA water, Olympus) and illuminated with a 488 nm wavelength laser source (3 W custom 488 nm laser; Genesis MX-488, Coherent, USA). We have adapted our setup to the specific excitation/emission properties of A6+ [6]. On our setup, 488 laser light is attenuated ten times between laser aperture and bath chamber and we use typically between 1.5 and 6.5 mW light for our experiments under the 60× objective. To increase the voltage signal, the emitted light is high-pass filtered at 561 nm (RazorEdge ultrasteep long pass edge filter, Semrock) and then imaged by a CMOS camera (MiCAM03, BrainVision). The slice in the recording chamber is constantly perfused with gassed and temperature-controlled SPS. Acquisitions are performed at 40 fps or 120 fps (256 × 256 pixel resolution).

Acquisition Steps

The steps taken during a full experiment are listed in the following.

1. Slice immobilization

    Fresh brainstem slices are immobilized with a "harp" or other immobilization methods (*see* **Note 8**). Complete slice immobility is absolutely essential for acquiring good quality signals. Vibrations and subtle drift in slice position might not be visible under 5× magnification, so we recommend to observe the slice with high magnification (40× or 60×) objective to be sure that the slice is indeed stable.

2. SPS perfusion

    Slices are completely immersed in a constant flow of SPS during the experiment.

3. Slice selection

    The condition and location of the region of interest is confirmed using low-magnification (such as 5×) visualization in DIC mode, including checking that the structure is not folded or "floating."

4. ROI selection

    After switching to fluorescence mode the A6+ staining can be visualized. Injection epicenters should be recognizable with low intensity illumination (less than 2 mW illumination with 10 ms exposure time with our CMOS camera and laser). While it is usually best that the A6+ injection is centered within the nucleus of interest (Figure 3a shows a successful targeting of IO dorsal nucleus), in some cases appropriate staining can be obtained by diffusion from an injection epicenter located outside of the structure (*see* **Note 9**).

Basic Illumination
and Acquisition Parameters

Optimal image acquisition parameters are found by balancing sampling rate and illumination intensity, with the ideal combination varying depending on the characteristics and strength of signals of interest. Briefly, while low sampling rates lead to better signal-to-noise ratio, they may result in unacceptable signal aliasing, and Nyquist laws must be taken into consideration [10, 21]. Conversely, high sampling rates (short exposure time) impair target visualization and lead to bad pixel histogram equalization. Similarly to choosing optimal sampling frequency, the illumination intensity selection involves balancing between stronger SNR and recording quality. Specifically, weak illumination may contain a significant component of laser noise that can occlude the neuronal signal. Importantly, some lasers are unstable when operated at very low power ranges; we recommend thoroughly familiarizing with the technical specifications provided by the manufacturer. In contrast, strong and prolonged laser illumination (we rarely use more than 20 mW laser power under objective) can lead to signal saturation,

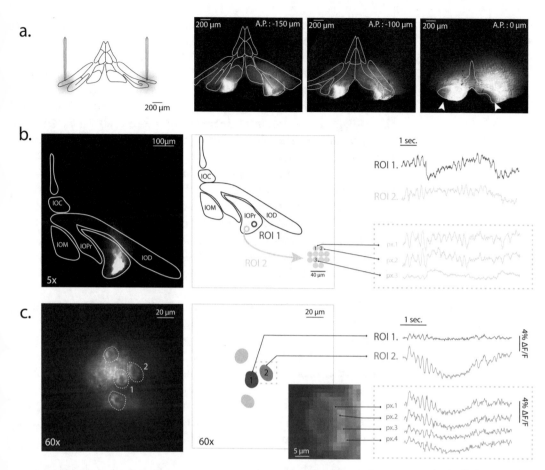

Fig. 3 Voltage imaging of IO STO with A6+ voltage reporter. (**a**) Confocal images of A6+ spread inside inferior olive a few hours after in vivo microinjection. A6+ is bilaterally targeted to the Principal and Dorsal nuclei of IO and travels hundreds of micrometers in the brain tissues. (**b**) Wide-field voltage imaging performed in vitro with low-magnification objective (5×). The compound is targeted to the principal nucleus of IO (IOPr, left). 2 different regions of interest display similar 3.5 Hz sinusoidal oscillation when A6+ is excited at 488 nm during 10 s recordings sampled at 80 fps. STO are also visualized within IO nuclei at the single pixel level (bottom inlay) (**c**) High magnification voltage imaging under 60× objective in vitro. IO neurons are stained by A6+ and can be identified individually on maximum intensity projection image (left image). Different regions of interest are defined by circling individual neurons (central panel and caption) and the average trace sampled at 60 fps shows typical 6 Hz STO obtained on the 2 different cells. 4 pixels of the cell membrane of one IO neuron show similar overlapping STO pattern (right panel)

bleaching, phototoxicity, or even physically damage of the tissue due to heating.

The optimal parameters are to be found through systematic measurements of signal-to-noise ratios and signal stability using the preparation that is to be used in experimentation, and regularly re-checked to avoid possible degradation of experimental quality due to equipment or dye performance changes.

Adjusting Signal Acquisition	First, trial acquisitions are performed with low illumination power (2–3 mW) under low-magnification 5× objective. It is advisable to start with a 5–10 s acquisitions at 100 Hz, and decrease the laser power in each trial until an appropriate value is reached to obtain a stable recording. In this condition, the strongest fluorescence signals originate at the injection center and falls off with distance. If the average signal intensity decreases by more than 10% during a 10-s acquisition, the illumination intensity needs to be decreased until the signal is stable. When lowering laser power, the exposure time must be increased (by decreasing sampling rate) to compensate for the decrease in signal strength. If the sampling rate must be decreased to the extent that it fails to correctly capture the temporal features of the signal of interest, the experiment will be useless and experimental conditions need to be adjusted (*see* **Note 10**). Conversely, if the staining is strong and signal saturation is evident, the sampling rate can be increased (sometimes up to 200 Hz).

Special Care for High Magnification Objectives

At 60× magnification, it is preferred to image stained neurons with low (around 2 mW to 5 mW) laser light (under objective) at 60 Hz sampling rate in order to avoid fast photobleaching. Also, IO cells are sensitive to 488 nm light. In our experimental conditions, we found out that powerful laser light suppresses STO while recorded in whole-cell patch-clamp configuration (*see* **Note 11**).

Examples of successful IO neuron staining and voltage imaging of STO is shown in Fig. 3. For 5× low-magnification (Fig. 3b) and 60× high magnification (Fig. 3c) objectives.

3.2.12 Tips for Improving the Quality of Recordings

Here are a few tips to optimize signal acquisition with air objectives.

Dust in Perfusion

Small particles (such as brain tissue parts, dust or other aggregates) create unwanted artefacts in wide-field signals. For wide-field imaging, we recommend an open perfusion system (clean SPS flowing through the chamber only once and disposed) using big quantities of ACSF (more than 0.5 l, for performing experiments for several hours) to avoid debris and dust accumulating in the perfusion system. If a closed perfusion system is necessary (such as when using expensive perfusion compounds), adverse effects of floating particles in wide-field fluorescence imaging can be limited by turning perfusion off for the acquisition periods if they last less than 10–20 s.

Bath Liquid Level

If the liquid level fluctuates during the recording, the focal point of the objective will be likewise unstable. This is why it is essential to minimize liquid vibrations during acquisition periods, and it is recommended to re-focus the objective before each acquisition to compensate for possible slow changes in liquid level.

Limiting Photobleaching	If badly adjusted, laser light can be a threat for brain tissue and can damage cells (overexposure can create overheating, photodamage, tissue section, blebbing). With the illumination parameters described here, photobleaching should be insignificant with $5\times$ objective. In our conditions, we can repeat 10-s acquisition trials for at least 30 times during 1 h without noticeable signal decrease (But *see* **Note 11**).
Slice Immobilization	Brain slices are commonly stabilized with "harps" that are often not well suited for small structures such as the IO. Harps often cut the structure of interest (the spacing between wires is small), impair visualization and wires are sometimes auto-fluorescent (*see* **Note 8**). Replacing harps by silver pellets (home-made from silver or platinum filament) can make visualization of the region of interest easier. Silver pellets are small but dense and heavy, and scattering them around the slice (but away from the region from interest) considerably improves the slice stability compared to harp use (*see* **Note 8**).

4   Notes

1. We cannot ensure that the parameters we quote for A6+ micro-injections in IO are optimal for all possible experimental conditions. Parameters that may lead to different outcomes include animal strain, age and sex, as well as technical features of the stereotactic device used. Furthermore, as described in our recent publication [25], it is recommended to try different times of post-injection incubation for best results.

2. Permanent marker pen is often used for labeling the drilling target on skull surface. However, pen can be imprecise and lead to off-target drills as pen tips are often too big, the ink stain spreads non-homogenously and the experimenter's hands can be shaky). For precise targeting, visual observation of the skull at the coordinates indicated and noting subtle imperfections (color contrasts, bone shapes, small blood capillaries) is essential for correct drilling.

3. On tip-filling the injection pipette with A6+. If the pipette is too small, the compound can block the tip. If this keeps on happening, adjust the pipette tip to be larger. Also, it is important to avoid air bubbles from going inside the pipette during tip-filling, as they form a "buffer" between the oil and A6+ fractions. This will disturb the smooth transmission of the pressure applied by the plunger to the liquids and impair dye injection.

4. After the pipette is completely out of the brain, application of slight pressure should result in appearance of a small droplet at

the tip of the pipette, confirming that the pipette did not get clogged during the injection. If the dye is still flowing through the tip during plunger advancement and the pipette collects dye correctly during plunger retraction, the same pipette can be used for contralateral side. If orange deposits are observed at the tip or the dye does not flow correctly, the pipette is to be discarded. If the dye level in the pipette didn't move towards the tip during injection, the procedure has to be done again on the same side with a new pipette.

5. For skin closure the wound has to be clean and dry before applying cyanoacrylate glue. We recommend to gluing the incision gradually from one end, as if closing a zipper. If the skin is wet, glue will polymerize and not form a strong bond with the skin. Finally, the wound closure has to be tight with no air allowed to enter between the skull and the skin.

6. When a new slice is being explored, the focus of objective and condenser, illumination diffraction angle and camera exposure settings need to be constantly adjusted for optimal DIC image visualization; note that each experimenter will develop their own style of viewing. The membrane of healthy cells is less visible in DIC than the membrane of dead cells, meaning that locating healthy neurons inherently depends on the experimenter's patience and expertise. Importantly, it should be repeated, especially to younger colleagues that it is better to spend an hour looking for a good sample location than record from ten poor cells and obtain useless data.

7. Slicing any brain region will inevitably induce massive neuronal death, and IO is relatively tricky due to heavy myelination at the brainstem. However, for various reasons we find that there are certain niches of the IO where neurons seem to be having a higher chance of surviving the trauma. These include the periphery of blood capillaries, regions close to slice external border (including ventral part of the principal and dorsal nuclei), and around the midline (medial part of medial nuclei or IOB, IOC).

8. On anchoring the slice to the recording chamber. Harps can be used for anchoring the slice inside the perfusion chamber (e.g. RC-27L harp, Warner Instruments, US) as well as mesh grids (e.g. HSG-MEA, ALA Science, US) or custom-made tools (e.g. fishing wires or fibers from nylon stockings glued on a bent piece of platinum wire). However, IO is a small structure and only few slices (less than 5) can be cut from a single adult brain. It often happens that the random positioning of the harp badly damages the only surviving neurons from the slice. To avoid this, we use 1–2 mm silver pellets positioned around the IO perimeter? They are heavy enough to maintain

brainstem anchored and keep intact the entire IO network. This method is optimal for wide-field imaging and small brain areas.

9. If the A6+ injection center is outside of IO, the dye diffusion can still stain the structure of interest. If the staining is strong enough, this "off-target" staining method may be advantageous as it prevents mechanical damage to the IO caused by the injection pipette.

10. Acquisition parameters are optimized for the visualization of biologically relevant signals and the experimenter needs to be aware of the potential sources of noise. First, the spatial localization of expected signal needs to match with coherent neurophysiological properties of signals (signals of interest are on membranes of healthy cells and not debris or bubbles). Strong signals are often artifacts: the acquisition parameters should not be adjusted to these noise sources (like harp, dye accumulation, auto-fluorescent structures). Also, perfusion flow can move small slice tissue fragments and generate artifacts. It is very tempting to select the brightest object for the acquisitions but the experimenter needs to dedicate time and energy searching to find the best ROI for recording biologically relevant signals.

11. The consequence of laser over-exposure are not always visible on a slice. Unlike blebbing or micro-sections caused by 2-photon illumination that are obviously damaging tissue, other damages can be invisible to the experimenter's eye. For example, during whole-cell current clamp experiments of IO neurons, laser light can inhibit IO STOs if power is stronger than 20 mW under objective. In this context, the tissue remains intact on DIC image (or fluorescence) while light alters electrophysiological properties (possibly by creating nanoscale perturbations of the membrane, or overheating).

References

1. Hodgkin AL, Huxley AF (1939) Action potentials recorded from inside a nerve fibre. Nature 44:710–711

2. Peterka PS, Takahashi H, Yuste R (2011) Imaging voltage in neurons. Neuron 69(1): 9–21. https://doi.org/10.1016/j.neuron.2010.12.010

3. Cohen LB et al (1974) Changes in axon fluorescence during activity: molecular probes of membrane potential. J Membr Biol 19(1): 1–36. https://doi.org/10.1007/BF01869968

4. Loew LM et al (1992) A naphthyl analog of the aminostyryl pyridinium class of potentiometric

membrane dyes shows consistent sensitivity in a variety of tissue, cell, and model membrane preparations. J Membr Biol 130(1):1–10. https://doi.org/10.1007/BF00233734

5. Hübener G, Lambacher A, Fromherz P (2003) Anellated hemicyanine dyes with large symmetrical solvatochromism of absorption and fluorescence. J Phys Chem B 107:7896–7902. https://doi.org/10.1021/jp0345809

6. Fromherz P et al (2008) ANNINE-6plus, a voltage-sensitive dye with good solubility, strong membrane binding and high sensitivity. Eur Biophys J 37(4):509–514. https://doi.org/10.1007/s00249-007-0210-y

7. Kuhn B, Roome CJ (2019) Primer to voltage imaging with ANNINE dyes and two-photon microscopy. Front Cell Neurosci 13:321. https://doi.org/10.3389/fncel.2019.00321

8. Knöpfel T (2012) Genetically encoded optical indicators for the analysis of neuronal circuits. Nat Rev Neurosci 13(10):687–700. https://doi.org/10.1038/nrn3293

9. Roome CJ, Kuhn B (2019) Voltage imaging with ANNINE dyes and two-photon microscopy. NeuroMethods 13:321. https://doi.org/10.1007/978-1-4939-9702-2\13

10. Nilsson J, Panizza M, Hallett M (1993) Principles of digital sampling of a physiologic signal. In: Electroencephalography and clinical neurophysiology/ evoked potentials. Wiley, New York, pp 348–358

11. Kuhn B, Denk W, Bruno RM (2008) In vivo two-photon voltage- sensitive dye imaging reveals top-down control of cortical layers 1 and 2 during wakefulness. Proc Natl Acad Sci U S A 105(21):7588–7593. https://doi.org/10.1073/pnas.0802462105

12. Dalphin N et al (2020) Voltage imaging of cortical oscillations in layer 1 with two-photon microscopy. eNeuro 7(3): ENEURO.0274-19.2020. https://doi.org/10.1523/ENEURO.0274-19.2020

13. De Zeeuw CI et al (1998) Microcircuitry and function of the inferior olive. Trends Neurosci 21(9):391–400. https://doi.org/10.1016/s0166-2236(98)01310-1

14. Llinás R, Baker R, Sotelo C (1974) Electrotonic coupling between neurons in cat inferior olive. J Neurophysiol 37(3):560–571. https://doi.org/10.1152/jn.1974.37.3.560

15. Lampl I, Y. Yarom Y. (1993) Subthreshold oscillations of the membrane potential: A functional synchronizing and timing device. J Physiol 70(5):2181–2186. https://doi.org/10.1152/jn.1993.70.5.2181

16. De Zeeuw CI et al (1990) Intracellular labeling of neurons in the medial accessory olive of the cat: II ultrastructure of dendritic spines and their gabaergic innervation. J Comp Neurol 300(4):478–494. https://doi.org/10.1002/cne.903000404

17. Leznik E, Makarenko V, Llinás R. (2002) Electrotonically mediated oscillatory patterns in neuronal Ensembles: An in vitro voltage-dependent dye-imaging study in the inferior olive. J Neurosci 22(7):2804–2815. https://doi.org/10.1523/jneurosci.22-07-02804.2002

18. Devor A, Yarom Y (2002) Coherence of subthreshold activity in coupled inferior Olivary neurons. Ann N Y Acad Sci 978:508. https://doi.org/10.1111/j.1749-6632.2002.tb07594.x

19. Leznik E, Llinás R. (2005) Role of gap junctions in synchronized neuronal oscillations in the inferior olive. J Neurophysiol 94(4): 2447–2456. https://doi.org/10.1152/jn.00353.2005

20. Dorgans K et al (2019) Short-term plasticity at cerebellar granule cell to molecular layer interneuron synapses expands information processing. Elife 8:e41586. https://doi.org/10.7554/eLife.41586

21. Candelas M et al (2019) Cav3.2 T-type calcium channels shape electrical firing in mouse lamina II neurons. Sci Rep 9(1):3112. https://doi.org/10.1038/s41598-019-39703-3

22. Eguchi K et al (2020) Advantages of acute brain slices prepared at physiological temperature in the characterization of synaptic functions. Front Cell Neurosci 14:63. https://doi.org/10.3389/fncel.2020.00063

23. Ankri L, Yarom Y, Uusisaari MY (2014) Slice it hot: acute adult brain slicing in physiological temperature. J Vis Exp 92:e52068. https://doi.org/10.3791/52068

24. Huang S, Uusisaari MY (2013) Elevated temperature during slicing enhances acute slice preparation quality. Front Cell Neurosci 7:48. https://doi.org/10.3389/fncel.2013.00048

25. Dorgans K, Kuhn B, Uusisaari MY (2020) Imaging subthreshold voltage oscillation with cellular resolution in the inferior olive in vitro. Front Cell Neurosci 14:440. https://doi.org/10.3389/fncel.2020.607843

Chapter 9

Mapping Synaptic Connectivity in the Cerebellar Cortex Using RuBi-Glutamate Uncaging

Ludovic Spaeth, Théo Gagneux, Kevin Dorgans, Antoine Valera, Izumi Sugihara, and Philippe Isope

Abstract

Synaptic functional organization of excitatory and inhibitory inputs in neuronal microcircuits organizes dynamics and underlies learning processes. Describing connectivity maps is, therefore, essential to bridge the gap between cellular neuroscience and behavior. Although in vivo extracellular recordings techniques can inform about neuronal discharge activity, patch-clamp recordings in brain slices enable synaptic mapping in principal neurons. Here, we describe a method allowing fast and reliable activation of presynaptic neurons on a wide area around the targeted cell and the establishment of excitatory synaptic connectivity maps. Specifically, we used glutamate uncaging (RuBi-glutamate) in combination with patterned light illumination to activate presynaptic neurons on acute cerebellar slices. Our goal is to describe a simple method to record, analyze, and reconstruct the functional synaptic organization of neuronal microcircuits in the cerebellum. This method may be applied in many other brain regions.

Key words Glutamate uncaging, RuBi-glutamate, Synaptic mapping, Cerebellum

---

1 Introduction

Information processing in neuronal networks relies on key cellular mechanisms. Sensory information such as stimulus intensity and duration is encoded into neuronal discharges that are translated into synaptic release onto principal cells of the cerebral cortex. Dendritic integration in principal cells governs the collective discharge of the local network and defines a population response—a barcode—of the network to a given stimulus [1–3]. If this population code produces a correct behavior, the dynamic of this neuronal network is then stabilized through long-term cellular learning mechanisms. Alternatively, undesirable responses (e.g. a failed or underperforming movement) lead to modification of network dynamics to improve the behavior. Decades of physiological experiments demonstrated that stabilization of network dynamics occurs through modification of synaptic transmission at individual

Roy V. Sillitoe (ed.), *Measuring Cerebellar Function*, Neuromethods, vol. 177, https://doi.org/10.1007/978-1-0716-2026-7_9,
© Springer Science+Business Media, LLC, part of Springer Nature 2022

synapses via potentiation, depression or silencing [4, 5]. Therefore, the functional synaptic organization, i.e. the connectivity maps defined by excitatory and inhibitory inputs onto a given neuron or group of neurons, underlies network dynamics [6]. The description of these maps in normal or pathological conditions should help understanding the etiology of neurological disorders.

While network dynamics and spike trains can be described using in vivo extracellular recordings or population imaging techniques, it remains difficult to infer the functional synaptic organization from these recordings. Glutamate uncaging methods combined with patch-clamp recordings of a target neuron on acute brain slices, developed in the 1990's [7, 8], allowed the systematic excitation of single or small groups of presynaptic cells [9–11] while monitoring the resulting synaptic inputs in the target cell soma. In these methods, presynaptic cell excitation relies on glutamate receptors located in the dendrites or the soma avoiding the direct excitation of neighboring axons as opposed to electrical or optogenetic stimulation. If glutamate uncaging is combined with a system enabling specific and localized illumination, the spatial organization of the presynaptic cells as well as their synaptic weights on the target cell can be reconstructed, defining synaptic connectivity maps. In the last two decades, this technique has been has been essentially developed using UV light caged glutamate (e.g. MNI-glutamate; for review see [12–15]. A considerable amount of information has been gathered in many areas of the nervous system (spinal cord, [16]; thalamus, [17]; cortical areas, [18–21]; cerebellum, [22]). Uncaging can be coupled with imaging technique allowing for mesoscale studies (e.g. in combination with autofluorescence [23, 24] or the study of postsynaptic transduction pathways such as calcium dynamics by coupling it with two-photon microscopy [25–31].

In all these applications, glutamate was uncaged with UV light and combined with laser scanning method. While we have used similar illumination methods in the past [32], in this chapter we present another method that does not rely on scanning devices and allow uncaging in the visible light and at a faster rate ([33]; Fig. 1; see also Note 4.1). We replaced the scanning components by a digital micromirror device (Mosaic, Andor, Oxford instruments, UK) that controls light source illumination (for technical considerations on patterned light illumination through DMDs, see [34]). This device contains an 800×600 array of individual micromirrors that can be toggle ON/OFF individually at high speed ($<\mu s$) by a computer driven software. The resulting patterned illumination update at a rate of up to 5 kHz update rate. Although commercial systems are available (e.g. ANDOR, Mightex) such device can be developed in-house using system derived from video projectors [35]. We took advantage of the then recently developed caged compound RuBi-glutamate, which can be excited with

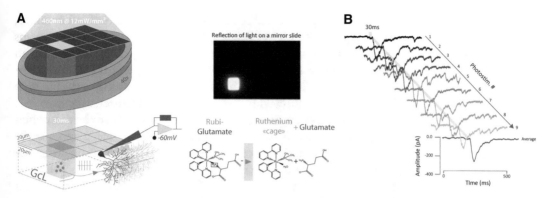

Fig. 1 Experimental setup for RuBi-glutamate uncaging using patterned illumination (see Notes 4.1 and 4.2). (**a**) A digital micromirror device (DMD, Mosaic, Andor Technology) allows illumination of well-delimitated squared regions (20 μm × 20 μm, see reflection of light on a mirror slide in the *upper right panel*) in the granule cell layer (in brown, GcL). Photostimulation of GC clusters (in green) may elicit excitatory postsynaptic currents (EPSCs) in selected Purkinje cell (in red) that are recorded via voltage clamp at −60 mV (whole-cell configuration). *Bottom right panel*, RuBi-glutamate. The ACSF perfusion contains 100 μM of RuBi-Glutamate. In its native form, this molecule is inactive as glutamate is covalently bound to a photochemical, ruthenium (Ru)-based, protecting group. A brief blue-light pulse (in our conditions, 460 nm) releases a functional molecule of glutamate that can activate glutamate receptor (AMPA/NMDA receptors on GC cell bodies and dendrites). (**b**) Glutamate uncaging on GCs elicit EPSCs in PCs. Example of a sequence of photostimulation (9 stimulations) on a 20 × 20 μm site. 30 ms blue-light flashes on GCs induce reliable photo-release of glutamate and corresponding EPSCs in the PC (traces #1 to #9 in shades of gray). In red, averaged EPSCs

one-photon visible light illumination (e.g. 460–470 nm, *see* [36, 37] and Notes 4.2). At these wavelengths, the same light sources (LED or diode-pumped solid-state laser, DPSS) can be conveniently used for both optogenetic stimulation and uncaging experiments.

We will describe here glutamate uncaging methods for the study of the functional synaptic organization of the cerebellar cortex. In this structure, the two major excitatory inputs, the climbing (CF) and the mossy fibers (MF), are topographically organized [38]. Groups of neighboring Purkinje cells (PC), the sole output of the cerebellar cortex, receive CF inputs conveying information from the same receptive fields defining functional units called microzones [39]. Microzones are parasagittal zones of the cerebellar cortex spanning across layers and composed of a group of PCs, the molecular layer above it, the granule cell layer underneath it, as well as their CF and MF inputs [38, 40, 41]. PCs from a given microzone receive MF information via a relay in the granule cell layer and project to specific zones of the cerebellar nuclei [42]. Therefore, microzones are considered as individual modules processing information from defined area of the body [38, 43–45]. How neighboring microzones are interconnected is largely unknown, but several cells (e.g. granule cells, Lugaro cells, molecular interneurons) can send processes across several of these parasagittal regions. Granule cells (GC) have a long axon, the parallel fibers, that transmits information to many different microzones, from the closest PCs to distant groups located hundreds of microns away from their somatic location [32, 46]. Since GC to PC synapses can be active or silent [47], this pathway controls the communication between microzones. Our recent findings using acute cerebellar slices in mice showed that this functional synaptic organization is conserved between animals, but can be modified by plasticity rules [32]. As opposed to the CF system, MF inputs originate in many precerebellar nuclei and carry sensorimotor information from the cerebral cortex (e.g. efferent copies of motor commands [48], the brainstem or the spinal cord (e.g. proprioceptive signals [49]. While MF inputs are topographically organized and partially match with CF receptive fields, they send the same information in several parasagittal area of the cerebellar cortex [50]. Therefore, microzones combine information from a single CF receptive field with a wide range of MF inputs from different receptive fields. Since CF inputs control synaptic plasticity at the GC-PC synapses [51, 52], MF and CF inputs synaptic processing organize PC discharges. Altogether, the operational rules underlying microzonal communication in the cerebellar cortex might be at the core of cerebellar computation[38]. Indeed clinical [53–55] and functional [56–58] studies have identified task-related microzones that can be selectively modified or altered leading to movement impairments in one part of the body and movement disorders. Many studies have also demonstrated that individual microzones have anatomical regional differences and diverse functional properties leading to a large diversity of information processing in individual microzones [59]. For example, a family of neurochemical markers, the Zebrins, which are expressed in Purkinje cells define parasagittal bands partially matching with CF topography [40, 60–62]. Therefore, an exhaustive functional description of the synaptic communication between microzones is necessary to understand how they differentially process information and influence motor coordination. In order to establish connectivity maps between microzones, we developed this method based on glutamate uncaging and described the synaptic weights at GC-PC synapses.

2   Materials

2.1   Mice

Cerebellar microzones can be observed through expression of neurochemical markers of the Zebrin family, in particular Aldolase C, also known as Zebrin II. The Zebrin II pattern is a reliable molecular tool to identify cerebellar functional units. Therefore, we took advantage of the knock-in mouse line expressing a reporter gene, Venus, fused to Aldolase C (ALDOC-Venus) [63] and bred under a CD1 background. These mice allow direct visualization of the Zebrin II pattern in vivo or in acute cerebellar slices (Fig. 2a).

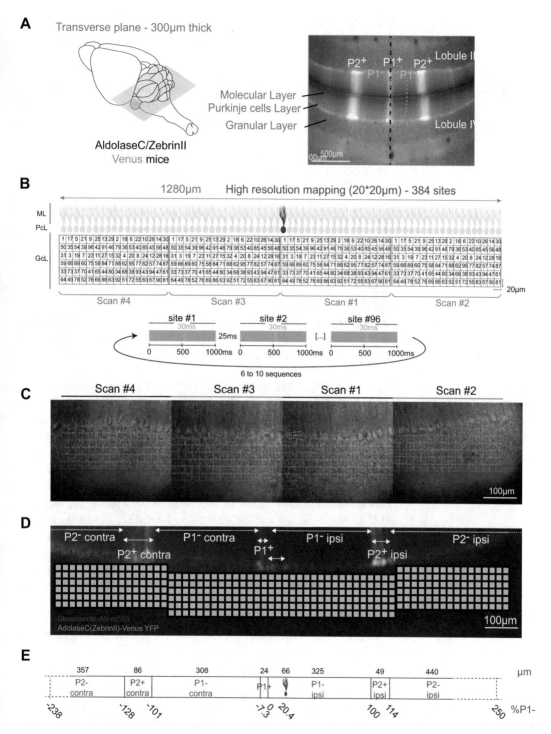

Fig. 2 Method for GC-PC synaptic mappings. (**a**) *left,* Transverse acute cerebellar slices from adult ALDOC-Venus mice are used to visualize of YFP-ZebrinII fluorescence during the experiment. *Right,* Region of interest in lobule III/IV of the anterior vermis. In our study, PCs are all located within a ~100 μm rostrocaudal stripe (in dashed blue) in the medial portion of P1⁻ Zebrin band. (**b**) *top,* Mapping protocol. The first field of view is positioned below the patched PC (in red, scan#1). A complete photostimulation sequence is performed when each site of the grid has been illuminated. The photostimulation sequence is repeated 6–10 times, then the

Since Venus fluorescence is present in both homo- and heterozygous individuals, this strain can be bred with other transgenic lines.

2.2 Setups for Photostimulation

Two different setups were used to develop connectivity synaptic mapping. We have initially used a confocal-based laser scanning method (FV-300, Olympus, Japan) mounted with a 473 nm DPSS laser. Light power measured under the objective (20× objective, Zeiss, Germany) was ~30 mW/mm². Now, our experiments are performed on a basic electrophysiological rig upgraded with a patterned illumination system (Mosaic2, Andor, Oxford Instruments, UK; see Notes 4.1) mounted with a collimated 460 nm ultra-high power LED (30 W, UHP-460, Prizmatix, Israel). The blue light is collimated and targeted to a chip composed of digitally controlled micromirrors (DMD) at the back of the illuminator of a microscope (BX51, Olympus, Japan). Light power measured under the objective (40× objective, Zeiss, Germany) is ~140 mW/mm² when all the mirrors are toggled ON. The Mosaic device is controlled by a native software, iQ Live Cell Imaging Software (version 3.0, Andor, Oxford Instruments, UK). iQ allows DMD calibration (*see* Subheading 3.2), setup configuration, photostimulation, and image acquisition through a cMOS camera (Zyla, Andor, UK).

2.3 Electrophysiological Setup

We use an Olympus microscope (BX51, Olympus, Japan) positioned on a XY movable platform and a LN junior 3-axis micromanipulator (Luigs & Neumann, Germany). We designed a closed and partially dark perfusion system controlled by a peristaltic pump (model PPS-2, MultiChannel Systems, Harvard Bioscience Inc., USA) in order to limit the volume of the perfusion to 10 mL. PCs are recorded using a Multiclamp 700B headstage (Molecular Devices, USA). The signal is recorded at 2–5 kHz, digitized at 20 kHz via a digital-analog interface (NI-6322, National Instruments, USA) and acquired by the WinWCP software written by John Dempster (John Dempster, University of Strathclyde, UK). The WinWCP software synchronizes recordings and photostimulation via TTL control of the light source and the Mosaic unit.

2.4 Pipettes

PC recording pipettes (3–4 MΩ) are pulled from borosilicate capillaries (ID = 1.17 mm and OD = 1.56 mm, Warner Instruments,

---

Fig. 2 (continued) field of view is moved along the mediolateral axis to extend the mapping. A complete map is composed of the concatenation of 2–4 scans. (**c**) Example of grids superimposed with the slice during an experiment. The recorded PC is circled in red, the tip of the patch pipette is represented by the white dashed lines. (**d**) Map reconstruction. Following experiments, slices are fixed in PFA in order to localize the recorded cell (in red) and to measure the size of the P2$^-$, P2$^+$, P1$^-$, and P1$^+$ Zebrin bands (in green). (**e**) Zebrin band measurements. The size of the Zebrin bands in the medial vermis and the distance between P1$^-$ ipsilateral medial edge and the recorded cell are measured in microns (upper values). Values are then normalized to the size of the P1$^-$ ipsilateral band (lower values) for each experiment

Harvard Bioscience Inc., USA) using a gravitational puller (model PC12, Narishige, Japan). Pipettes for loose cell-attached on GCs are pulled to reach 6–8 MΩ resistance with our internal solution.

2.5 Epifluorescence System

For post-hoc reconstruction (*see* Subheading 3.4), pictures are acquired using an epifluorescence microscope (Axio Imager M2, Zeiss, Germany) mounted with motorized platform.

3 Methods

3.1 Slice Preparation and Solutions

Acute cerebellar slices are prepared from P21 to P90 male ALDOC-Venus mice (*see* [33]). Mice are anesthetized by inhalation of iso-flurane 4% (Verflurane, Virbac, France) and then killed by decapitation. The cerebellum is kept in an Artificial Cerebro-Spinal Fluid (ACSF) containing (in mM): NaCl (120), KCl (3), NaHCO$_3$ (26), NaH$_2$PO$_4$ (1.25), CaCl$_2$ (2), MgCl$_2$ (1), C$_6$H$_{12}$O$_6$ (10), minocycline (0.00005) (Sigma Aldrich, USA) at 4 °C and bubbled with carbogene (95% O$_2$ and 5% CO$_2$). 300 μm thickness horizontal slices of cerebellum are made with a vibratome (HM 650 V, Microm, Germany), in a low-sodium solution containing (in mM): *N*-methyl D-glucamine (93), KCl (2.5), NaH$_2$PO$_4$ (1.2), NaHCO$_3$ (30), 4-(2-hydroxyethyl)-1-piperazineethanesulfonic acid (HEPES) (20), C$_6$H$_{12}$O$_6$ (25), Sodium Ascorbate (5), Sodium pyruvate (3), *N*-acetylcysteine (1), Kynurenic acid (1), MgCl$_2$ (10), CaCl$_2$ (0.5). After slicing, slices are maintained in a bubbled ACSF at 34 °C for 45 min and then at room temperature. During recording sessions, 10 mL of ACSF is perfused in a closed system and supplemented with (in mM): picrotoxin 0.1, D-AP5 0.05 (Ascent Scientific, Abcam Inc., UK), DPCPX 0.0005, AM251 0.001, CGP52432 0.001 and JNJ16259685 0.002 (Tocris-Cookson, UK) in order to prevent inhibitory transmission and modulatory activity from NMDA, adenosine, CB-1, GABA-B, and mGluR1 receptors on EPSCs amplitudes. The RuBi-glutamate is added just before the experiment in the closed perfusion system. RuBi-glutamate powder (Abcam, UK) is diluted at a stock dilution of 20 mM in ultrapure water. This stock solution can be stored in the dark at −20 °C.

For whole-cell PC recordings, pipette internal solution contained (in mM): CsMeSO$_4$ (135), NaCl (6), HEPES (10), MgATP (4), Na$_2$GTP (0.4). Biocytin (Sigma Aldrich, USA) or neurobiotin (Vector Lab, USA) were added (1 mg/mL each) for later cell labeling. For loose cell-attached GC recordings, pipette solution contained (in mM): NaCl (120), KCl (3), HEPES (10), NaH$_2$PO$_4$ (1.25), CaCl$_2$ (2), MgCl$_2$ (1), and glucose (10, Sigma Aldrich, USA). All solutions used are titrated at pH = 7.3–7.4 and osmolarity = 300 ± 10 mosmol.

3.2 Calibration of Photostimulation

To ensure homogeneous and precise photostimulation of cerebellar slices, a calibration procedure was developed in iQ software. This step is critical to ensure a reliable illumination of small areas (20 × 20 µm). The iQ software is set in acquisition mode from the camera while the photostimulation light source is turned ON. The slice is replaced by a mirror slide positioned at the top of the recording chamber and the objective is focused to image the surface of the mirror. Then in iQ the calibration setup wizard interacts with the camera to position perfectly a selection of pixels that can be observed on the mirror slide. At the end of this procedure all the micromirrors are toggled ON and we evaluate the available area for photostimulation. In our conditions, the 800 × 600 micromirrors illuminate an area of 320 × 262 µm. To select the photostimulated area, ROIs are drawn directly on the acquired pictures. In our experiments, grids of 96 squares (20 × 20 µm) were designed and each square is photostimulated in a pseudo-random mode.

3.3 Purkinje Cell Recording and Synaptic Mappings

PC are recorded at a holding potential (Vh) of −60 mV. Junction potential is not corrected and estimated at ∼ −10 mV yielding an actual Vh around −70 mV. Series resistance is monitored and compensated (60–80%) throughout the experiment using a square voltage step (30 ms at −10 mV from Vh) positioned at the beginning of each episode. Recordings are performed at room temperature.

Synaptic maps are built as follows (Fig. 2b, c): one file contains a collection of 96 episodes during which a single photostimulation (30 ms duration) is performed while the resulting EPSC is recorded in the PC. Therefore, a single file contains one map of a given field of view (320 × 120 µm). For each slice, the same map is recorded 6–10 times and the averaged map is calculated (Fig. 1c). Up to four field of views are concatenated to build one complete connectivity map in a given lobule (Total number of sites in a connectivity map = 384 sites, Fig. 2b, c).

3.4 Slice Reconstruction

After recordings (one PC per slice), slices are fixed in ACSF + paraformaldehyde (4%) solution for 24 h at 4 °C, then washed (3× 10 min with agitation) in a Tris–HCl (TBS), normal goat serum (1%, NGS), and Triton 100-X (1%) solution. Nonspecific epitopes are blocked by soaking slices overnight at 4 °C with agitation in TBS, NGS (10%), Triton100X (0.4%), supplemented with Serum Albumin (0.1%). The following day, slices are washed (3× 10 min with agitation) in TBS, NGS (1%), Triton (1%) and incubated in a TBS, NGS (1%),Triton 100-X (1%) solution supplemented with Alexa555-Streptavidin (1/1000, Molecular Probes, Thermo Fisher Scientific, USA) for 3 h at room temperature with agitation in order to label biocytin or neurobiotin. After a final wash (3× 10 min with agitation) in TBS, slices are mounted using Vectashield (Vector laboratories, USA).

Pictures of the slices are taken using a basic epifluorescence microscope and two channels were imaged: one to visualize the Zebrin band pattern via Venus fluorescence and a second one to visualize biocytin/neurobiotin via the Alexa555 fluorescence (Fig. 2d). Using ImageJ software (Rasband WS, NIH, USA), we measure the size of the zebrin bands and the relative position of the recorded PC along the mediolateral axis. In our conditions in CD1 mice, recorded PCs belong to the $P1^-$ zebrin band, in anterior lobules III or IV. Their position is determined by measuring the absolute distance between the recorded PC and the medial edge of $P1^-$ (i.e. boundary with P1+ zebrin band at the midline). In adult CD1 mice, averaged Zebrin band lengths in lobule III/IV were (in $\mu m \pm SD$): $P2^-$ contralateral, 417 ± 71; $P2^+$ contralateral, 72 ± 25; $P1^-$, contralateral 321 ± 63; $P1^+$, 35 ± 1 6; $P1^-$ ipsilateral, 320 ± 61; $P2^+$ ipsilateral, 70 ± 22; $P2^-$ ipsilateral, 438 ± 64. Since Zebrin bands are conserved across mice [40], PC positions are normalized according to the size of the ipsilateral $P1^-$ zebrin band (i.e. the band containing the recorded PC). Relative measures enable comparisons of synaptic maps between animals using a single system of coordinates. In figures, negative vs positive positions represent contralateral vs ipsilateral GC sites and Zebrin bands.

3.5 Connectivity Map Establishment

All subsequent analyses are performed using Python-based routines available on demand. For each recorded PC, the averaged synaptic map is built and the amplitude or the charge following photostimulation is measured in each site of the grid. In each average map, the distribution of the noise is estimated and used to calculate a z-score of amplitudes/charges (Fig. 3). A z-score above 1.96 indicates that the amplitude/charge of a given site is 2 standard deviations above the mean noise level yielding the description of connectivity maps related to zebrin band patterning. Figure 3c, d illustrates the patchiness of the GC-PC connectivity with silent sites and strongly connected GC areas.

4   Additional Recommendations

4.1 A Simplified Setup

Historically, photostimulation was performed using laser scanning methods as it was the only combination (laser and galvos) allowing for both localized light illumination and fast mapping in a small area of the tissue. Most of the time, these experiments were performed using confocal or two-photon microscopes which are quite expensive setups. New generation of diode-pumped solid-state lasers (DPSS) are now affordable and could be easily focalized for one-photon photostimulation. Moreover, the reinvestigation of digital micromirror devices initially developed for video-projection

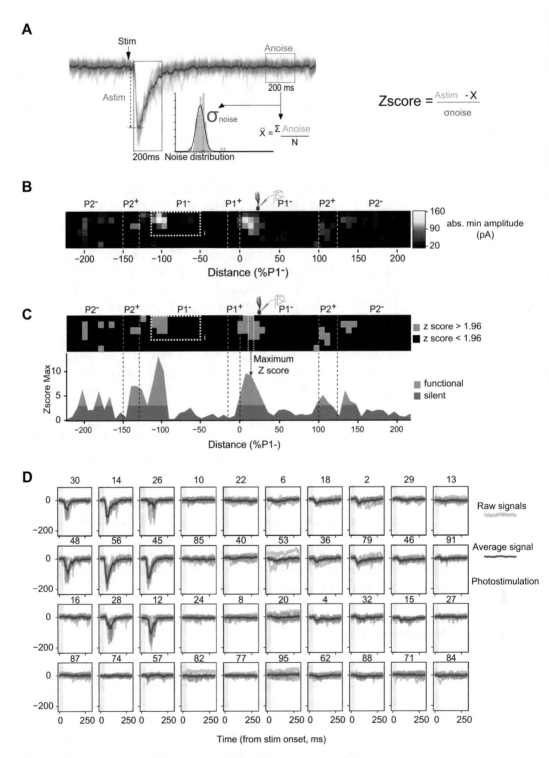

Fig. 3 Building a synaptic GC-PC connectivity map. (**a**) example of EPSCs evoked by illumination (30 ms, blue rectangle) in individual sites of the grid. Average signal for each site (in red) is superimposed to raw signals (in light gray). (**b**) synaptic amplitudes measurements and conversion into z scores. For a given site, the average (in red) amplitude (Astim) recorded in PCs is measured in a 200 ms window following onset of illumination. Corresponding background noise (Anoise) is measured in a 200 ms time window outside the

allowed for a precise one-photon illumination and can be used with similar spatial resolution with collimated LEDs or DPSS lasers [34, 35]. Toggling ON/OFF small micromirrors on a DMD chip in the optical path enabled simultaneous illumination of pixel-like area of the tissue simultaneously. This tool is cheaper and faster than laser scanning methods and an electrophysiological rig can then be upgraded for a few tens thousands of dollars. DMD based illumination can be very fast and meet important criteria for the generation of large mappings, notably the possibility to illuminate different areas of the slice with no delay (i.e. real patterned illumination). Activation of different groups of GCs could then be performed to study spatial and temporal summation at the GC-PC connection.

4.2 Controls for RuBi-Glutamate

We used RuBi-glutamate instead of MNI-glutamate or other UV excited compounds for three main reasons. First our DMD system for patterned light illumination blocks UV light (in our case the lower limit bound is around 400 nm, but recent version may have solved this technical limitation). Second, blue visible light penetrates deeper in brain slices with less damages [64–66]. Third, blue visible light allows to combine uncaging with optogenetic stimulation. Since RuBi-glutamate at 300 µM partially blocks GABAergic currents in cortical neurons [36], GC inhibitory transmission was not studied in our first series of experiments. However, for one-photon uncaging the concentration of RuBi-Glutamate can be decreased to 100 µM (and even below if using DPSS lasers) as opposed to 300 µM for to two-photon uncaging. At this concentration, our preliminary experiments suggest that inhibitory pathway is poorly affected in the cerebellar cortex. Indeed, we found that neither the amplitude nor the frequency of inhibitory miniature inputs on PCs are significantly affected by the perfusion of RuBi-glutamate at 100 µM (Amplitude: baseline, 29.25 ± 10.40 pA; RuBi-glutamate perfusion, 32.18 ± 7.94 pA; $p = 0.254355$, Friedman's test, $n = 9$; frequency: baseline, 7.73 ± 4.49 Hz; RuBi-glutamate perfusion, 5 ± 2.31 Hz; $p = 0.114899$, Friedman's test, $n = 9$). These findings suggest that this method may be applied to establish the connectivity maps for inhibition.

Fig. 3 (continued) illumination period (typically after the EPSC). The distribution of the noise is estimated (mean, X) and the half-width at half-maximum value of the Gaussian fit of the noise (σ) is used to calculate the z-score. (**c**) Map of the evoked amplitudes (Astim measured in A). (**d**) *Top*, binary z-score map. GCs eliciting significant synaptic amplitudes (i.e. z-score > 1.96) in recorded PC (in red) are shown in red; non connected GCs (i.e. z-score < 1.96) are shown in black. *Bottom*, the maximum z-score value in each column of GC layer is selected in order to build individual GC-PC input pattern. Significant z scores are in red, non-significant in dark gray

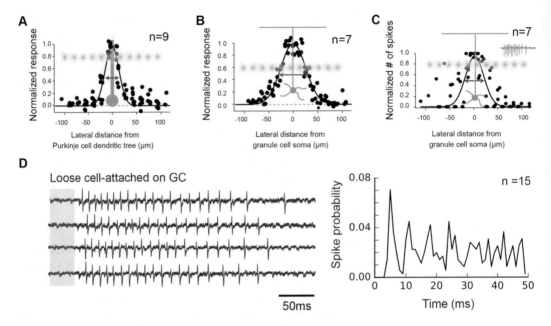

Fig. 4 Controls for RuBi-glutamate photostimulation. Data illustrated in (**a**, **b** and **c**) were recorded using laser scanning and DPSS laser illumination. Data illustrated in D were collected using patterned light photostimulation and LED illumination. (**a**) Normalized direct glutamatergic current recorded in a PC when the point scan is moved across the dendritic tree (half-width $= 33.0 \pm 1.8$ μm, $n = 9$, see text). (**b**) Normalized direct glutamatergic current recorded in a GC when the point scan is moved across the cell (half-width $= 59.6 \pm 2.1$ μm, $n = 7$, see text). (**c**) Normalized number of action potential elicited in GCs when the point scan is moved across the cell (half-width $= 19.6 \pm 6.6$, $n = 7$). (**d**) Action potentials are reliably elicited in GCs. *Left*, example traces of GC discharge following photostimulation. *Right*, timecourse of spike probability in the GCs recorded in left panel. (Adapted from 32, 33)

We previously assessed the diffusion of glutamate after RuBi-glutamate photoactivation by recording AMPA receptor currents evoked in PC dendrites when the spot of light (15–20 μm diameter, point scan using a DPSS laser) was moved by steps of 20 μm steps (Fig. 4a). We also estimated the extension of the activation of a given GC by the same spot of light by recording GCs both in loose cell-attached mode (half-width of evoked action potentials: 45.5 ± 10.9 μm, $n = 7$) and in whole-cell mode (half-width of direct stimulation: 59.6 ± 2.1 μm, $n = 7$) while uncaging glutamate in 10 μm steps (Fig. 4b, c). The average maximum number of evoked action potentials at the center of the spot was 19.6 ± 6.6 ($n = 7$) at near physiological temperature (32 °C). This experiment has been replicated at room temperature using pattern light photostimulation (Mosaic, Andor) with a LED source (UHP LED, Prizmatix, Israel). Similar activation was observed (Fig. 4d; mean spike number $= 22.6 \pm 13.5$ spikes, $n = 10$).

In our previous experiments [32], point scan photostimulation at a given site could elicit EPSCs of up to 10 pC. Since the synaptic weight of a single GC-PC connection is approximately 0.1 pC [47], tens of active GC connections are therefore stimulated simultaneously, attesting to the efficacy of the photostimulation protocol. In a recent study using patterned light illumination (MOSAIC, Andor) and LED illumination we demonstrated that a small number of GCs could be elicited yielding the study of unitary GC-MLI connections [33].

In conclusion, this glutamate uncaging method is a very versatile method that can be used to quickly and reliably map synaptic connectivity in many brain areas.

Acknowledgements

This work was supported by the Centre National pour la Recherche Scientifique (CNRS), Strasbourg University, the Agence Nationale pour la Recherche (ANR-2015-CeMod, ANR-2019-MultiMod, ANR-2019-NetOnTime) and by the Fondation pour la Recherche Medicale to PI (# DEQ20140329514) and LS.

References

1. Shepherd G, Grillner S (2010) Handbook of brain microcircuits, first edit. Oxford University Press, USA

2. Grillner S et al (2005) Microcircuits in action--from CPGs to neocortex. Trends Neurosci 28: 525–533

3. Grün S, Rotter S (2010) Analysis of parallel spike trains. Springer US, Boston, MA

4. Mack S et al (2013) Principles of neural science, 5th edn. McGraw-Hill Education, New Yor

5. Abbott LF, Nelson SB (2000) Synaptic plasticity: taming the beast. Nat Neurosci 3: 1178–1183

6. Buonomano DV, Merzenich MM (1998) Cortical plasticity: from synapses to maps. Annu Rev Neurosci 21:149–186

7. Callaway EM, Katz LC (1993) Photostimulation using caged glutamate reveals functional circuitry in living brain slices. Proc Natl Acad Sci U S A 90:7661–7665

8. Katz LC, Dalva MB (1994) Scanning laser photostimulation: a new approach for analyzing brain circuits. J Neurosci Methods 54: 205–218

9. Shepherd GMG (2012) Circuit mapping by ultraviolet uncaging of glutamate. Cold Spring Harb Protoc 2012:998–1004

10. Callaway EM (2002) Cell type specificity of local cortical connections. J Neurocytol 31: 231–237

11. Callaway EM, Yuste R (2002) Stimulating neurons with light. Curr Opin Neurobiol 12: 587–592

12. Ellis-Davies GCR (2019) Two-photon uncaging of glutamate. Front Synaptic Neurosci 10:48

13. Tran-Van-Minh AN et al (2019) Two-photon neurotransmitter uncaging for the study of dendritic integration, Neuromethods. Humana Press Inc, Totowa, New Jersey

14. Canepari M, De Waard M, Ogden D (2013) Combining Ca^{2+} imaging with L-glutamate photorelease. Cold Spring Harb Protoc 2013: 1165–1168

15. Specht A et al (2013) Characterization of one- and two-photon photochemical uncaging efficiency. Methods Mol Biol 995:79–87

16. Kato G et al (2007) Differential wiring of local excitatory and inhibitory synaptic inputs to islet cells in rat spinal lamina II demonstrated by laser scanning photostimulation. J Physiol 580:815–833

17. Deleuze C, Huguenard JR (2006) Distinct electrical and chemical connectivity maps in the thalamic reticular nucleus: potential roles in synchronization and sensation. J Neurosci 26:8633–8645

18. Shepherd GMG et al (2005) Geometric and functional organization of cortical circuits. Nat Neurosci 8:782–790

19. Shepherd GMG, Pologruto TA, Svoboda K (2003) Circuit analysis of experience-

dependent plasticity in the developing rat barrel cortex. Neuron 38:277–289

20. Fino E, Yuste R (2011) Dense inhibitory connectivity in neocortex. Neuron 69(6):1188–1203

21. Camon J et al (2018) The timing of sensory-guided behavioral response is represented in the mouse primary somatosensory cortex. Cereb Cortex 7:1–14

22. Walter JT, Dizon MJ, Khodakhah K (2009) The functional equivalence of ascending and parallel fiber inputs in cerebellar computation. J Neurosci 29:8462–8473

23. Lam YW, Sherman SM (2005) Mapping by laser photostimulation of connections between the thalamic reticular and ventral posterior lateral nuclei in the rat. J Neurophysiol 94: 2472–2483

24. Llano DA et al (2009) Rapid and sensitive mapping of long-range connections in vitro using flavoprotein autofluorescence imaging combined with laser photostimulation. J Neurophysiol 101:3325–3340

25. Carter BC, Jahr CE (2016) Postsynaptic, not presynaptic NMDA receptors are required for spike-timing-dependent LTD induction. Nat Neurosci 19:1218–1224

26. Abrahamsson T et al (2012) Thin dendrites of cerebellar interneurons confer sublinear synaptic integration and a gradient of short-term plasticity. Neuron 73:1159–1172

27. Nikolenko V, Poskanzer KE, Yuste R (2007) Two-photon photostimulation and imaging of neural circuits. Nat Methods 4:943–950

28. Sobczyk A, Scheuss V, Svoboda K (2005) NMDA receptor subunit-dependent $[Ca^{2+}]$ signaling in individual hippocampal dendritic spines. J Neurosci 25:6037–6046

29. Zhang YP, Holbro N, Oertner TG (2008) Optical induction of plasticity at single synapses reveals input-specific accumulation of αCaMKII. Proc Natl Acad Sci U S A 105: 12039–12044

30. Ngo-Anh TJ et al (2005) SK channels and NMDA receptors form a Ca^{2+}−mediated feedback loop in dendritic spines. Nat Neurosci 8: 642–649

31. Lovett-Barron M et al (2012) Regulation of neuronal input transformations by tunable dendritic inhibition. Nat. Neurosci 15: 423–430

32. Valera AM et al (2016) Stereotyped spatial patterns of functional synaptic connectivity in the cerebellar cortex. elife 5:e09862

33. Dorgans K et al (2019) Short-term plasticity at cerebellar granule cell to molecular layer interneuron synapses expands information processing. elife 8:e41586

34. Anselmi F, Banerjee A, Florin AD (2015) Patterned Photostimulation in the brain. Springer international Publishing, Cham, Switzerland

35. Dhawale AK et al (2010) Non-redundant odor coding by sister mitral cells revealed by light addressable glomeruli in the mouse. Nat Neurosci 13:1404–1412

36. Fino E et al (2009) RuBi-glutamate: two-photon and visible-light Photoactivation of neurons and dendritic spines. Front Neural Circuits 3:2

37. Amatrudo JM et al (2015) Caged compounds for multichromic optical interrogation of neural systems. Eur J Neurosci 41:5–16

38. Apps R et al (2018) Cerebellar modules and their role as operational cerebellar processing units. Cerebellum 17:654–682

39. Oscarsson O (1979) Functional units of the cerebellum - sagittal zones and microzones. Trends Neurosci 2:143–145

40. Apps R, Hawkes R (2009) Cerebellar cortical organization: a one-map hypothesis. Nat Rev Neurosci 10:670–681

41. Ruigrok TJH (2011) Ins and outs of cerebellar modules. Cerebellum 10:464–474

42. Sugihara I et al (2009) Projection of reconstructed single Purkinje cell axons in relation to the cortical and nuclear aldolase C compartments of the rat cerebellum. J Comp Neurol 512:282–304

43. Shambes GM, Gibson JM, Welker W (1978) Fractured somatotopy in granule cell tactile areas of rat cerebellar hemispheres revealed by micromapping. Brain Behav Evol 15:94–140

44. Apps R, Garwicz M (2005) Anatomical and physiological foundations of cerebellar information processing. Nat Rev Neurosci 6: 297–311

45. Ito M (1984) The cerebellum and neural control. Raven Press, New York

46. Pichitpornchai C, Rawson JA, Rees S (1994) Morphology of parallel fibres in the cerebellar cortex of the rat: an experimental light and electron microscopic study with biocytin. J Comp Neurol 342:206–220

47. Isope P, Barbour B (2002) Properties of unitary granule {cell→Purkinje} cell synapses in adult rat cerebellar slices. J Neurosci 22: 9668–9678

48. Kratochwil CF, Maheshwari U, Rijli FM (2017) The long journey of pontine nuclei neurons: from rhombic lip to Cortico-Ponto-cerebellar circuitry. Front Neural Circuits 11: 33

49. Sengul G et al (2015) Spinal cord projections to the cerebellum in the mouse. Brain Struct Funct 220:2997–3009

50. Shinoda Y et al (2000) The entire trajectory of single climbing and mossy fibers in the cerebellar nuclei and cortex. Prog Brain Res 124: 173–186

51. Ito M (2006) Cerebellar circuitry as a neuronal machine. Prog Neurobiol 78:272–303

52. Jörntell H, Hansel C (2006) Synaptic memories upside down: bidirectional plasticity at cerebellar parallel fiber-Purkinje cell synapses. Neuron 52:227–238

53. Martin TA et al (1996) Throwing while looking through prisms. I. Focal olivocerebellar lesions impair adaptation. Brain 119: 1183–1198

54. Delmaire C et al (2007) Structural abnormalities in the cerebellum and sensorimotor circuit in writer's cramp. Neurology 69:376–380

55. Patel VR, Zee DS (2015) The cerebellum in eye movement control: nystagmus, coordinate frames and disconjugacy. Eye (Basingstoke) 29: 191–195

56. Horn KM, Pong M, Gibson AR (2010) Functional relations of cerebellar modules of the cat. J Neurosci 30:9411–9423

57. Pijpers A et al (2008) Selective impairment of the cerebellar C1 module involved in rat hind limb control reduces step-dependent modulation of cutaneous reflexes. J Neurosci 28: 2179–2189

58. Darmohray DM et al (2019) Spatial and temporal locomotor learning in mouse cerebellum. Neuron 102:217–231

59. Cerminara NL et al (2015) Redefining the cerebellar cortex as an assembly of non-uniform Purkinje cell microcircuits. Nat Rev Neurosci 16:79–93

60. Sillitoe RV, Joyner AL (2007) Morphology, molecular codes, and circuitry produce the three-dimensional complexity of the cerebellum. Annu Rev Cell Dev Biol 23:549–577

61. Ahn AH et al (1994) The cloning of zebrin II reveals its identity with aldolase C. Development 120:2081–2090

62. Sugihara I, Shinoda Y (2004) Molecular, topographic, and functional organization of the cerebellar cortex: a study with combined aldolase C and olivocerebellar labeling. J Neurosci 24: 8771–8785

63. Fujita H et al (2014) Detailed expression pattern of aldolase C (Aldoc) in the cerebellum, retina and other areas of the CNS studied in Aldoc-Venus knock-in mice. PLoS One 9: e86679

64. Hernandez O et al (2018) Optogenetic stimulation of complex spatio-temporal activity patterns by acousto-optic light steering probes cerebellar granular layer integrative properties. Sci Rep 8:1–1

65. Chaumont J et al (2013) Clusters of cerebellar Purkinje cells control their afferent climbing fiber discharge. Proc Natl Acad Sci U S A 110:16223–16228

66. Al-Juboori SI et al (2013) Light scattering properties vary across different regions of the adult mouse brain. PLoS One 8:e67626

Chapter 10

Multi-site Extracellular Electrode Neuronal Recordings in the Rodent Cerebellar Cortex and Nuclei

Thibault Tarpin, Victor Llobet, Guillaume Dugué, Zuzanna Piwkowska, Andrés P. Varani, Andrei Khilkevich, David DiGregorio, Daniela Popa, and Clément Léna

Abstract

Multi-site electrodes have become essential tools for simultaneous recording of multiple neurons in vivo. Here, we aim to cover the basic knowledge needed to set up such recordings in the cerebellar cortex and nuclei, from the physical principles governing the extracellular signal to the practical protocols for recording and processing neurophysiological data. We discuss the different types of electrodes and the criteria for choosing a recording system. We then provide protocols for building simple and cost-effective fixed implants or for building custom microdrives and performing brain implant surgeries. Finally, we provide guidance for spike sorting, including identification of Purkinje cell complex spikes.

Key words Extracellular recordings, Current dipoles, Wire bundles, Tetrodes, Silicon probes, Microdrive, Implant surgeries, Spike sorting, Cerebellum, Purkinje cells, Complex spikes

1 Introduction

Extracellular electrophysiological recording is a powerful approach to measure neuronal action potential discharges with high temporal accuracy, while minimizing perturbations of neuronal physiology by the recording system. In contrast, intracellular recordings require piercing the membrane, while optical recordings require filling the intracellular medium or the membrane with organic dyes or genetically expressed fluorescent reporter protein. Extracellular recordings are also less sensitive than intracellular recordings to fluctuations over time in the distance between the electrode and the recorded neurons, and are therefore more amenable to in vivo recordings in awake and even in freely moving animals. Single electrode extracellular recordings have a long history in the cerebellum, but in this chapter we will focus on the use of multi-site electrodes in the cerebellum of rodents, which has been more

Roy V. Sillitoe (ed.), *Measuring Cerebellar Function*, Neuromethods, vol. 177, https://doi.org/10.1007/978-1-0716-2026-7_10,
© Springer Science+Business Media, LLC, part of Springer Nature 2022

recently developed (for example, *see* [1–4]). The main motivation for the use of multi-site electrodes is to increase the yield of experiments and provide better access to network dynamics by recording several cells at the same time. Another benefit is that generally each cell is recorded on multiple channels, with the spike amplitude on each channel reflecting the distance between the neuron and the recording site [5]; this redundant knowledge improves the signal-to-noise ratio in the detection of spikes and allows the separation of nearby neurons using spike sorting methods.

The aim of this chapter is to provide a simple guide to facilitate an experimenter's first multi-site extracellular recordings. We will provide a basic set of methodologies, which is for the most part similar to methods used in any region of the brain, but we will also attempt to provide guidance on the choices and adaptations that can or should be made to suit the objectives being pursued and provide examples of data that can be obtained in rodent cerebellum. For a recent description of multi-site recordings in the primate cerebellum, *see* [6, 7] for the issues of deep multielectrode recording in large animals.

2   General Considerations

2.1   The Extracellular Signal

When planning to record cerebellar neurons, it is important to keep in mind that the identification of cell types by the sole inspection of extracellular spike waveforms is not very reliable.

The recorded extracellular potential changes are due to electric fields generated by currents flowing in the extracellular space between current "sources" and current "sinks" (Fig. 1). Extracellular currents are caused by transmembrane neuronal currents that trigger a reorganization of the ions in and around the neuron. To schematize the process, negative deviations of tens to hundreds of microvolts occur at a short distance from the neuronal surface where an inward current takes place (this represents a current sink in the extracellular medium), for example, around the sodium channels during a spike or around groups of excitatory synapses activated at the same time (e.g. in Purkinje cell dendrites following climbing fiber activation). However, the depolarization of the membrane caused by an inward current is not limited to the immediate vicinity of the channels: in the parts of the neuron remote from where inward currents take place (Fig. 1c, d), the neuronal membrane operates as a capacitor and the depolarization of the neuron is associated with a discharge of the capacitor, which produces a current moving away from the membrane (i.e. a current source), and a positive deviation of the extracellular potential; since the inward and outward currents balance each other and are far away from each other, we speak of current dipoles. The simplified diagram of Fig. 1c, d shall indeed be completed since the change

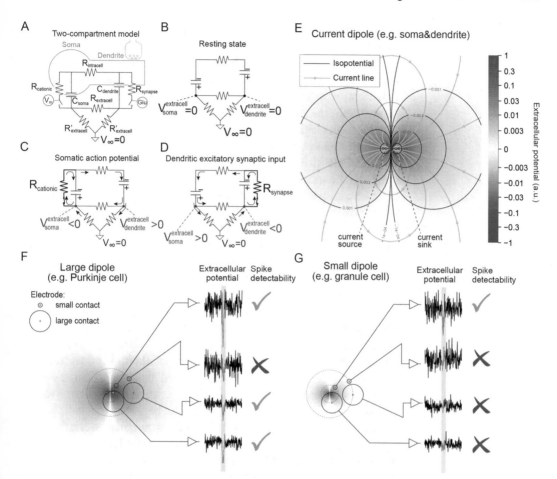

Fig. 1 Schematic principles of extracellular recordings. (**a**) Schematic of a simplified two-compartment model of neuron with only a cationic somatic conductance (represented by the resistance $R_{cationic}$) gated by the transmembrane potential (V_m) and dendritic synaptic conductance ($R_{synapse}$) gated by glutamate (Glu). The soma and dendrite have distinct membrane capacitance (C_{soma} and $C_{dendrite}$); the intracellular dendrite is resistively coupled to the soma ($R_{intracell}$), and the extracellular medium is reduced to three points: a point close to the dendrite, another close to the soma, and the last one far from the cell (with an electrical potential V_∞ null by convention); these points are coupled by resistances ($R_{extracell}$). (**b**) Resting state for the simple model: the somatic and dendritic capacitances are charged, but there is no current flow. The extracellular potentials are null. (**c**) Impact of a somatic cationic conductance (e.g. opening of sodium channels): the capacitances are discharging through the somatic conductance and yield a negative potential near the soma and a positive potential near the dendrite. (**d**) Impact of an excitatory dendritic input: the capacitances are discharging through the dendritic conductance yielding a positive potential near the soma and a negative potential near the dendrite. (**e**) Spatial distribution of the extracellular currents (green lines) and extracellular potential around a pair of current source/sink ("dipole") in a homogeneous, isotropic resistive medium. Note the logarithm scale of color: rapidly decreases in space (as the inverse of the squared distance to the dipole). NB, the electrical potential is also proportional to the dipole size and thus to the intensity of the current source/sink, and the distance between the source and sink. (**f**) Schematic of detection of a transient current dipole (e.g. extracellular spike) by small or large electrodes, as a function of distance. Small electrodes tend to have larger noise, but if positioned close to the cell can produce large spikes. Large electrodes can record the spike provided enough of their surface is exposed to a region of with large deflection of the extracellular potential during the spike. (**g**) Same as (**f**) for the same current source/sink intensity as in (**f**), but ten times closer to another. Note that the spatial region undergoing large extracellular potential deflection is much

in extracellular potential caused by a current dipole in a resistive medium depends from the distance to the dipole (Fig. 1e).

The size and the morphology of the neurons determine the current flow around the cells when they produce an action potential. First, a spike in a small cell is associated with a smaller total transmembrane current and thus generates a smaller extracellular current than a spike in a large cell (Fig. 1f, g). Second, the extracellular currents are determined by the structure of the neurites: large and long dendrites will produce extracellular currents in a wider region than short and thin dendrites. Third, current dipoles produce an extracellular potential signal that decreases as the squared inverse distance to the dipole; the size of the dipole depends on two factors: the intensity of the current and the distance between the source and sink of current. Hence, small cells are doubly penalized since their current sources and sinks are both small and closer to one another. The cerebellar cortex hosts a spectrum of cell sizes and morphologies, from the tiny granule cells with only four dendrites or the slightly larger mono-dendritic unipolar brush cells to the Purkinje cells with their large soma and huge dendritic tree. Similarly, the size of the neurons varies in the cerebellar nuclei. Due to the resulting variety in extracellular potential amplitudes, the sampling of neurons may be strongly biased depending on the electrode, as different electrodes will differ in their capability to detect the smallest, focal, signals (Fig. 1f, g, see below).

In real neurons, when a spike occurs, negative deviations are usually larger because the inward current is more focal (for example, stronger at the axon hillock), while the capacitive outward current is distributed over a larger part of the neuronal surface. A clear exception to this rule of thumb in the cerebellum is provided by Purkinje cells. Indeed, Purkinje cell simple spikes trigger detectable positive deflections in the proximal layer (*see* Fig. 3b, PC2). Moreover, during the complex spike, which results from climbing fiber activation, a very large synaptic current, associated with calcium spike(s) takes place in the dendrites. When recorded close to the Purkinje dendrites in the proximal molecular layer, the complex spikes are recorded as broad negative deflections of the extracellular potential ("fat spikes") corresponding to the dendritic inward currents (e.g. Fig. 3b, PC3), while when recorded close to the soma of Purkinje cells, the complex spikes are associated with a positive deflection of the extracellular potential due to the dendritic events (together with brief negative deflections due to the somatic sodium currents). Complex spikes can, therefore, be studied in isolation if recorded in the molecular layer (*see* for example [1, 8]).

Fig. 1 (continued) smaller. The extracellular signal is only detected by the small electrode provided that it is positioned close enough. For the large electrode, even if it is close, the portion of the surface exposed to a significant deflection of potential is too small to be detected

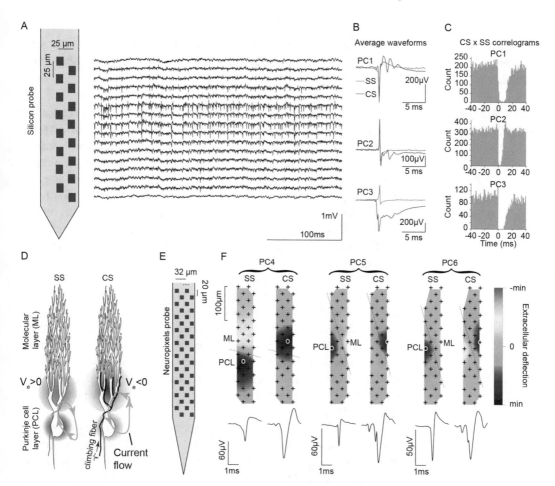

Fig. 3 Silicon probe recordings in the cerebellar cortex of awake mice. (**a–c**) Recordings with a Cambridge NeuroTech P4 probe. (**a**) Schematic of the electrodes and unfiltered extracellular traces. (**b**) Extracellular waveforms of pairs of simple and complex spikes recorded from the same channel. The spike sorting was performed with MountainSort4 (parameters for the complex spikes: "clip_size": 50, "freq_min": 150, "freq_-max": 2000, "detect_sign": 0, "detect_threshold": 3, "adjacency_radius": 51, "detect_interval": 10, "curate": "false"; parameter for the simple spikes: "clip_size": 40, "freq_min": 600, "freq_max": 6000, "detect_sign": 0, "detect_threshold": 3, "adjacency_radius": 51, "detect_interval": 10, "curate": "false"). Note the variability of the polarity of the fast and slow components of the spikes (displayed for the channel with the largest deflection). (**c**) crosscorrelograms between the complex spikes (CS) and simple spikes (SS) shown on panel C; a "pause" of simple spikes is observed after the occurrence of complex spikes. (**d**) Schematic of the deflection of the extracellular potential (Ve) during the simple spikes (left), when the main current inflow takes place in the soma and axon hillock, and during the complex spikes (right), when a strong current inflow due to synaptic activity takes place in the dendrites. Red and blue indicate positive and negative deflection of Ve. (**e**) Schematic of a Neuropixels probe. (**f**) Map of the extracellular potential at the peak deflection of simple spikes and complex spikes for three Purkinje cells. Each recording site is marked by a "+" sign. The site of the electrode which recorded the peak deflection of each type of spike is indicated with a white circle, and the average filtered waveform is indicated at the bottom. The color represents a linear interpolation of the extracellular potential recorded at the time of the peak of each type of spike. The boundary (dashed lines) between the Purkinje cell layer (PCL) and molecular layer (ML) is hypothesized to follow the sites of inversion of the potential. The spikes were sorted using Kilosort 2 with default parameters, followed by manual curation of the automatically detected clusters with Phy2 (merging/splitting and selection of good units)

2.2 Choosing (or Designing) an Electrode

Several factors should be considered when choosing an electrode or even designing a custom electrode; we only discuss here metal electrodes used for action potential recordings.

2.2.1 Size/Impedance/ Surface Treatment

The first degree of freedom is the size/impedance of the contacts of the electrode with the extracellular medium. Since the thermal noise increases with the impedance, the value of impedance impact on the signal/noise ratio; moreover, the combined impedance of the electrode and input impedance of the pre-amplifier stage of recording may form a voltage divider (*see* discussion in [9]). A low impedance may seem better. However, the impedance also reflects multiple factors which have a strong impact on signal detection, notably the size of the electrical contact of the electrode and the surface of the contact.

There is no ionic/electronic current flow across the interface between the electrode and the extracellular medium, and the electric coupling between the electrode and the extracellular medium is thus essentially capacitive; the impedance of the electrode is thus a complex impedance and depends on the frequency; by convention, it is usually reported at 1 kHz (which is close to the peak of the spectral power of extracellular potential). The impedance of the electrode decreases when the surface area of the contacts (hence the capacitance) increases.

Extracellular electrodes do not record the electrical potential at a single point; instead the electrical potential of the electrodes roughly corresponds to the average of the extracellular potential of the extracellular medium facing the electrode (this is in reality far more complicated since the electrodes disrupt the extracellular current flow). Therefore, large electrodes (Fig. 1f, bigger capacitance, lower impedance) will have a good signal-to-noise ratio, but will fail to capture efficiently focal deflections of extracellular potential at a scale smaller than their size (Fig. 1g); hence they will be more effective for large cells. In contrast, small electrodes will have higher noise but they may capture focal deflections of the extracellular potential if located at the right place (Fig. 1g).

Contact dimensions are not the sole determinants of the impedance. For a given electrode size, surface treatments may improve the impedance by increasing the surface of contact between the electrodes and the extracellular medium, without changing the section of the electrode (by creating a villous surface by material depositions or etching). Surface treatments of the electrodes with carbon fibers or conductive polymer (e.g. [10]) may also improve the biocompatibility, limit the inflammation caused by the electrode in the tissue, and prevent the glial ensheatment that can insulate the electrode from the extracellular medium.

The cerebellum hosts the largest spectrum of neuronal size from the tiny granule cells/unipolar brush cells to the Purkinje cells. Most commercially available electrodes have contacts dimensions in the 10–20 μm range and a relatively low impedance (<1 MΩ down to ~50 kΩ). These electrodes have a good signal-to-noise ratio but will work mostly in the cerebellum for targets that produce a significant extracellular signal over the surface of the recording site: Purkinje cells, Golgi cells, and glutamatergic projection neurons in the cerebellar nuclei are probably the easiest to record with such electrodes. Other interneurons will be harder to record: the range of electrode positions relative to the targeted cell yielding enough signal will be smaller, and electrodes on the lower end of contact size will be more likely to provide satisfactory detection of the neurons. Even the discharge of the smallest cells and compartments of the cerebellar cortex (granule cells, unipolar brush cells, and mossy fiber terminals) produce significant deflections of the extracellular potential, but only at very short distances; yet electrodes with small contact size (~5 μm, e.g. Cambridge Neurotech H1 electrodes) may capture this signal in the immediate vicinity of these targets.

2.2.2 Mechanical Stress and Trauma

The second factor to consider is the disruption induced by the electrode: all multi-site electrodes will have to push aside and tear the neuropil when inserted in the brain. This factor particularly matters in the cerebellum. Notably, the molecular layer of the cerebellar cortex is like a 3D fabric of interwoven parallel fibers perpendicular to sheets of dendrites, and if a 100 μm wide electrode passes through the bundle of parallel fiber afferents of a Purkinje cell, it may disrupt a large proportion of the inputs (there may be less disruptions for other parts of the cerebellum where afferents come from all directions). The mechanical disruption may also trigger a local inflammation and some gliosis which will degrade the recordings. A sign of disruption is that neuronal discharge is lost along the track when the electrode moves upward after a descent. During the penetration, limiting the lateral stress by applying a movement of the electrode as collinear as possible to the main axis of the electrode may reduce the damage induced by the electrode.

2.2.3 Stability, Head-Fixed vs. Freely Moving Animals

The third factor to consider is linked to the recording conditions: freely moving or head-fixed animal, chronic or acute recordings (electrodes left in place or not). These conditions have different requirements in terms of mechanical stability: the movements of the brain relative to the skull are minimal in anesthetized animals and low in (properly trained) head-fixed animals; in contrast in freely moving animals, the cerebellum tends to move relative to the skull, notably when the animal moves the neck. Rigid electrodes tightly fastened to the skull will provide unstable recordings (the brain will move relative to the electrodes on the timescale of minutes or hours) and will tend to produce a stronger mechanical stress on the tissue (e.g. [11]); in contrast electrodes with a loose attachment to the skull (and connected to the recording system through a flexible cable) may become "floating electrodes," that is, they will move with the brain rather than with the skull because of the friction between the brain and the electrodes and the weakness of constraints arising from the connections with the skull.

In our experience, the amount of electrical activity in the cerebellar cortex and nuclei seems stronger in freely moving animals than in head-fixed animals, and this shows up in the recordings as medium and high frequency, indistinct background activity (whether this is due to differences in recording conditions, e.g. contamination by electromyographic signal from the neck muscle- or differences in sensory inputs and behavioral outputs in typical head-fixed vs. freely moving condition is not clear); in this situation, electrodes in the smaller size range may be more susceptible of providing larger spikes, if correctly positioned, and thus a better signal/(noise+background) ratio.

2.3 Multi-site Electrode Typology

There are several classes of electrodes to choose from: wire bundles, glass-insulated multi-site electrodes, and micro-fabricated "silicon probes."

2.3.1 Wire Bundles

Wire bundles can easily be home-made, are extremely cost-effective and are flexible enough to become floating electrodes with excellent stability; they tend to be harder to advance, and to produce more damage in the rat cerebellar cortex than other types of multi-site electrode [1]), but they can provide satisfactory results in the cerebellar cortex [12]) and cerebellar nuclei (Fig. 2, [13]). We provide below a description of the manufacture of wire bundles (for a movie see also [14]).

2.3.2 Glass-Insulated Multi-site Electrodes

Glass-insulated multi-site electrodes, as commercialized by Thomas Recordings, are also effective in the cerebellar cortex [2, 6] and cerebellar nuclei (Fig. 2). Their morphology makes the adjustment smooth, and they induce minimal damage on the way as assessed by the possibility to record cells when pulling back the electrode along its track and by post-mortem histology. Their main drawback is their rigidity, which prevents them from behaving as floating electrodes in the mouse or rat brain.

2.3.3 Silicon Probes

Commercial silicon probes exist from multiple suppliers and in many variants, with single or multiple shanks, and with multiple distributions and sizes of contacts available. The main limitation for these electrodes is also their stiffness, but some models of electrodes with flexible connectors can be positioned in the brain at the

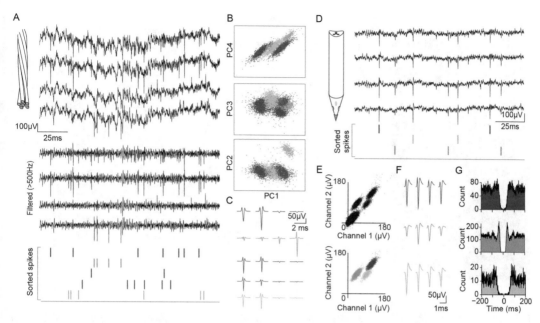

Fig. 2 Multi-site electrode recordings in the cerebellar nuclei. (**a–c**) Wire bundle recording in the dentate nucleus of a freely moving mouse. (**a**) Unfiltered (top 4 traces) and filtered (bottom 4 traces) extracellular potential. The raster at the very bottom identifies the spikes from 5 different units: each line color corresponds to a unit. (**b**) Spike sorting is performed with Tridesclous: the signal was bandpassed between 500 Hz and 7500 Hz and then subtracted from the common noise and a threshold of departure from more than 8 negative median absolute deviation was applied to select the waveforms. The clustering was then performed by the "pruning shears" method after a dimensional reduction keeping only the first three components of a PCA performed for each channel. To visualize the cluster separation, the three plots of clusters correspond to three 2D projections of the 12 PCA components considered (each dot corresponds to a spike). C: average unfiltered waveforms for the 5 units found. Same color code as for the units indicated in panels (**a**, **b**). (**d**, **e**) Thomas Recordings glass-insulated tetrode recording in the interposed nucleus of an anesthetized rat. (**d**) Unfiltered extracellular potential recorded from the 4 channels of the tetrode. The raster at the bottom indicates the sorting of spikes across units. (**e**) Visualization of the clusters before (top) and after (bottom) clustering using xclust2 (as in [2]). Three units are distinguished by the colors. (**f**) Average unfiltered waveforms for the three isolated units. (**g**) Autocorrelograms of the three units, showing a clear refractory period for each of them

time of surgery and not adjusted afterwards, leaving to chance to decide which contacts will end-up close to neurons. The silicon probes may provide nice recordings in the cerebellar cortex (Fig. 3; [15]) and in the cerebellar nuclei [16, 17]. Silicon probes may be implanted at a fixed position defined at the time of surgery, but if the density/number of channels is not very high, it may be of interest to mount the electrode on a microdrive. Such drives are commercially available, but we provide below a simple solution for building one by 3D-printing (Fig. 4).

The high-density high-count (hundreds of contacts) electrodes such as Neuropixels (384 simultaneous recording sites) provide multi-site recordings yielding many units in a single session or a series of closely-spaced sessions (within about a week), at the cost of

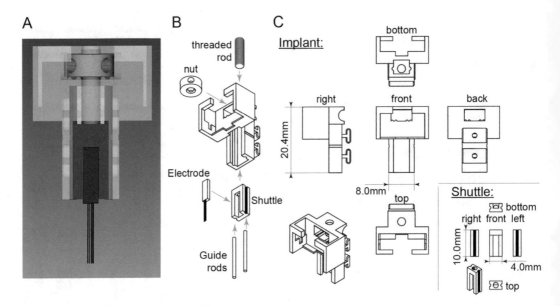

Fig. 4 plan for a 3D printed Microdrive. *See* Subheading 2.2 Microdrive for description of the parts. A resolution of 25 μm is needed. Nut: Thorlabs M2 × 0.20 LN2M20. Threaded rod: Thorlabs M2 × 0.20 F2ES10

more difficult chronic implantability. This dense data also provides insights into the spatial structure of the spikes (Fig. 3). These electrodes work well in the cerebellum [15] and are well suited for recording from different cellular layers simultaneously (granule cell layer, Purkinje cell layer and molecular layer with both inter-neurons and complex spikes) and even from several lobules at the same time.

2.4 The Future of Multi-site Electrodes

The combination of the advantages of floating wire bundles with the wide sampling of silicon probes is found in flexible multi-site electrodes. At the time of writing, micro-fabricated flexible electro-des have already been experimented and they hold promises of long-lasting stable recordings in freely moving animals [18–20]; these approaches will certainly be very useful for multi-site record-ings in the cerebellum during unconstrained behavior. The improvement of biocompatibility or even the engineering of the biological response to the presence of the electrode [21] may also considerably improve the quality and long-term stability of the recordings in the future.

2.4.1 Acquisition Hardware and Software

Nowadays, many commercial and open-source (Open-Ephys) solu-tions for amplification, digitization and treatment of extracellular recordings are available for purchase, and they will all provide good to excellent recordings for a wide range of cost and quality of support. Multiple criteria may need to be taken into account to choose the recording system. First, for freely moving animals, it is important to pay attention to the size and weight of the pre-amplifier/amplifier which will be placed on the head and thus may hinder the natural movements of the animal. A mouse can carry a (very) few grams on the head and an adult rat can carry up to 20 g without major hindrance, but the lighter the implant+headstage the better. Second, the integration of the neurophysiological recordings with other elements required by the experiments (e.g. video flux, behavioral apparatus, accelerometer, electrical and optical stimulations, etc.) may be variably easy, depending on the system and the instrumentation skills of the experimenter. Closed-loop experiments may be more demanding in terms of features of the recording system, and this is particularly true if online spike sorting is required. For experiments in the cerebellar cortex, spike sorting may be challenging due to the very high activity (many units firing above 100 Hz and thus generating temporal overlaps of spike waveforms require specific handling) and the presence of complex waveforms (Purkinje cells' complex spikes); it may thus be easier to collect continuous recordings of the extracellular potential, rather than spikes isolated by the recording system, for offline analysis. Finally, the spikes of the cerebellar units tend to be very brief compared to other brain regions (hippocampus or cortex), and a high sampling rate will improve the restitution of waveforms: 20 kHz or even 25 kHz sampling rates are less likely to capture the peaks of the spikes.

3   Experimental Procedures

3.1   Electrodes and Implants

3.1.1   Protocol for Bundles

Compared to commercially available electrodes, the use of wire bundles is extremely cost-effective, and allows the experimenter to easily record from multiple regions in a single mouse, combined with one or several optical fibers, etc., and generally to rapidly adapt the design to the changing constraints from one experiment to another. The main steps of the bundle construction are (1) to assemble wires into a bundle, (2) to place guide cannulas in space to target defined stereotaxic coordinates, and (3) to load the bundles and connect the wires to a connector to form the implant.

3.1.2   Bundle Creation

Each bundle is created from several loops of a Kanthal Nichrome (80/20) wire with polymide insulation (12 μm diameter). The wire is very fragile, so sharp tension should be avoided at all times to avoid micro-breaks in the wire which will degrade the signal. The wires are assembled into a bundle by twisting them together: this will notably stiffen the bundle enough to allow for the penetration in the brain. The twisting is achieved by fixing one point of the loops to an elevated fixed horizontal rod, and another point (2–3 cm below) to a clamp freely hanging on the wires. Another 2–3 cm of wires should be left untwisted above and below these points, for attaching the connectors later. The clamp is coupled via small magnets to a rotating magnet placed underneath and driven by a motor at approximately 60 rpm. Rotation should continue until all the wires are tightly bundled (a few tens of turns should suffice). A brief heating with a heat-gun along the bundled wires then fuses the polymide insulation to keep them permanently together. Heating for ~1 s along each point of the bundle is typically enough; the top and bottom part of the wires should not be heated since each individual wire will have to be separated from the others and connected separately. Finally, to obtain two bundles ready for use, a scissor cut is performed in the middle of the twisted part of the bundle. The loops of wire at the end of each bundle are then cut in the middle in order to separate each individual wire of the bundle using a soft tool (toothpick, carbide tweezer, etc.). The parts of the wire that seem damaged (e.g. with a kink) should be discarded.

3.1.3 Preparation of the Guide Cannulas

To position the bundles at the desired recording sites once in the brain, they are held in place by cannulas targeted to the sites of interest. The coordinates of the sites are determined from a stereotaxic atlas and/or test injections or test acute implantations of a DiI-coated metal needle. Guide cannulas are made of 30G stainless steel tubing and are initially positioned in space using a mold with holes at the required 3D positions; the mold can be produced by 3D printing or simply by micro-drilling an acrylic cube at the appropriate position and depth. The diameter of the holes in the mold receiving the cannulas must be adjusted as close as possible to the diameter of the cannulas, to ensure the accuracy and reproducibility of the implants. All cannulas must be parallel to avoid damage during the penetration. The extremities of the cannulas must be cleaned and rounded to avoid damage of the bundle when it is inserted.

3.1.4 Assembly of the Implant

The implant is made by assembling the cannulas to the connector, inserting the bundles in the cannulas and connecting the wires to the connector. In order to achieve the required level of precision during this step, all the manipulations should be performed on a stereotaxic frame under a binocular magnifying glass.

First, the cannulas, positioned in the mold, are glued with cyanoacrylate to the rim of the connector's circuit board (Neuralynx 16 or 32 EIB). Moreover, an isolated wire, which will be connected to the skull, must be soldered to the pin of the EIB corresponding to the ground (this pin will depend on the acquisition system). The mold is then removed, the EIB is held in the air with a manipulator. The bundles are then slid from top to bottom into the cannula, leaving only the 2–3 cm of untwisted wires at the top. Then, using surgical forceps, the wires must be individually inserted in the contact holes of the EIB, and Electrode Attachment

Pins from Neuralynx must be inserted into each contact hole after the wire to hold it in place and ensure good electrical contact between the wire and the board. Once all the contacts have been made, both sides of the EIB are covered with dental cement to protect from mechanical stress all parts of the wires which will not enter the brain. Finally, the twisted parts of the bundles are cut with precision scissors at about 1 mm from the cannula exit (the length may be adjusted to limit the amount of cannula entering the brain). The cut of the bundle should be clean and the tip of the bundle should be recut if it looks "pinched" or if the insulation appears torn at the tip.

3.1.5 Impedance Adjustment

A last step consists in adjusting the impedance of each wire. The impedance is tested at 1 kHz. It is reduced by electrodeposition on the wires with a gold solution (i.e. gold plating of the wire's end). For this purpose, the tips of the bundles are immersed in a cyanid-free gold solution (Sifco) and an electrical circuit is established with a current generator to run current pulses (3 μA during 1 s, the current running from the solution to the wire) until the impedance of each wire reaches ~200 kΩ. The presence of shorts between the wires must be tested at the end of the process. In case of shorts, the current can be reversed to attempt to remove the deposit and resume the plating. Beware that the interface crossing, when the bundles are pulled out of the solution, may occasionally rip the plating from the wire (this can be circumvented by testing that the impedance is unchanged after pulling out and returning the bundle to the solution). This step is best performed on the day of the surgery.

3.2 Microdrive

Self-production of microdrives for silicon probes by 3D printing is fairly easy and cost-effective. Moreover, it offers the possibility to personalize the design to suit electrode arrangements, addition of an angle, choice of steric hindrance, adaptation to a specific head fixation system, addition of parts facilitating the recycling of electrodes, etc. Stereolithography 3D printing technology, for example Forms3, is well suited for the drive construction in order to obtain high-precision resistant parts, with a precision of the order of 25 μm. The parts are first designed with a CAD software such as SolidWorks, which is very powerful (but expensive), or an open-source solution, such as FreeCad.

We present on Fig. 4 a simple design of a microdrive used for chronic cerebellar recording. The electrode is attached to a shuttle that moves under the action of a screw on a threaded rod collinear to the electrode's shuttle; the lateral slack is limited by metal guide rods which are inserted in grooves collinear to the movement on the outer sides of the shuttle and inner sides of the implant. The shuttle is moved up or down by the threaded rod under the effect of a captive nut, blocked in translation, but not in rotation, within a cavity. The upper part of the implant also comprises a cavity designed to host and protect a connector, held in place by lateral grooves. The protrusions on the back of the implant are adapted for a head fixation system (not drawn here).

The complete assembly is done before surgery. The electrode is manipulated with a clamp supported by stereotaxic arms, using a small piece previously glued to the surface of the electrode. The electrode is placed in the shuttle and glued to it. It is possible to use polyethylene glycol in order to recover the electrode later. The connector is then carefully placed in its cavity and glued with dental cement. The remaining length of the flexible cable is slid under the connector. A last piece (not shown) is finally added to protect the junction between the flexible cable and the connector.

This design provides a satisfactory stability of recordings and provides steps of ~25 µm in the translation of the electrode, corresponding to quarter of turns of the nut, in order to optimize the depth of the electrode during the course of the experiment.

3.3 Surgery

3.3.1 Before the Surgery

The microdrive and electrode assembly is prepared ahead of the surgery. To facilitate the histological reconstruction of electrode tracks, the shanks of electrodes can be covered with DiI (prepared by diluting DiI crystals in ethanol). Alternatively, sulfonated derivatives of DiI and DiO, or CM-DiI, can be used to keep the dye after fixation. Surgical tools are sterilized using a dry sterilizer or preferably by autoclaving. A thermostatically-controlled heating blanket fitted with a rectal temperature probe is placed in the stereotaxic apparatus.

3.3.2 Pain Management

Preemptive analgesia can be obtained by subcutaneous administration of buprenorphine (0.05 mg/kg for rats and 0.1 mg/kg for mice) 20 min before the surgery. A good analgesia reduces physiological stress responses to the surgery (e.g. the production of corticosterone) and increases the potency of inhaled anesthetic gases (e.g. isoflurane), thus reducing their consumption. Local anesthetics (lidocaine, xylocaine) are also used during the surgical procedure to prevent priming the pain system (see below).

3.3.3 Preparing the Animal

1. When using isoflurane, rapid induction is obtained by placing the animal in a dedicated box receiving 3–4% isoflurane in pure oxygen or air at 2–4 L/min. After induction, the animal's nose is either placed directly in the mask attached to the head holder of the stereotaxic frame or in a side mask. When using a stereotaxic apparatus with optical encoders, the use of a side mask for shaving the animal's head is preferable as hair debris might alter the functioning of the encoders; shaving can otherwise be done after having placed the animal's head in the stereotaxic frame.

2. Head fixation is performed by inserting ear bars in the external ear canals while maintaining the nose in the mask. Xylocaine gel can be applied on the tip of the bars before insertion; this will produce an analgesic effect at the level of the canal epithelium and will facilitate the insertion of the bars by lubricating the canals. When both bars have been inserted, they can be jointly slid while maintaining a gentle pressure against the canals in order to align the animal's head with the center of the stereotaxic frame. After having secured the bars in position, the tooth bar is placed behind the upper incisor teeth such that it exerts a gentle pull in the rostral direction, the head holder is tightened in position and the snout is locked in place using a nose bar or by sliding the mask toward the animal, depending on the model of head holder.

3. Insert the rectal temperature probe and turn on the heating blanket control system to ensure that body temperature is maintained at 37 °C. The concentration of isoflurane and gas flow can then be progressively lowered to reach a maintenance level (generally around 1% at 0.3–0.6 L/min depending on the species and size of the animal). Pure oxygen is preferable over air for long surgeries (>1 h) to compensate for the respiratory depression induced by isoflurane. At this point and at regular intervals during the surgery, consider providing warm subcutaneous injections of saline to keep the animal hydrated and prevent hypovolemia and hypotension which might occur from bleeding. Finally, apply ophthalmic gel on both eyes to prevent drying of the corneal surface.

3.3.4 Preparing the Skull

1. After having shaved the head from the eyes to the back of the head (see above), scrub the scalp with povidone-iodine and then with a 70% ethanol solution. Perform a pre-incision infiltration with 0.5% lidocaine in the subcutaneous space along the midline. After a minute or two, incise the scalp along the midline from the eyes to the back of the head using a #10 scalpel blade and move the borders of the skin to the sides.

2. Gently scrape the periosteum off the skull bone using the scalpel blade and clean the skull using saline. A thorough removal of soft tissues and cleaning is crucial to guarantee an optimal bond between the bone and the adhesive (see below). Small pieces of hemostatic sponge soaked in saline can be placed over bleeding spots until bleeding stops. Perform an additional cleaning step by applying hydrogen peroxide using a Q-tip or microsponge, making sure not to touch soft tissues around the skull. The bone should now be clean and whitish.

3. The bregma and lambda reference points must then be adjusted to the same height. Rather than checking the height of each point sequentially, a quicker approach is to use a

dedicated tool comprising two vertically-aligned conical pins separated by the expected bregma-lambda distance. When lowering the tool over the skull, the two pins should touch the bone at the same time when the head is properly leveled. If necessary, the head pitch angle is corrected using the vertical adjustment knob on the head holder. Beware that this procedure might increase the pulling force exerted on the incisor teeth by the tooth bar; it might thus be necessary to untighten the head holder to release the pulling force while adjusting the head pitch angle.

4. Mark the position of skull screws and of the craniotomy using an extra fine tip marker. For rats, consider using a minimum of 3 pairs of screws: two pairs in the parietal bone and one pair in the frontal or interparietal bone. Given that the craniotomy is almost systematically localized in the interparietal bone for electrode implantation in the cerebellum, adding one or two screws in this part of the skull might be tricky. These screws will however provide an excellent anchoring point to the skull as the interparietal bone is particularly thick compared to other bone plates. The next step is to drill burr holes using an air driven or electric handpiece or a handheld drill bit holder. Skull screws (stainless steel 0-80 or M1.6 screws for rats, 00-96 or M1.2 for mice, cleaned in 70% ethanol or, preferably, sterilized by autoclaving) are then inserted into the holes using the appropriate screwdriver. One of the screws, connected to a thin wire, will serve as a ground for the electrodes. Because soldering stainless steel requires a specific type of flux and some practice, an easier solution is to glue the wire onto the screw using a conductive epoxy. As minor bleeding can occur during the procedure, do a final check of the skull surface once all screws are in place and clean up any trace of dried blood using a fine tip (e.g. microsponge) soaked in distilled water. Once the skull is perfectly dry and clean, apply a layer of Superbond on the bone and at the base of the screws, avoiding the area where the craniotomy will be performed. If a precise positioning of the electrode(s) relative to bregma is important, also leave this point exposed.

5. Variant for head-fixed, acute recordings in mice (e.g. when using Neuropixels): either a small ground wire or screw is implanted (alternatively during each recording session a silver ground wire is placed in the bath above the craniotomy). Two other steps of skull preparation are also performed:

   (a) Implantation of a rigid but light-weight head fixation system such as a metal bar or metal plate over the front of the skull, using dental cement. This device will be used on recording days to attach the animal's head in the electrophysiology rig. Prior to the first recording day, the mouse is habituated to this head fixation in the rig over several days.

(b) Creation of a recording chamber over the location of the desired craniotomy. Dental cement can be used on its own or in combination with a plastic ring to build this chamber. A large chamber with good access to the skull is preferable. The bone at the bottom of the chamber needs to be protected for example with a thin layer of transparent dental cement.

3.3.5 Craniotomy

1. Drill the craniotomy hole using an air driven or electric handpiece fitted with an appropriate dental drill bit. The cerebellum is most often accessed through the interparietal bone, which is a thick skull plate and has a tendency to produce significant bleeding. Keep in mind that bleeding during the drilling procedure will likely occur when reaching the interparietal bone marrow and will require hemostatic measures (e.g. the use of hemostatic sponges). If possible, reduce the speed of the drill as the bone gets thin and brittle. Remove bone pieces using fine forceps to reveal the underlying dura.

2. In rats, a durotomy is preferable to facilitate electrode insertion. In contrast, in mice, electrodes with a fine tip can pierce the dura without excessive bending or damage to the electrode. If necessary, silicon probes can be sharpened using a custom polishing system made of a polishing pad glued to the upper platter of a recycled hard drive. Efficient tools for performing a durotomy can be fabricated using two single-use insulin syringes with a fixed 29G needle. One needle is bent to approach the dura bevel up and tangentially; the other is gently tapped against the sterilized bench surface to obtain a fine hook. The first needle is used to open a slot in the dura using a gentle shear motion while the second is used to grab one side of the dura. Once the durotomy is performed and the underlying brain surface is exposed, place a piece of hemostatic sponge soaked in saline on the craniotomy.

3. (optional) For head-fixed, acute recordings using Neuropixels probes in mice, the craniotomy is not performed during the first surgery, but rather during a second brief surgery. At least several hours of recovery after the surgery are allowed before the start of the experiment. In the case of multiple recording days, the craniotomy is protected between sessions with a sealant such as Kwik-Cast. An alternative approach is to use a 3D printed chamber with a closable lid, *see*, for example, https://github.com/blinklab/recording-chamber. In this case the craniotomy should be covered with Dura-Gel (Cambridge NeuroTech), which does not need to be replaced on subsequent recording sessions.

3.3.6 Electrode Insertion and Sealing

1. Attach the implant (microdrive and electrode assembly) on one of the stereotaxic arms. To make the insertion easier, the assembly can be mounted onto a motorized micromanipulator. Note that the electrode should be perfectly aligned with the axial movement of the arm or manipulator.

2. If necessary, place the electrode close to bregma to calculate the exact coordinates of the penetration point. Remove the sponge protecting the craniotomy and place the electrode over the insertion point to check whether it falls in the durotomy and that the electrode course will not hit a superficial vein. If necessary, perform minor corrections of the insertion point. Insert the electrode at a speed of 10–20 µm per second. Electrode insertion can be made easier by using a motorized micromanipulator.

3. (For chronic implantation) Once the electrode has reached its final position, cover the craniotomy with a dural sealant (e.g. Dura-Gel). If the upper part of the electrode and the microdrive are to be embedded in dental cement, cover the electrode shank and eventually the microdrive shuttle mechanism (if not within a protective casing) using vaseline melted with a low temperature cautery or a low melting point agarose gel dispensed via a syringe. Position the electrode interface board and connector (e.g. Omnetics connector) at the front using, for example, the second arm of the stereotaxic frame. If necessary, perform final connection steps (e.g. of the ground wire or electrode wires to the interface board).

4. Apply dental cement over the screw heads and Superbond layer, covering all parts of the implant which need to be sealed in place (microdrive, connector, interface board for chronic recordings, head-fixing system for acute recordings). It is important to obtain a smooth surface to avoid itching at the contact with the skin borders and potential injury or discomfort during self-grooming.

5. (For acute recordings, e.g. using Neuropixels probes in mice) The electrode is inserted acutely using a motorized micromanipulator at a slow speed (10–30 µm/s) at each recording session. The craniotomy is covered in ACSF and periodically inspected to prevent drying during the experiment. The ground silver wire connected to the probe is inserted in the ACSF as well. Alternatively, if the ground wire was implanted during the surgery, connect it to the probe ground. In this case ACSF is not needed and the craniotomy should stay covered with Dura-Gel. After the recording session, the Neuropixels probe is carefully rinsed as fast as possible in PBS and is then placed for several hours (typically overnight) in a solution of 1% Tergazyme detergent to remove tissue debris. Before the next use it is briefly rinsed in deionized water and propanol.

3.3.7 Final Steps and Postoperative Management

1. Stitch the skin in the front and at the rear of the implant such that the borders of the skin are pressed against the cement without excessive tension. To facilitate healing, an antiseptic veterinary wound powder can be applied onto the skin. Alternatively, a topical antibiotic such as fusidic acid can be used.

2. After having turned off the isoflurane vaporizer and removed the animal from the stereotaxic frame, inject a final dose of warm saline subcutaneously. At this point, it can be wise to also administer a nonsteroidal anti-inflammatory drug (e.g. carprofen or meloxicam).

3. Place the animal on a dedicated hot water blanket, electric heating blanket or under a heat lamp and monitor recovery from anesthesia every 10 min until the animal is awake. Put the animal back to its home cage with nutritional support (e.g. concentrated milk).

4. In the hours and days following the surgery, monitor food and water intake and keep providing nutritional support and pain killers (buprenorphine) if necessary. The animal should be kept separated from its congeners for at least 5 days.

3.4 Spike Sorting

Spike sorting is a crucial stage that conditions the possibility of exploiting the recorded data, so it must be approached with care and upon proper training and/or validation. Commercial spike sorting solutions are sometimes bundled with the recording equipment. However, the academic field of spike sorting is very active and many open-source solutions are available and often under continuous development (e.g. Kilosort2 https://github.com/MouseLand/Kilosort2, Klusta https://klusta.readthedocs.io/en/latest/, MountainSort https://github.com/flatironinstitute/mountainsort_examples/blob/master/README.md, Spiking-Circus https://spyking-circus.readthedocs.io/en/latest/#, Tridesclous https://github.com/tridesclous/tridesclous, YASS https://github.com/paninski-lab/yass...). It is, however, quite likely that a blind, uneducated use of any of these software tools will fail to produce reliable results (https://www.biorxiv.org/content/10.1101/796599v1 [22]. An exploration of the various existing solutions is, therefore, necessary in order to find the one best suited to your situation and programming skills. The python project spikeInterface (https://github.com/SpikeInterface) offers a unified programming interface to launch several of the algorithms available, compare their outputs, and manipulate and explore the data.

Here, we provide examples of sorting with Tridesclous (Fig. 1a–c), a custom software (Fig. 1d–g), MountainSort (Fig. 2a–c), and Kilosort 2 followed by curation with Phy2 (https://github.com/cortex-lab/phy) (Fig. 2e, f).

3.4.1 Spike Sorting in the Cerebellum

Cerebellar recordings are plagued with the same general dangers of spike sorting: instabilities (some units disappear or appear, transiently or suddenly), difficulty of separating neighboring units (yielding spike-trains composed from spikes from more than one neuron), over-separation of the spikes (spikes from a neuron distributed in two groups; this happens notably for bursting neurons, where the first and last spikes of bursts may look different enough to be sorted as different units, but also for very short spikes where a low sampling rate will generate different shapes).

Some features of cerebellar activity have also to be taken into account. First, many cells have very brief spikes, so a low sampling rate (20 kHz) and/or the corresponding low-pass filter setting (e.g. 5 kHz) may distort the spikes. Second, the level of activity is often very high, and the presence of several fast-firing units (>100 Hz) around the electrode will produce many temporal overlaps of spikes which will yield combined waveforms; this situation is a challenge for any spike sorting software so care must be taken either to treat the overlapping spikes or not to draw scientific conclusions which would be sensitive to the mis-sorting of these temporal collision events (e.g. in synchrony analysis). Third, the complex spikes require separate, specific procedures.

3.4.2 Sorting Complex Spikes

The detection of complex spikes in Purkinje cells is a challenge. Indeed, these spikes combine a fast somatic spike with strong dendritic currents lasting several milliseconds and a variable number of dendritic calcium spikes in a given cell [23]. These features make them difficult to detect with traditional methods (brief windows, high-pass filtering). The somatic spike is associated with a (classical) negative deflection of extracellular potential near the cell (in and around the Purkinje cell layer), but it also associated with a positive extracellular potential deflection in the proximal layer. The dendritic spike produces a negative deflection in the proximal molecular layer, but a negative deflection in the Purkinje cell layer. The combination of the somatic spike(s) and dendritic currents will produce a wide variety of waveforms as a function of the position of the recording site relative to the cell, and of the Purkinje cell morphology (Fig. 2).

In order to optimize the detection of complex spikes, it is useful to perform a specifically dedicated analysis, by filtering with a lower frequency band (150–2000 Hz) and extracting the waveform with wider windows adapted to their low frequency component (e.g. [2]). Several studies have attempted to improve the detection of complex spikes [24, 25]), but none of them provides a perfectly satisfactory solution for the moment. Among the algorithms mentioned above, we found that MountainSort4 is particularly effective in detecting complex spikes (Fig. 2a–c). If the detection of complex spikes is crucial for your experiments, a thorough exploration of the problem remains necessary.

3.4.3 Cell Identification

Once the spike sorting step has been carried out, it is necessary to identify the recorded cell types. Purkinje cells are most easily identified if the complex spikes are detected; the correspondence of the simple and complex spike from the same unit is made based on two factors: the spatial proximity of the two events (*see* Fig. 3b: shape of simple and complex spike pairs on the same channel, and Fig. 3f: shapes on the channels with largest negative amplitude) and the "pause" in the simple spikes which follows the occurrence of complex spikes, as shown in complex spike*simple spike crosscorrelograms (Fig. 3c, this pause results from the resetting of the Purkinje cell's pacemaker by the complex spike, combined with a transient hyperpolarization of the cell).

The identification of the other cell types is more complicated. In the cerebellar cortex, the layers may be identified by the features of their neuronal activity. The Purkinje cell layer corresponds to the area where many very active neurons are continuously firing. In this layer, the complex spikes consist of a fast downward component, similar to the Purkinje cell's single spikes, and a slow positive component (Fig. 3b). The molecular layer has rather low extracellular background activity during resting periods, and close to the Purkinje cell layer, "fat spikes" are occurring at low rates (~1 Hz) ([1], Fig. 3f). "Fat spikes" correspond to the dendritic currents during the complex spike when the recording takes place close to the proximal part of the Purkinje cells' dendritic tree; since these "fat spikes" rarely show oscillatory structure, they more likely reflect the synaptic currents evoked by the climbing fiber rather than the bursts of dendritic calcium spikes evoked by these currents. Using these criteria, it is relatively easy to determine the relative position of the cerebellar layers.

Several attempts to classify the cell types of the cerebellar cortex have been proposed on the basis of discharge rate [26–28]. However, these categories derive mostly from anesthetized animals and their validity in behaving animals remains uncertain. Waveforms and first order statistics (i.e. slow versus fast firing, bursting or not) may offer an alternative way to identify the cerebellar interneurons [29–33]. In the cerebellar nuclei, the width of the spike's waveform should also help in distinguishing excitatory from inhibitory neurons [34]. It is regrettable however that in most cases the spike waveforms published in the literature are high-pass filtered; since filtering strongly modifies the waveform, and since filter types and parameters differ between laboratories, this complicates comparisons between studies. Yet, it is likely that the advent of massively parallel recordings in the cerebellar cortex will provide in the coming years atlases of waveforms for each cell type, which will help to unequivocally identify the units.

References

1. Gao H, De Solages C, Lena C (2012) Tetrode recordings in the cerebellar cortex. J Physiol Paris 106:128–136

2. De Solages C, Szapiro G, Brunel N et al (2008) High-frequency organization and synchrony of activity in the purkinje cell layer of the cerebellum. Neuron 58:775–788

3. Soteropoulos DS, Baker SN (2008) Bilateral representation in the deep cerebellar nuclei. J Physiol 586:1117–1136

4. Delescluse M, Pouzat C (2006) Efficient spike-sorting of multi-state neurons using inter-spike intervals information. J Neurosci Methods 150:16–29

5. Mechler F, Victor JD, Ohiorhenuan I et al (2011) Three-dimensional localization of neurons in cortical tetrode recordings. J Neurophysiol 106:828–848

6. Sedaghat-Nejad E, Herzfeld DJ, Hage P et al (2019) Behavioral training of marmosets and electrophysiological recording from the cerebellum. J Neurophysiol 122:1502–1517

7. Ulyanova AV, Cottone C, Adam CD et al (2019) Multichannel silicon probes for awake hippocampal recordings in large animals. Front Neurosci 13:397

8. Tang T, Blenkinsop TA, Lang EJ (2019) Complex spike synchrony dependent modulation of rat deep cerebellar nuclear activity. eLife 8: e40101

9. Neto JP, Baiao P, Lopes G et al (2018) Does impedance matter when recording spikes with Polytrodes? Front Neurosci 12:715

10. Rivnay J, Wang H, Fenno L et al (2017) Next-generation probes, particles, and proteins for neural interfacing. Sci Adv 3:e1601649

11. Nguyen JK, Park DJ, Skousen JL et al (2014) Mechanically-compliant intracortical implants reduce the neuroinflammatory response. J Neural Eng 11:056014

12. Halverson HE, Khilkevich A, Mauk MD (2015) Relating cerebellar purkinje cell activity to the timing and amplitude of conditioned eyelid responses. J Neurosci 35:7813–7832

13. Campolattaro MM, Kashef A, Lee I et al (2011) Neuronal correlates of cross-modal transfer in the cerebellum and pontine nuclei. J Neurosci 31:4051–4062

14. Nguyen DP, Layton SP, Hale G et al (2009) Micro-drive array for chronic in vivo recording: tetrode assembly. J Vis Exp 26:1098

15. Kostadinov D, Beau M, Blanco-Pozo M et al (2019) Predictive and reactive reward signals conveyed by climbing fiber inputs to cerebellar Purkinje cells. Nat Neurosci 22:950–962

16. Chabrol FP, Blot A, Mrsic-Flogel TD (2019) Cerebellar contribution to preparatory activity in motor neocortex. Neuron 103: 506–519 e504

17. Gao Z, Davis C, Thomas AM et al (2018) A cortico-cerebellar loop for motor planning. Nature 563:113–116

18. Chung JE, Joo HR, Fan JL et al (2019) High-density, long-lasting, and multi-region electrophysiological recordings using polymer electrode arrays. Neuron 101:21–31 e25

19. Musk E, Neuralink (2019) An integrated brain-machine Interface platform with thousands of channels. J Med Internet Res 21:e16194

20. Fu TM, Hong G, Zhou T et al (2016) Stable long-term chronic brain mapping at the single-neuron level. Nat Methods 13:875–882

21. Adewole DO, Serruya MD, Wolf JA et al (2019) Bioactive Neuroelectronic interfaces. Front Neurosci 13:269

22. Buccino AP, Hurwitz CL, Garcia S, Magland J, Siegle JH, Hurwitz R, Hennig MH. SpikeInterface, a unified framework for spike sorting. Elife. 2020 Nov 10;9:e61834. doi: 10.7554/ eLife.61834

23. Yang Y, Lisberger SG (2014) Purkinje-cell plasticity and cerebellar motor learning are graded by complex-spike duration. Nature 510:529–532

24. Zur G, Joshua M (2019) Using extracellular low frequency signals to improve the spike sorting of cerebellar complex spikes. J Neurosci Methods 328:108423

25. Markanday A, Bellet J, Bellet ME et al (2020) Using deep neural networks to detect complex spikes of cerebellar Purkinje cells. J Neurophysiol 123:2217–2234

26. Hensbroek RA, Belton T, Van Beugen BJ et al (2014) Identifying Purkinje cells using only their spontaneous simple spike activity. J Neurosci Methods 232:173–180

27. Ruigrok TJ, Hensbroek RA, Simpson JI (2011) Spontaneous activity signatures of morphologically identified interneurons in the vestibulocerebellum. J Neurosci 31:712–724

28. Van Dijck G, Van Hulle MM, Heiney SA et al (2013) Probabilistic identification of cerebellar cortical neurones across species. PLoS One 8: e57669

29. Blot A, De Solages C, Ostojic S et al (2016) Time-invariant feed-forward inhibition of

Purkinje cells in the cerebellar cortex in vivo. J Physiol 594:2729–2749

30. Prsa M, Dash S, Catz N et al (2009) Characteristics of responses of Golgi cells and mossy fibers to eye saccades and saccadic adaptation recorded from the posterior vermis of the cerebellum. J Neurosci 29:250–262

31. Barmack NH, Yakhnitsa V (2008) Functions of interneurons in mouse cerebellum. J Neurosci 28:1140–1152

32. Holtzman T, Rajapaksa T, Mostofi A et al (2006) Different responses of rat cerebellar Purkinje cells and Golgi cells evoked by widespread convergent sensory inputs. J Physiol 574:491–507

33. Vos BP, Maex R, Volny-Luraghi A et al (1999) Parallel fibers synchronize spontaneous activity in cerebellar Golgi cells. J Neurosci 19:RC6

34. Ozcan OO, Wang X, Binda F et al (2020) Differential coding strategies in glutamatergic and GABAergic neurons in the medial cerebellar nucleus. J Neurosci 40:159–170

Chapter 11

Investigating Cerebrocerebellar Neuronal Interactions in Freely Moving Mice Using Multi-electrode, Multi-site Recordings

Yu Liu, Brittany L. Correia, Mia B. Fox, and Detlef H. Heck

Abstract

The function of any brain structure is to a large extent determined by its interactions with other brain areas or the sensory inputs it receives. The cerebellum has massive reciprocal connections with the cerebral cortex via the thalamus and pontine nuclei. Functional magnetic resonance imaging studies used measures of correlated BOLD signals to demonstrate functional connectivity between virtually all cerebral cortical areas and the cerebellum. Those studies ultimately led to a crucial revision of the long-standing belief of a predominance of sensorimotor areas being connected with the cerebellum. Instead, the results revealed that much of the cerebellar cortical surface is functionally connected with association areas of the cerebral cortex, providing a neurophysiological basis of cerebellar cognitive function. Studies of cerebellar involvement in cognitive function must take those interactions into account by observing neuronal activity in functionally connected areas during relevant behaviors. Here we describe a multi-site extracellular recording approach we developed to simultaneously record neuronal activity in the medial prefrontal cortex, dorsal hippocampus, and cerebellar cortex. Our focus is on the role of coherence of neuronal oscillations as a means of controlling neuronal communication between cerebral cortical areas and the proposed role of the cerebellum in coordinating the task specific modulation of cerebral cortical coherence.

Key words In vivo electrophysiology, Cerebellum, Oscillations, Cerebellar cognitive function, Spatial working memory, Coherence, Instantaneous phase, Cognition, Multi-site recordings, Freely moving electrophysiology, Cerebrocerebellar communication

1 Introduction

Traditional approaches mostly viewed cerebellar neuronal activity in isolation and electrophysiological experiments were designed to evaluate neuronal responses to sensory stimuli, motor events, or learning paradigms. It is now widely accepted that the cerebellum is involved in a multitude of tasks that include cognitive and affective functions [1]. Imaging studies revealed far more extensive interconnections between the cerebellum and non-motor cortical structures than previously thought [2, 3]. Cerebellar contribution to any

Roy V. Sillitoe (ed.), *Measuring Cerebellar Function*, Neuromethods, vol. 177, https://doi.org/10.1007/978-1-0716-2026-7_11,
© Springer Science+Business Media, LLC, part of Springer Nature 2022

function that involves cerebral cortical structures is mediated via the pathways connecting the cerebellum and cerebral cortex. Clearly, understanding cerebellar function must take this interaction into account and include the investigation of cerebellar communication with cerebral cortical structures. In electrophysiological experiments this requires recordings in at least one cerebellar site and one or more sites in the cerebral cortex. The efficacy of this approach was demonstrated by an intriguing observation by Popa and colleagues who recorded in the rat primary motor and sensory areas while pharmacologically manipulating cerebellar neuronal output from the interposed nucleus [4]. The key finding was that inhibition of the interposed nucleus starkly reduced the coherence of gamma oscillations between sensory and motor cortex but left gamma oscillations within each structure intact [4]. Coherence of neuronal oscillations is believed to be a mechanism for dynamically modulating neuronal communication between cerebral cortical areas in a task dependent manner [5, 6]. Popa et al.'s findings indicated that the cerebellum might be involved in controlling or coordinating that communication via coherence. To investigate this further it becomes necessary to have at least two recording sites in the cerebral cortex and one in the cerebellum, with all three in areas involved in controlling the same task. Here we describe an experimental approach we developed to simultaneously record neuronal activity in the medial prefrontal cortex (mPFC), the dorsal hippocampal CA1 area (dCA1), and cerebellar lobulus simplex (LS) or Crus I in head-fixed mice. The choice of the three recording sites was based on existing literature which documents a crucial role for the mPFC and dCA1 in spatial working memory in rodents and humans [7–10], with imaging studies in humans also implicating the cerebellar LS/Crus I region [11–13]. Recordings of interareal neuronal communication at rest can be made in head-fixed conditions, an approach we used to investigate the representation of phase information of oscillations in the mPFC and dCA1 in LS and Crus I Purkinje cell (PC) simple spike activity [14]. These recordings required measuring local field potentials (LFPs) in the mPFC and dCA1 and spiking activity in the cerebellum and will be described in the first half of the chapter.

The ultimate evaluation of functional interaction between brain areas requires observation during behavior and the behavioral task should require interaction between the observed structures. Spatial working memory in mice can be tested by evaluating spontaneous alternations in a Y maze or plus maze [15]. In the second half of the chapter, we describe techniques for recording from freely moving mice, detailing the benefits and drawbacks of the two recording systems.

2   Materials

2.1   Surgical Procedures

Surgical anesthesia was initiated by exposing mice to 3–4% isoflurane in oxygen in an induction chamber (*see* **Note 1**). Sedation is confirmed by lack of a righting reflex when the chamber is tilted. Transfer mouse to a stereotaxic head frame once sedated, placing the nose in the nose cone first to ensure sedation is maintained while stabilizing head in stereotaxis. During surgery, anesthesia is maintained using an Ohio isoflurane vaporizer (Highland Medical Equipment; Deerfield, IL, USA) at 1–2.5% isoflurane in oxygen, or at the lowest concentration at which mice fail to show the toe-pinch reflexive withdrawal. Maintain core body temperature at 37–38 °C using a servo-controlled heat blanket (FHC; Bowdoinham, ME, USA) monitored by a rectal thermometer. Drop lubricant ointment (Retaine PM, OCuSOFT Inc.; Rosenberg, Tx, USA) on each eye to prevent drying. Surgical preparation begins with removing the hair on the skull with a depilatory (Nair, Church & Dwight UK Limited; Folkestone, Kent, UK), followed by alternating application of alcohol wipes and iodine solution on the hair-free skin. At the beginning of each surgery, after mice were anesthetized but before the first incision, mice received a single subcutaneous injection of the analgesic Meloxicam SR (4 mg/kg, 0.06 ml) to alleviate pain.

A Microtorque II dental drill (Ram products, Inc.; Dayton, NJ, USA) with a bur carbide RA ½ drill bit (0.7 mm round drill bit) (SS White Dental; Lakewood, NJ, USA) is required to create holes for 1/16 SL PAN Skull screws (Antrin Miniature Specialties, Inc.; Fallbrook, CA, USA) and craniotomy sites for electrode placement. Craniotomy placement was based off target brain regions for the specified experiment, all coordinates were found in *The Mouse Brain in Stereotaxic Coordinates* [16].

Cyanoacrylate gel glue (Loctite, Henkel Corporation; Rocky Hill, CT, USA) is used for a variety of surgical procedures for both head-fixed and freely moving recording surgeries; it was used in creating an isoflurane chamber and in gluing the skin to the head cap for head-fixed recordings. In the freely moving surgery, it is required to fix the Omnetics Connector (Omnetics Connector Corporation; Minneapolis, MN, USA) to the headpost, also referred to as a head-fixation bar. In either surgery, acrylic cement (Palavit G, Heraeus Kulzrer GmbH; Hanau, Germany) is used to cover the skull, surrounding the drinking straws, with the effect of creating a secure barrier around the electrodes. Acrylic cement is required to secure the head-fixation bar in either surgery and the optical fiber in freely moving recordings.

All animal experimental procedures described here adhered to guidelines approved by the University of Tennessee Health Science Center Animal Care and Use Committee. Principles of laboratory animal care (NIH publication No. 86-23, rev. 1996) were followed.

Mice were housed in a breeding colony with 12-h light/dark cycles in standard cages with ad libitum access to food and water. All experiments were performed during the light cycle, between 12:00 noon and 17:00 h.

2.2 Electrophysiological Recordings

For head-fixed experiments, collect data with a 7-electrode manipulator (System Eckhorn, Thomas Recording; Germany) threaded with insulated glass tungsten electrodes (80 μm outer diameter, 3–5 MΩ). Use Spike2 software (Cambridge Electronic Design Limited; Cambridge, England, UK) for data acquisition, which has been digitized by Power 1401 (Cambridge Electronic Design Limited; Cambridge, England, UK).

The key component to freely moving recordings is a custom-built electrode micro-drive connected to a wireless head-stage (W2100-HS16, Multichannel Systems; Germany). It consists of a driving screw and two electrode guiding tubes, with glass insulated tungsten/platinum electrodes (80 μm diameter; impedance: 3.5–5.0 MΩ). A custom holder is required for holding and positioning the micro-drive during the surgery. LEDs, emitting 465 nm light, mounted on a 5 mm long optical fiber (200 μm diameter, Doric Lenses Inc.; Quebec, Canada) were used for optical stimulation. Two miniature male gold pins were used to deliver power to the LEDs via a thin wire (2 m, 40 AWG solid nickel).

2.3 Behavioral Tasks

For freely moving recordings, behavioral tasks were completed in an elevated plus maze and recorded using the *Viewer* video system (Biobserve GmbH; Bonn, Germany). Data was saved to a hard disk (W2100-HS16, Multichannel Systems; Germany).

3 Methods

3.1 Triple Site Recordings in Head-Fixed Mice

A major decision to make before embarking on a multi-site recording experiment is to determine whether recordings are to be made in head-fixed or freely moving conditions. Each approach has specific advantages and disadvantages. With head-fixed recordings there is no concern about the weight of equipment, such as amplifiers and connectors, the mice would have to carry on their head. Typically, only recording chambers and a head-fixation bar are fixed to the head, which adds minimal weight. Key advantages of head-fixed recordings are greater recording stability and the ability to change recording sites during the experiment. The latter is crucial if the ideal site is not yet known and needs to be mapped out. Also, movement artifacts are typically less severe since the animals usually move far less after adapting to head fixation and do not bump the amplifier and connector into maze walls etc., as they do in a freely moving setup.

3.1.1 Surgical Procedures

1. Once the surgical site is prepared as described in Subheading 2, make a single rostro-caudal incision in the skin down the midline. Push the skin to the side and clean the skull with sterile cotton swabs. Use 4% hydrogen peroxide to remove any remaining tissue attached to bone.

2. Insert three skull screws to anchor the head-fixation bar and recording chamber. Holes are drilled with a dental drill and round drill bit (*see* **Note 2**). Placement of the screws on the skull depends on what the target recording sites are. In our case recordings are from the left mPFC, left dCA1, and right cerebellum LS/Crus I. Accordingly, we place our anchoring screws above the left cerebellum and right frontal area, with a third screw placed over the left visual cortex (*see* **Note 3**).

3. Carefully remove the skull overlying the recording sites. Use the same drill bit to prepare these openings by drilling an open circle around the stereotaxic marking of the recording site, always being careful to avoid damaging the dura (*see* **Note 4**). Eventually the piece in the center of the circle will detach and can be carefully removed with tweezers (*see* **Note 5**). The exposed dura should immediately be covered with triple antibiotic (Walgreens; Deerfield, IL), which serves as a mechanical protection from acrylic cement entering and prevents it from drying out.

4. Create a recording chamber with acrylic cement to protect the recording site. Make a barrier around the recording sites using a sterilized plastic drinking straw cut to approximately 5 mm in length (*see* **Note 6**). Once the straw barriers are established, their interiors are filled with triple antibiotics (*see* **Notes 7 and 8**). Next, we apply a thin layer of acrylic cement to cover the skull and flow around the screws and chambers. Then the head-fixation bar is put in place, held by a stereotaxic manipulator, and attached to the skull with acrylic cement (*see* **Notes 9 and 10**). After the cement has hardened a cement cap is build up slowly to completely cover the screws, the sidewalls of the recording chambers, and the center part of the head-fixation bar. Once the cement cap is finished and hardened, the skin is carefully pulled over the acrylic cap and spot-glued to the acrylic using cyanoacrylate glue (*see* **Note 11**). Mice should recover in their home cages, singly housed, for 3 to 4 days prior to recording.

3.1.2 Electro-physiological Recordings

In order to simultaneously insert extracellular electrodes in all three recordings sites we use a 7-electrode system that allows for independent control of the penetration depth of all 7 electrodes at micrometer precision. Electrode movements are controlled with step motors via a remote control. Each electrode is threaded

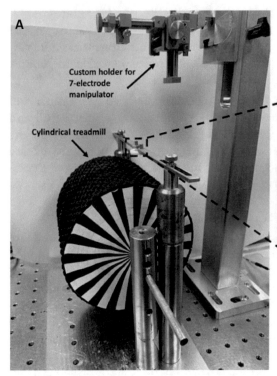
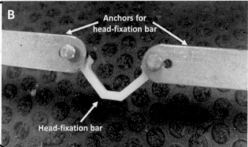

Fig. 1 Recording apparatus for head-fixed in vivo electrophysiology. (a) Picture of the head-fixed recording apparatus, including a cylindrical treadmill placed under the head-fixation bar to allow for mouse movement if necessary. Once the mouse is secured in the apparatus, attach the 7-electrode manipulator to the custom holder, which can then be lowered to the correct position just above the recording sites. (b) A close-up of the head-fixation bar and how the animal is anchored into the recording apparatus. In an experiment, the head-fixation bar itself would be secured to the mouse's skull using acrylic cement as described in Subheading 3.1.1

through a different guiding tube, whose positions can be manipulated to fit the parameters of your experiment (*see* **Note 12**).

1. Head-fix the mouse on the recording setup (Fig. 1). Remove the triple antibiotic ointment from the chambers and rinse multiple times with saline. Next, position the 7-electrode manipulator over the mouse's head using a custom-made holder and lower the manipulator until the guiding tubes are touching the brain (*see* **Note 13**). Once electrode positions have been established, fill the chambers halfway with saline (*see* **Note 14**).

2. Advance hippocampal recording electrodes until sharp wave ripples (SWRs), a high-frequency LFP oscillation characteristic for the hippocampus [17], are detected (Fig. 2b). Advance electrodes in the cerebellum until Purkinje cells (PC) can be identified by the presence of complex spikes [18]. Advance prefrontal cortex recording electrodes until the target depth is

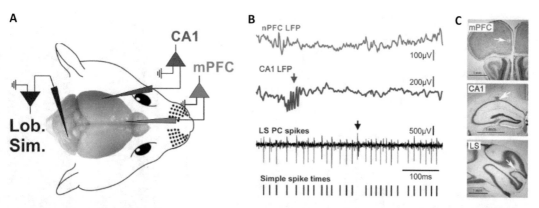

Fig. 2 Illustration of experimental approach, examples of raw LFP and single-unit spike signals and anatomical reconstructions of recording sites. (a) Illustration of recordings sites for simultaneous recordings of LFP activity in the mPFC and dCA1 and PC spike activity in cerebellar LS. (b) Raw LFPs recorded in the mPFC (mPFC, blue trace) and dCA1 (dCA1 LFP, red trace), and raw single-unit PC spike activity recorded in the cerebellar LS (LS PC spikes, black trace). The arrow in the LFP recording from the dCA1 region points to a sharp wave ripple event, a brief high-frequency oscillation characteristic for the hippocampus. The presence of sharp wave ripples in the LFP recordings was used for online verification of electrode tip placement within dCA1. The arrow in the trace of raw PC spike activity marks the occurrence of a complex spike, which reflects input from the inferior olive. Complex spikes are characteristic of PCs and were used for online identification of PC activity. The bottom trace shows tick marks representing the time sequence of simple spikes extracted from the raw trace and used for subsequent analysis. (c) Photomicrographs of Nissl-stained sections of the mPFC, dCA1, and cerebellar LS showing micro lesions at recording sites (white arrows). (**b** and **c** from: McAfee et al. [14])

reached according to predetermined stereotaxic coordinates, which are verified post-mortem anatomically using small electrolytic lesions.

3. Separate LFPs and spike signals by band-pass filtering at 0.1 to 200 Hz and at 200 Hz to 8 kHz, respectively, digitizing (sampling rate: >20 kHz for action potentials and >1 kHz for LFPs) and storing on hard disk, using a 16-bit A/D converter and Spike2 software.

4. Repeat recording sessions on three successive days for each mouse. After completion of each recording session, rinse the chambers and re-fill with triple antibiotic ointment.

3.2   Triple Site Recordings in Freely Moving Mice

The major argument for freely moving recordings is the ability to record neuronal activity while the mice perform more complex behaviors that require free movement and that the freely moving condition might be overall less stressful for the mice. In this section, we describe a procedure for recording from the mPFC, dCA1, and cerebellar cortex while mice are exploring a plus maze. We also describe how to incorporate PC optogenetic stimulation into this paradigm. New experimental techniques utilizing virtual environments in rodents provide a hybrid approach between head-fixed and freely moving approaches [19, 20].

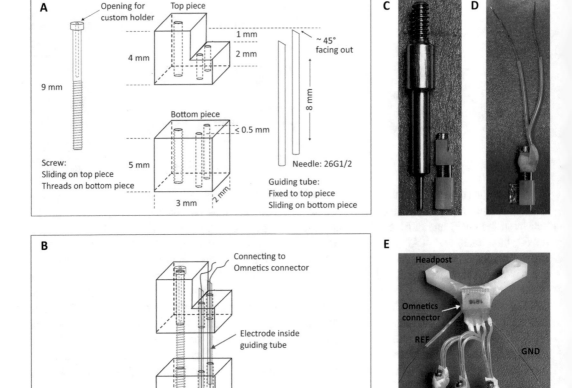

Fig. 3 Electrode micro-drive. (**a**) Schematic drawing of micro-drive parts and (**b**) an assembled micro-drive. (**c**) Photos of a micro-drive holder (left) and a micro-drive (right). (**d**) Photo of a micro-drive with two electrodes inserted and fixed to the top piece. A small piece of thin aluminum (4 × 4 mm) was attached to the side of the bottom piece for holding the micro-drive during the electrode preparation. It can be removed or kept in place if necessary. (**e**) Photo of a set of three micro-drives (6 electrodes) prepared for surgery. A piece of yellow paper between the screw and the electrode wires is to prevent acrylic cement from coming into the top piece and around the screw. A small amount of clay (blue) can be applied around the screw to hold the top piece up to the top end of the screw. *REF* reference, *GND* ground

3.2.1 Micro-Drive Assembly

A key component of our freely moving recording technique is an electrode micro-drive that was recently developed in our laboratory (Fig. 3). The micro-drive consists of a top piece (acrylic; without screw thread), a bottom piece (acrylic; with screw thread), a driving screw (only threaded on the bottom half), and two electrode guiding tubes (Fig. 3a). Once assembled, the driving screw catches the bottom piece, aligning the two pieces with the screw. While turning the screw clockwise brings the top and bottom pieces closer, a counterclockwise turn of the screw does not necessarily change the distance between the two pieces because the top piece is not connected to the threaded aspect of the screw (Fig. 3b). The two guiding tubes are fixed with the top piece and slide into the bottom piece. When the inserted electrodes are fixed with the top piece (at the top ends of the guiding tubes), pushing the two pieces closer together using clockwise turns advances the electrodes (Fig. 3b). Each micro-drive carries two electrodes through two guiding tubes (Fig. 3b, d). Using three micro-drives with such precision (*see* **Note 15**), we simultaneously record LFPs and single-unit activity from the mPFC, dCA1, and cerebellar cortex (or deliver optogenetic stimulation to the cerebellar cortex) in mice during resting conditions or a behavioral task. The driving screw has a hole (4 mm deep) at the top end. A custom holder, which can be inserted into the hole, is used for holding and positioning the micro-drive during surgery (Fig. 3c, left).

1. Prepare a set of electrodes prior to surgery. Attach an Omnetics connector to an acrylic headpost using cyanoacrylate gel glue. Then solder the electrodes, reference, and silver ground wires to the corresponding pins of the connector. Ground unused connector pins.

2. Apply acrylic cement around the connector/pins to stabilize the electrode ensemble (Fig. 3e) (*see* **Note 16**).

3.2.2 Surgical Procedures

1. Prepare the surgical site as described in materials, then make a single rostro-caudal incision in the skin down the midline. Prepare two round openings (1.0 mm diameter) in the skull overlying the left mPFC and left hippocampus (*see* **Note 4**) using a dental drill, leaving the underlying dura intact (Fig. 4a).

2. For mice receiving optical stimulation with an optical fiber-coupled LED (*see* **Note 17**), prepare a small opening (1.5 × 2 mm) overlying the right LS (*see* **Note 4**) and secure the fiber in place to touch but not penetrate the dura.

3. Center two extracellular recording electrodes attached to the custom-made micro-drive over of the mPFC and dCA1 skull openings and fix the micro-drives to the skull using acrylic cement (*see* **Note 18**). Repeat this step for an additional recording site over the right LS if it is not being used for optogenetic manipulation. Use two electrodes at each recording site.

4. The acrylic headpost, which is not used to affix the mouse to an external structure, provides a handle to manually stabilize the head while connecting and disconnecting the wireless head-stage. Embed the micro-drives, headpost, and wire connectors in acrylic cement and anchor to the skull via the skull screws. Finally, allow for a recovery period of 3–4 days before beginning any electrophysiological experiments.

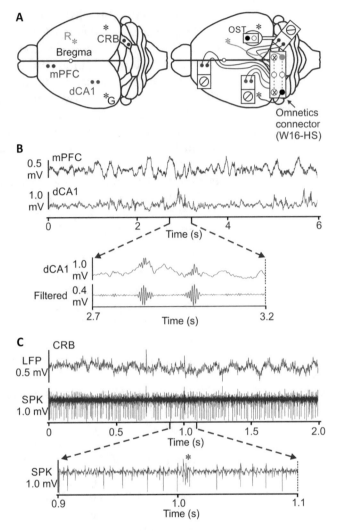

Fig. 4 Illustration of recording locations, electrode wiring, and example data. (**a**) Schematic drawing of the top view of a mouse brain. Left: LFPs and/or unit activities were recorded from the left mPFC, left dCA1, and right cerebellum (CRB). The asterisks mark the locations of three skull screws for stabilizing the implant. Two screws are also used for connecting the reference (R; AP -1 mm; ML 3 mm) and ground (G; AP -4 mm; ML 4 mm) wires. Right: Schematic drawing of wire connections to the Omnetics connector. An LED along with its connector is installed for optical stimulation (OST) of the right cerebellum. (**b**) Top traces are LFP signals recorded in mPFC and dCA1. The two bottom panels show enlarged sections of dCA1 LFPs around sharp wave ripple activities (upper traces) and band-pass filtered versions (130–200 Hz). (**c**) Top two traces are LFP signals and Purkinje cell (PC) activities recorded in the right LS. The bottom panel shows an enlarged section of PC activities with a complex spike (asterisk)

3.2.3 Electro-physiological Recordings

1. To begin a recording session, plug a wireless head-stage into the Omnetics Connector. To reduce weight, do not plug the battery directly into the head-stage (*see* **Note 19**). Instead, supply power by connecting the battery and head-stage with two highly flexible thin nickel wires.

2. Perform recordings on five consecutive days. On the first day, manually advance the electrodes into the left mPFC, dCA1, and LS while the animals are in their home cage (*see* **Note 15**). Use the presence of SWRs to determine electrode tip placement in dCA1 [17] (Fig. 4b). In the cerebellum, try to isolate single-unit activity (Fig. 4c). Once electrode positions have been established, perform recordings during plus maze exploration (see below). Only alter electrode positions between sessions if signals are lost.

3. Digitize broadband voltage signals (0.1–8 kHz) at 20 kHz and save to a hard disk. After completion of the final experiment, disconnect the head-stage and the LED power wires and mark the recording sites by making small electrolytic lesions (10 μA DC; 12 s). Deliver the lesion current through a needle that was inserted into the corresponding channel of the Omnetics Connector.

3.2.4 Behavioral Task

We use the plus maze task to quantify spatial working memory (SWM) using counts of spontaneous alternations [15]. During spontaneous maze exploration, healthy mice tend to avoid entering recently visited arms and generate arm entry sequences without repetition at above chance level. Such repeat free sequences are called spontaneous alternations [15]. In a four-arm maze, like the plus maze used here, random arm visits would result in 22.5% spontaneous alternations. Healthy mice generate significantly higher numbers of spontaneous alternations and a decrease in spontaneous alternations is interpreted as a deficit in SWM [21]. In our experiments, mice explore the plus maze for 12 min while their arm entry sequences are automatically tracked with a video system (30 frames/s). Arm entry sequences and resulting spontaneous alternations are analyzed offline.

Each mouse performs the plus maze test five times, once per day, on five successive days. Use the first session for habituation to let the mouse freely explore the maze environment. On the four subsequent days, collect electrophysiological recordings during each session and track mouse movements by video. This paradigm can also incorporate optogenetic manipulation, which we apply when mice are passing through the center zone of the maze.

3.2.5 Optical Stimulation of LS PCs

Manually apply photostimulation during the decision-making phases of plus-maze trials, which begins the moment the mouse's nose enters the center area of the plus maze. Photostimuli consisted of a 1 s light stimulus (sinusoidally modulated illumination, 100 Hz or 120 Hz) (*see* **Note 20**). The sinewave voltage used to control LED illumination was digitized (2 kHz) and recorded simultaneously with the electrophysiological data. For complete findings, *see* [22].

3.3 Data Analysis

3.3.1 Histology

At the end of the last recording session, create electrolytic lesions at the final recording sites in the mPFC, dCA1, and cerebellum by passing a small current (5 μA/10 s) through the recording electrodes (Fig. 2c). Within 12 h of creating the lesion, euthanize the mice and then transcardially perfuse them. After perfusion, remove the brain and post-fix it in 4% formaldehyde for 24 h. Cut brain tissue into 50 μm thick coronal sections and then mount and stain with cresyl violet. Confirm lesion sites using a stereotaxic atlas of the mouse brain.

3.3.2 Data Preprocessing

Use Spike2 software to perform band-pass filtering to generate LFP (0.1–200 Hz) and spike (200–8000 Hz) data for each brain region. Apply a hum-removal algorithm on Spike2 to diminish 60 Hz noise while maintaining power spectrum of LFP at 60 Hz (*see* **Note 21**). Only include sections of data with stable single-unit spike isolation and LFP recordings, which are free of movement artifacts. dCA1 recording sites were included only if they showed clear SWR (Fig. 2b, dCA1 LFP).

For freely moving data collected in the plus maze, align all LFP signals at the time point that marked the moment the mice had entered a new arm and hence had completed their choice. This time point was determined by analyzing the video offline and was defined as the time all four paws were first placed inside the newly chosen arm. For further data analysis, 10 s of LFP data centered on these time points were selected and stored to a hard disk (at 2 kHz).

3.3.3 Analysis of PC Activity

Identify PCs based on location, firing characteristics, and the presence of complex spikes [23] (Fig. 2a, b). We identify simple spike activity offline using a shape-based spike-sorting algorithm in Spike2 software. PCs simple spike firing rates ranged from 36.5 to 157.9 Hz, with a mean of 89.1 Hz, consistent with previously reported rates in awake mice [24] and primates [23]. We typically remove complex spikes from our analysis because their low frequency (∼1 Hz) results in an insufficient number of incidents for a meaningful evaluation.

3.3.4 Estimation of LFP Phase and Phase Relationships

In order to determine phase and phase relationships of LFP oscillations in mPFC and dCA1 over time, band-pass filter LFPs for frequencies between 0.5 and 100 Hz. Apply FIR filters in frequency steps of 0.25 Hz with bandwidth ranging from 0.5 to 10 Hz with increasing frequency. Filter order was determined to be the number of samples within 5 cycles of the center frequency. Apply the

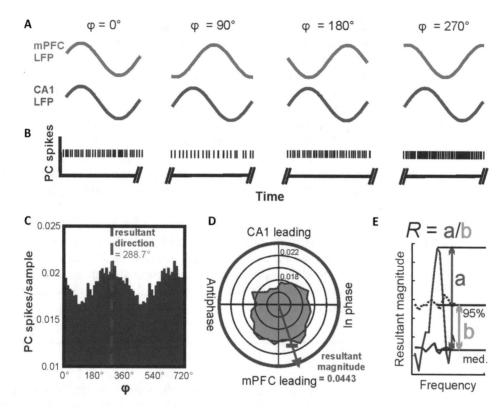

Fig. 5 Conceptual illustration of the data analysis applied to determine the correlation between phase differences between mPFC-CA1 oscillatory LFP activity and PC simple spike activity in LS. (**a**) Illustration of hypothetical oscillations in the mPFC (blue traces) and CA1 (red traces) at different phase relationships (φ). (**b**) Hypothetical PC spikes recorded simultaneously with the LFP oscillations. Here, PC activity significantly increased when the phase difference reached around 270°. (**c**) Phase histogram of real PC simple spike activity. The histogram shows spike activity as a function of mPFC-CA1 phase differences at 11 Hz, with a preferred phase difference of 288.7°. (**d**) Same data as in C represented in polar coordinates. The resultant vector magnitude was used to quantify the degree of modulation and tested against surrogate results for statistical significance. (**e**) Quantitative evaluation of the strength of representation (R) of mPFC-CA1 phase difference in PC spike trains. The solid blue line represents resultant vector magnitude plotted as a function of frequency. The solid black line ("med.") and dotted black line ("95%") represent the bootstrap-statistics derived median and 95th percentile boundary of the surrogate distribution, respectively. Resultant magnitude peak values were expressed as the ratio (R) of the peak resultant magnitude value minus surrogate median (red double arrow "a"), divided by the difference between the 95th percentile and median values (green double arrow "b") for each frequency. (From: McAfee et al. [14])

Hilbert transform on each band-pass filtered signal, and get the instantaneous phase values, which are defined as the angular value derived from the complex-valued analytic signal. Finally, subtract the instantaneous mPFC phase from the dCA1 phase to find the frequency specific phase relationship between mPFC and dCA1 oscillations. *See* Fig. 5 in [14] for more complete methods.

3.3.5 Time-Resolved Coherence Analysis of LFP Data

Of the two LFP signals available from two recording sites in each the mPFC and in the dCA1, one signal was chosen for further analysis. In the dCA1 we chose the LFP with the highest amplitude SWRs for further analysis, unless there was no difference, in which case one signal was chosen at random. In the mPFC, if LFP signals were equal in quality and one was chosen at random. In both the dCA1 and the mPFC recordings, if one signal showed increased line noise or artifacts, the other was used. All LFP data were first z-scored (Matlab function code: *z score*). Time-resolved coherence was calculated in Matlab using custom scripts (Matlab R2019b; function code: *mscohere*; sliding window: 1 s; step size: 20 ms) [22].

3.3.6 Time-Frequency Analysis of LFP Data

To examine time-frequency aspects of LFP activity in the mPFC and dCA1 region of the hippocampus, the same z-scored LFP data that were used for coherence analyses were used to conduct a time-frequency analysis, using custom scripts based on the Matlab function code: *pspectrum*. The temporal resolution of the time-resolved analysis was 0.2 s.

4   Notes

1. The induction chamber was built from a plastic, Tupperware-like box with a tightly closing lid. Tube connections for gas in- and outflow were cut into the sidewalls and plastic connector pieces were glued in, using a cyanoacrylate glue.

2. To keep the pilot hole at the exact diameter of the drill bit it is important to drill the hole in one go and keep a steady hand. It requires some practice to learn to stop advancing the drill as soon as there is no more resistance, which indicates that the bit is touching the dura.

3. Use of three screws is in our experience necessary and sufficient. Using only one or two anchoring screws is not recommended.

4. Stereotaxic left mPFC: lat. 0.5 mm, ant. 1.8 mm, left dCA1: lat. 2.0 mm, ant. −2.3 mm, right cerebellar LS: lat. 2.0 mm, ant. −5.52 mm, and right Crus I: 2.5 mm, ant. −6.25 mm. AP positions relative to bregma.

5. As the bone is removed for the craniotomies, occasional bleeding is normal and can be readily stopped with pressure applied with a sterile cotton swab. It is advised to rinse with sterile saline throughout drilling, especially as one approaches the dura, which should not become dry.

6. The bottom of the straw can be shaped with fine scissors to match the curvature of the skull.

7. Since the mPFC and dCA1 recording sites are close together we place one large chamber over both sites and a second chamber over the cerebellar LS/Crus 1 area.

8. When you have cut the straw pieces to their correct sizes, dip the edges meeting the skull in triple antibiotic ointment before placing them in their permanent position. This holds the straw in place and upright and the ointment also prevents acrylic cement from seeping underneath the straw barrier into the craniotomy.

9. Anchoring the bar to the skull is achieved using acrylic cement that is in turn anchored to the skull via four skull screws. In order to solidly anchor the screws to the skull bone the diameter of the pilot hole is crucial to ensure that the thread will cut into the bone when the screw is inserted.

10. Alternative methods use UV-curable cement applied directly to the skull bone, avoiding the use of screws [25].

11. It is important to not completely seal the skin around the cap to allow for fluid drainage.

12. We used a custom-made set of guiding tubes that directed 2 electrodes to each of our three recording sites (mPFC, dCA1, cerebellar LS/Crus I). The seventh channel was not used.

13. At the beginning of this process, the electrode tips should be just barely retracted within the guiding tubes. Ideally, once the 7-electrode manipulator is lowered to its final position, the guiding tubes should almost be touching the brain. In situations where this is not possible, do the best you can and then extend the electrode tips until they touch the brain tissue.

14. In the Thomas Recording system the metal guiding tubes serve as the electrical reference for the extracellular recordings. It is thus important to have the guiding tube tips immersed in saline that is filling the recording chambers, to establish electrical contact between the guiding tube ground and the brain tissue.

15. During recording, we usually make about 100 turns (or steps) of the driving screw to advance the electrodes for 1 mm that is about 10 μm per step.

16. We recommend covering the top of the Omnetics connector with a small piece of paper tape before applying acrylic cement.

17. We also tried optogenetic stimulation with the Chip-LED (1.25 × 2 mm, Lighthouse LEDs, LLC; Washington, USA), but this method generated large artifacts in the electrophysiological signals.

18. It is recommended to connect the reference and ground wires (2–3 loops around the skull screw) before positioning the electrodes (Fig. 4a).

19. In our experience, we have found that connecting the battery directly to the head-stage makes the ensemble too heavy for the mouse to carry. To adjust for this, we place the battery outside the recording environment and connect it to the head-stage with a long, thin cable.

20. Sinusoidal modulation of light was chosen to avoid the possibility of a depolarization block to occur with sustained DC illumination. The light modulation frequency was chosen to fall outside the frequency range of interest for coherence analysis and elevated from 100 to 120 Hz when high gamma oscillations between 80 and 90 Hz were identified as the frequency band with significant changes in decision related coherence.

21. We prefer to use a spectrum interpolation method instead of a notch filter to eliminate line noise. In our experience, a notch filter distorts our data by altering the power spectrum of the signal at 60 Hz.

References

1. Schmahmann JD (2019) The cerebellum and cognition. Neurosci Lett 688:62–75

2. Buckner RL (2013) The cerebellum and cognitive function: 25 years of insight from anatomy and neuroimaging. Neuron 80(3):807–815

3. Buckner RL et al (2011) The organization of the human cerebellum estimated by intrinsic functional connectivity. J Neurophysiol 106(5):2322–2345

4. Popa D et al (2013) Functional role of the cerebellum in gamma-band synchronization of the sensory and motor cortices. J Neurosci 33(15):6552–6556

5. Fries P (2015) Rhythms for cognition: communication through coherence. Neuron 88(1):220–135

6. Fries P (2005) A mechanism for cognitive dynamics: neuronal communication through neuronal coherence. Trends Cogn Sci 9(10):474–480

7. Gordon JA (2011) Oscillations and hippocampal-prefrontal synchrony. Curr Opin Neurobiol 21(3):486–491

8. Churchwell JC, Kesner RP (2011) Hippocampal-prefrontal dynamics in spatial working memory: interactions and independent parallel processing. Behav Brain Res 225(2):389–395

9. Tort AB et al (2008) Dynamic cross-frequency couplings of local field potential oscillations in rat striatum and hippocampus during performance of a T-maze task. Proc Natl Acad Sci U S A 105(51):20517–20522

10. Jones MW, Wilson MA (2005) Theta rhythms coordinate hippocampal-prefrontal interactions in a spatial memory task. PLoS Biol 3(12):e402

11. Brissenden JA et al (2018) Topographic Cortico-cerebellar networks revealed by visual attention and working memory. Curr Biol 28(21):3364–3372. e5

12. Lefort JM et al (2015) Cerebellar contribution to spatial navigation: new insights into potential mechanisms. Cerebellum 14(1):59–62

13. Tomlinson SP et al (2014) Cerebellar contributions to spatial memory. Neurosci Lett 578: 182–186

14. McAfee S et al (2019) Cerebellar lobulus simplex and crus I differentially represent phase and phase difference of prefrontal cortical and hippocampal oscillations. Cell Rep 27: 2328–2334

15. Richman CL, Dember WN, Kim P (1987) Spontaneous alternation behavior in animals: a review. Curr Psychol Res Rev 5(4):358–391

16. Paxinos G, Franklin KBJ (2001) The mouse brain in stereotaxic coordinates, vol 2. Academic Press, San Diego

17. Buzsaki G (2015) Hippocampal sharp wave-ripple: a cognitive biomarker for episodic memory and planning. Hippocampus 25(10):1073–1188

18. Thach WT (1968) Discharge of cerebellar Purkinje and nuclear neurons during rapidly alternating arm movements in the monkey. J. Neurophysiol 31:785–797

19. Sofroniew NJ et al (2014) Natural whisker-guided behavior by head-fixed mice in tactile virtual reality. J Neurosci 34(29):9537–9550

20. Vishniakou I, Ploger PG, Seelig JD (2019) Virtual reality for animal navigation with camera-based optical flow tracking. J Neurosci Methods 327:108403

21. Lalonde R (2002) The neurobiological basis of spontaneous alternation. Neurosci Biobehav Rev 26:91–104

22. Liu Y, McAfee SS, Van Der Heijden ME, Dhamala M, Sillitoe RV, Heck DH (2020) Causal evidence for a cerebellar role in prefrontal-hippocampal interaction in spatial working memory decision-making. bioRxiv 2020.03.16.994541; https://doi.org/10.1101/2020.03.16.994541

23. Thach WT (1970) Discharge of cerebellar neurons related to two maintained postures and two prompt movements. II. Purkinje cell output and input. J Neurophysiol 33:537–547

24. Cao Y et al (2012) Behavior related pauses in simple spike activity of mouse Purkinje cells are linked to spike rate modulation. J Neurosci 32(25):8678–8685

25. Kros L et al (2017) Synchronicity and rhythmicity of Purkinje cell firing during generalized spike-and-wave discharges in a natural mouse model of absence epilepsy. Front Cell Neurosci 11:346

Chapter 12

In Vivo Optical Detection of Membrane Potentials in the Cerebellum: Voltage Imaging of Zebrafish

Kanae Hiyoshi, Narumi Fukuda, Asuka Shiraishi, and Sachiko Tsuda

Abstract

Recent advances in our understanding of the brain function have been achieved using light to detect neuronal activity. Optical measurement of membrane potentials via voltage imaging is a desirable approach, enabling fast, direct, and simultaneous detection of membrane potentials in a population of neurons. Its high speed and directness can help detect synaptic and action potentials and hyperpolarization. Identifying various information that is encoded in these signals is essential to understand brain function, particularly in the cerebellum. Years of effort have yielded two types of voltage sensors: voltage-sensitive dyes (VSDs) and genetically encoded voltage indicators (GEVIs). In vivo application of these voltage sensors further promotes the optical interrogation of neuronal populations, leading to a deeper understanding of information processing in brain circuits. This chapter provides an overview of voltage imaging procedures and their application in circuit analysis. We focus on the cerebellum of zebrafish, a powerful animal model in the fields of developmental biology and neuroscience. The advantages of zebrafish for optical techniques enable the application of voltage imaging to various behavioral modalities, thereby providing essential information on brain function.

Key words Genetically encoded voltage indicator, Voltage-sensitive dye, Cerebellum, Purkinje cell, Electrophysiology, Zebrafish

1 Introduction

Unraveling the function of brain circuitry is fundamental to understanding how the brain works during various behavioral conditions. One of the challenges is to define how populations of neurons within brain circuits communicate with each other to yield a network function. Recent advances in optical techniques provide powerful tools for analyzing organizing principles and the development of neuronal circuits. Prominent examples include optical measurements and control of neuronal activity known as *optogenetics* [1–3]. These approaches have expanded the functional analysis of neuronal populations, previously analyzed using mostly electrophysiological recordings. For example, compared to electrophysio-

Roy V. Sillitoe (ed.), *Measuring Cerebellar Function*, Neuromethods, vol. 177, https://doi.org/10.1007/978-1-0716-2026-7_12,
© Springer Science+Business Media, LLC, part of Springer Nature 2022

logical detection, where a limited number of neurons are recorded, optical approaches enable the analysis of spatiotemporal dynamics from a large number of neurons.

Optical measurement of the membrane potential by voltage imaging is an attractive approach that enables fast, direct, and simultaneous detection of membrane potentials from a population of neurons, providing a desirable tool for the functional analysis of neuronal circuits [4–10]. Its high speed and directness allow for the detection of action potentials, hyperpolarization, and subthreshold responses, which are difficult to analyze by calcium indicators, due in part to the slow kinetics of the indicators and the dynamics of calcium [4, 11]. This is particularly important for analyzing cerebellar function, where inhibition is pivotal, and various information is encoded in the firing properties, such as simple spikes and complex spikes of Purkinje cells [12, 13]. Voltage-sensitive dyes (VSDs) and genetically encoded voltage indicators (GEVIs) detect changes in the membrane potentials. VSDs were initially identified to detect neuronal excitation in the 1970s [5–7]. Since then, extensive efforts have yielded a collection of VSDs with various colors, speed, brightness, and sensitivity [14–16]. These VSDs have been used for circuit analysis of the various brain regions, including the cerebellum [17–22]. Among these are red-shifted VSDs that can be combined with optogenetic control of neuronal activity [14, 18, 23]. VSDs indicate membrane potential changes by several mechanisms, such as a shift in their absorption/emission spectra upon depolarization or hyperpolarization [9, 24]. Previous studies have shown the advantages of VSD imaging in analyzing neuronal functions, such as the dendritic computation and inhibition, in several brain regions including the cerebellum (Di4-ANEPPS, ANNINE-6plus [17–19, 25, 26]). Although VSDs are useful for analyzing brain function, one limitation of VSD imaging is the lack of cell-type specificity. This is critical for the detailed analysis of neuronal circuitry by separating populations of neurons. To overcome this problem, considerable efforts have been made to develop GEVIs. Recent advances have produced increasing numbers of GEVIs with faster kinetics and brighter fluorescence, overcoming prior weaknesses in both speed and signal-to-noise ratio [4, 27–30]. Regarding the structure of GEVIs, there are a few classes: one is a mosaic protein that utilizes the voltage-sensitive domain of voltage-sensitive phosphatase originating from Ciona and other vertebrate species (VSFP-Butterfly 1.2 [27], ArcLights [30], and ASAPs [29]). Another class of GEVIs utilizes the voltage sensitivity of microbial rhodopsins. These include QuasArs (red-shifted GEVI, [31]), Ace2N-4AAmNeon [32], Archon [33], and Voltron [34]. Comparative evaluation of recently developed GEVIs is available elsewhere [8, 28]. The development of the improvement of voltage sensors has occurred in parallel with advances in optics, which have advanced the field of voltage imaging [35]. Regarding the application of GEVIs in studying brain function, more research is now underway on various brain regions, including the olfactory bulb, hippocampus, and cerebellum [36–38].

To maximize the optical readout of brain activity, observation targets are also critical. A prominent example is zebrafish, a small vertebrate that is widely used in many research fields, including developmental biology, neuroscience, and medical sciences [39, 40]. Its transparent and relatively small brain, together with thousands of transgenic lines available, provides an excellent system to analyze the functional organization of the brain in various behavioral modalities by whole-brain application of optical techniques in circuit analysis [39, 41–44]. As for the cerebellum, zebrafish has cerebellar circuits that resemble mammalian ones in their structure and function [45, 46]. An increasing number of studies have used voltage imaging in zebrafish [33, 34, 38, 47, 48]. Recently, some GEVIs (cf. ASAPs, Vortron, Archon) have successfully been used for voltage imaging of the zebrafish brain—including the cerebellum—by establishing transgenic fish lines [33, 34, 38].

Furthermore, combining voltage imaging with other techniques, such as optogenetic control or detection of neuronal activity via red-shifted optical probes [49, 50], behavioral tests, and various disease models, can deepen our understanding of the functional organization of the brain and its development.

Here, we describe the procedures of voltage imaging with VSDs and GEVIs, and the application of these methods for circuit analysis, focusing on the zebrafish cerebellum.

2 Materials

2.1 Staining of the Zebrafish Cerebellum by Voltage-Sensitive Dye

1. Di-4-ANEPPS, Invitrogen, D-1199.
2. Agarose, low gelling temperature, Sigma-Aldrich, A9414.
3. Puller, Narishige, PC-100.
4. Glass pipettes, Narishige, G-1.
5. Micromanipulator, Sutter Instrument, MPC-385.
6. Syringe, Teromo, SS-05SZ.
7. Tricaine, Sigma-Aldrich, A5040.
8. N-phenylthiourea (PTU), Nacalai Tesque, 27429-22.

2.2 Establishment of Transgenic Zebrafish Expressing GEVIs

1. Micromanipulator, Narishige, MN-153.
2. Stereo microscope, Olympus, SZ61/Leica, M205FA.
3. Confocal microscopes, Olympus, FV1000/Nikon, A1R.

2.3 Voltage Imaging of the Zebrafish Cerebellum and Spinal Cord

1. Tungsten wire, Nisshin EM, φ0.1 m/m, φ0.2 m/m.
2. 35 mm Glass Base Dish, Glass 12φ, No.1-S, Iwaki, 3971-035.
3. Surgical blade, Feather, No.11.
4. Agarose, low gelling temperature, Sigma-Aldrich, A9414.
5. Tubocurarine hydrochloride pentahydrate, Sigma, T2379.
6. Fluorescence stereo microscope, Leica, MZFL III.
7. Upright microscope, Nikon, FN.
8. Image acquisition software, Nikon, NIS-AR.
9. cMOS camera, Hamamatsu photonics, ORCA-Flash4.0.

2.4 Drug Treatments

1. Tetrodotoxin citrate, Tocris, 1069.
2. Picrotoxin, Abcam, ab120315.
3. Tubocurarine hydrochloride pentahydrate, Sigma, T2379.
4. Tube clamper, Narishige, CAT-1.
5. Peristaltic pump, World Precision Instruments, Peri-Star Pro 4 WPI PERIPRO-4LS.
6. Vacuum pump, Ulvac, DAP-6D.
7. Filter, WHATMAN, VACU-GUARD™.

2.5 Electrical Stimulation, Electrophysiology

1. Glass pipettes, 1.5 mm OD ×4 length with fila, World Precision Instrument, IB150F-4.
2. Millex GV syringe Driven Filter Unit (0.22 μm), Millipore, SLGVR04NL.
3. Microloader, MicroFil 34 Gauge 67 mm long, World Precision Instrument, MF34G-5.
4. Micromanipulator, Narishige, NMN-25/Sutter Instrument, MPC-385.
5. Electronic stimulator, Nihon Kohden.
6. Stimulus isolator, Nihon Kohden, SS-201J.
7. Amplifier, Nihon Kohden, CEZ-2400.
8. DAQ device, National Instruments, BNC-6012-USB.
9. Electrophysiology Software, University of Strathclyde, WinWCP.
10. Alexa Fluor™ 594 Hydrazide, Invitrogen, A10438.

2.6 Analysis

1. NIS-Elements (NIKON).
2. ImageJ programs, NIH.
3. Origin Pro, Lightstone.
4. HCImage software, Hamamatsu Photonics.
5. HSR software, Hamamatsu Photonics.

3   Methods

3.1   Zebrafish Lines

Wild-type zebrafish *(Danio rerio)* with a Riken wild-type genetic background are mainly used. Some fish are crossed with a nacre line [51] to increase their transparency. Adult zebrafish are maintained in a 14-h light/10-h dark cycle. For most experiments, embryos and larvae are incubated at 28 °C. For live imaging, zebrafish embryos are treated with 0.005% phenylthiourea, as needed (*see* **Note 1**). All procedures were performed in accordance with a protocol approved by the Saitama University Committee on Animal Research.

3.2   Staining of the Zebrafish Cerebellum by Voltage-Sensitive Dye

Larvae from 5 to 9 days post-fertilization (dpf) are paralyzed with 0.02% tricaine and mounted on 1% low melting-point agarose (*see* **Note 2**). The Di-4-ANEPPS solution (343.5 μM) is prepared according to a previously reported method [48, 52, 53] and loaded into the cerebellum using glass pipettes with a manipulator (MPC-385, Sutter Instrument) (Fig. 1a). The tip of the glass pipettes is cut using a surgical blade, and the dye solution is back filled into the pipettes. After an hour of incubation, the embryos are washed with an external solution (in mM: 134 NaCl, 2.9 KCl, 2.1 $CaCl_2$, 1.2 $MgCl_2$, 10 HEPES, and 10 glucose, adjusted to pH 7.8 with NaOH).

For imaging, embryos stained with Di-4-ANEPPS are paralyzed with 0.02% tricaine and tubocurarine (0.1 mM) and transferred to an extracellular solution containing 0.02 mM tubocurarine. Next, the skin above the cerebellum is carefully removed using fine forceps to expose the brain.

Although Di-4-ANEPPS is expected to be distributed across the cellular membrane, it is desirable to confirm whether Di4-ANEPPS is distributed properly after the staining step. For example, the localization of Di-4-ANEPPS on the cellular membrane can be confirmed using confocal microscopy (Fig. 1b, [48]). If there is an ectopic distribution or an aggregated signal, it is better to adjust the concentration of the dye, amount of dye injected, and/or washing process. In addition to the cerebellum, Di-4-ANEPPS fluorescence can also be observed in other parts of the brain, such as the optic tectum and hindbrain (Fig. 1b).

3.3   Establishment of Transgenic Zebrafish Expressing GEVIs

One of the advantages of using genetically encoded voltage indicators (GEVIs) is that it enables the optical detection of the membrane potential in specific cell types. The GAL4-UAS system is a powerful tool to achieve cell-type-specific expression of certain genes in zebrafish and other animals [54, 55]. For example, fish can express a GEVI in a specific cell type by crossing a UAS:GEVI fish line with a Gal4 fish line that expresses *gal4* in those cells. Continuous efforts in the zebrafish research community have

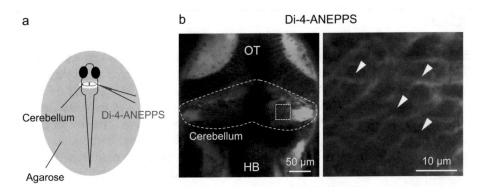

Fig. 1 Staining of Di-4-ANEPPS (VSD) and its membrane localization in the zebrafish cerebellum. (**a**) Schematic diagram of the Di-4-ANEPPS staining. (**b**) Confocal images of the zebrafish cerebellum stained with Di-4-ANEPPS at 7 dpf. Dorsal views are shown. Higher magnification images are shown in the right panel. Membranous distribution of Di-4-ANEPSS was observed (arrowheads). OT: optic tectum, HB: hindbrain. (Modified from Okumura et al. [48])

yielded thousands of Gal4 and UAS transgenic lines, most of which are available worldwide through scientific organizations such as the Zebrafish International Resource Center (ZIRC) in the USA and the National BioResource Project in Japan. For example, transgenic zebrafish lines that express *gal4* or other genes, like optogenetic probes, specifically in cerebellar neurons are available and widely used [38, 56]. These neurons include Purkinje cells, granule cells, inferior olivary neurons, and eurydendroid cells (the equivalent of the neurons in the mammalian deep cerebellar nuclei).

To establish transgenic zebrafish lines with GEVIs, the Tol2 transposon system is efficient and widely used [54]. First, a DNA construct that possesses UAS and GEVI sequences is generated or obtained as described previously (Fig. 2a, [38]). The critical step is to investigate whether GEVIs can be expressed in zebrafish tissues and whether they are correctly distributed in cellular membranes. This information can easily be obtained by examining fish with transient expression of GEVIs following co-injection of *gal4* mRNA and GEVI plasmid DNA containing transposase, UAS, and GEVI sequences at the one-cell stage. The expression of GEVI can be checked at a relatively early stage of development (cf. 1 to 2 dpf) using fluorescence stereomicroscopy (*see* **Note 3**).

If GEVIs are confirmed to be expressed and correctly localized to the cellular membrane in zebrafish, the next step is to obtain fish that have a GEVI sequence inserted in their genome. The widely used approach is the co-injection of the GEVI plasmid DNA and *transposase* mRNA into one-cell stage embryos. Screening of the fish possessing the transgene is performed via genotyping of the injected fish. The germline transmission is then tested. After mating candidate fish with *wild-type* fish or Gal4 fish, their offspring are

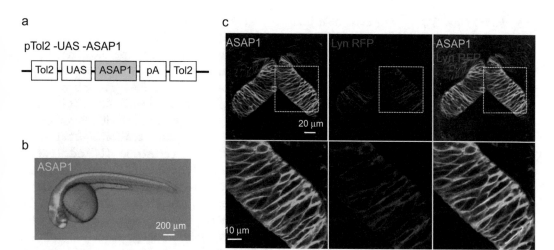

Fig. 2 Transgenic zebrafish showed membrane-localized ASAP1 (GEVI) in the neural tube. (**a**) Schematic diagram of plasmid construct for ASAP1. (**b**) ASAP1 was distributed widely in the neural tube of *Tg(elavl3: GAL4-VP16;UAS:ASAP1)* fish at 1 dpf. (**c**) Dorsal view of the neural tube of *Tg(elavl3:GAL4-VP16;UAS:ASAP1; UAS:lyn-RFP)* embryos at 1 dpf. ASAP1 (green) was co-localized with Lyn-RFP (red: cell membranes) in the neural tube. Higher magnification images are shown in the lower panel. (Modified from Miyazawa et al. [38])

investigated using genotyping or fluorescence observation, respectively (Fig. 2b). We use *Tg(elavl3:GAL4-VP16)* to examine GEVI signals in neurons, whereas cerebellar gal4 lines are used to drive the GEVI expression specifically in the cerebellar neurons [38].

For a detailed evaluation of GEVI expression and distribution in GEVI fish, confocal microscopic observation is preferable (*see* **Note 3**). Mating GEVI fish with transgenic fish, which express membrane-bound RFP (Lyn-RFP, Red Fluorescence Protein), is useful for confirming the membrane localization of the GEVI protein (Fig. 2c, [38]). If the GEVI is co-localized with Lyn-RFP, we can conclude that the GEVI is properly distributed to the cellular membranes. When using red-shifted versions of GEVI, it is appropriate to use membrane-bound GFP instead. It is important to note that GEVIs can also be distributed to the membranes of intracellular organelles. To evaluate GEVI expression in the cerebellum, larvae of approximately 5 to 7 dpf are preferable. This is because cerebellar neurons, such as Purkinje cells, are known to start proliferation from 3 dpf, and the basic structure of the cerebellar circuits is formed by 5 to 7 dpf [45, 57]. For confocal observation of zebrafish, fish are paralyzed in 0.02% tricaine and mounted on 2% low melting-point agarose (dorsal side up for upright microscopes, ventral side up for inverted microscopes). Optical sectioning is performed using a confocal microscope (Olympus FV1000 or NIKON A1R).

3.4 Voltage Imaging of Zebrafish Cerebellum and Spinal Cord

For cerebellar imaging, zebrafish larvae of 5 to 9 dpf are paralyzed using 0.02% tricaine and 0.01 mM tubocurarine (15 min) and transferred to an extracellular solution containing 0.01 mM tubocurarine. Larvae are then mounted on 2% low melting-point agarose dorsal side up. It is important to use 2% agarose as this can reduce the body movement of the fish (Figs. 3, 4, 5) (*see* **Notes 2** and **4**). For high-speed voltage imaging, wide-field microscopy is mainly used because it enables faster image acquisition than scanning microscopy. A fluorescence microscope (Nikon, FN-1) equipped with a cMOS camera (Hamamatsu Photonics, ORCA-Flash4.0) is used with a 25×/1.1 NA or 40×/0.8 NA water-immersion lens. The objective lens was chosen according to the purpose of the imaging. In general, a lower magnification lens with higher NA is preferable as this enables voltage detection across a wider area in the cerebellum or other brain regions with higher spatial resolution. Fluorescence images are acquired using HCImage or HSR software (Hamamatsu Photonics) at around 100 Hz (excitation filter: 470/40, emission filter: 535/50). We usually use cropping and binning functions to increase the acquisition speed and brightness, although the binning process reduces the spatial resolution of the images. It is essential to carefully set the acquisition parameters, such as exposure time, speed, and crop/binning of the imaging to optimize the voltage detection. The details are described in the following section (*see* **Note 5**).

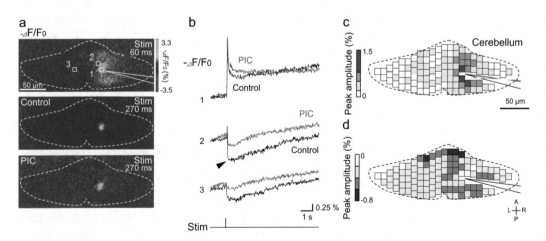

Fig. 3 Inhibitory signals in the cerebellum detected by Di-4-ANEPPS imaging. (**a**) Images of changes in Di-4-ANEPPS fluorescence 60 (top) and 270 ms (middle) after the electrical stimulation. (Bottom) Images of changes in Di-4-ANEPPS fluorescence after treatment with picrotoxin (PIC). Changes in fluorescence ($-\Delta F/F_0$) are indicated by the pseudocolor scale shown at the right. (**b**) Di-4-ANEPPS signals elicited by electrical stimulation of the cerebellum before and after treatment with PIC. Traces indicate signals detected at the three numbered locations indicated in (Fig. 3a). (**c, d**) Response map of the excitation (**c**) and inhibition (**d**) in the cerebellum. The peak amplitudes of the each response are encoded by the pseudocolor scale shown at the left. (Modified from Okumura et al. [48])

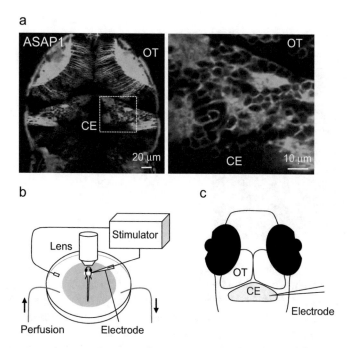

Fig. 4 Voltage imaging and electrical stimulation in the zebrafish cerebellum. (**a**) Membrane-localized ASAP1 in the cerebellum and optic tectum. Dorsal view of *Tg(elavl3:GAL4-VP16;UAS:ASAP1)* fish brain at 6 dpf. Higher magnification images are shown to the right. *CE* cerebellum, *OT* optic tectum. (**b, c**) Schematic diagram of the system for voltage imaging and electrical stimulation. (**a** and **b** was modified from Miyazawa et al. [38])

One disadvantage of wide-field microscopy is the lack of spatial resolution in the Z-axis. When required, voltage imaging using confocal microscopy is also useful. For this purpose, high-speed scanning is performed using a confocal microscope equipped with a resonant scanner (NIKON A1R). After acquisition of voltage imaging by either wide-field microscopy or confocal microscopy, it is preferable to obtain high-resolution images of the recorded specimen using confocal microscopy. The combination of high-temporal resolution data and high-spatial resolution data can help interpret voltage signals, such as identifying the sources of signals.

The spinal cord is known to develop quickly and exhibit spontaneous activity from around 17 hpf (0 dpf, [58, 59]), providing a useful system for in vivo evaluation of voltage sensors. This is due to the fact that spinal cord neurons can easily be observed at the single-cell level based on their large size, characteristic shape, and position (basally located) (Fig. 6). We usually perform voltage imaging of the spinal cord region first to evaluate voltage sensors in zebrafish neuronal tissues. For spinal code imaging, embryos are raised at 28 °C or 23 °C. Embryos of 20–23 hpf (22–28 somites, dechorionized) are paralyzed using tubocurarine (0.5 mM in

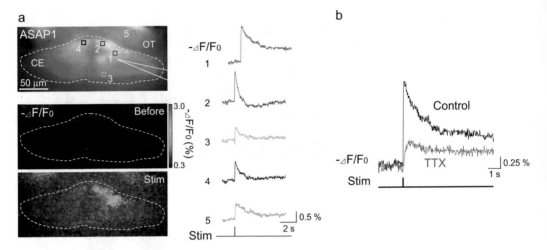

Fig. 5 Neuronal activity in the cerebellum was detected by ASAP1 imaging. (**a**) The evoked depolarization in the cerebellum was detected by ASAP1 imaging. (Top left) A fluorescence image of the cerebellum of *Tg (elavl3:GAL4-VP16;UAS:ASAP1)* fish. (Bottom left) Images of changes in ASAP1 fluorescence before and 0.14 s after the stimulation. Changes in fluorescence ($-\Delta F/F_0$) are indicated by the pseudocolor scale shown at right. (Right) ASAP1 signals produced by electrical stimulation of the cerebellum. Traces indicate signals detected at the 5 numbered locations indicated in the top-left image. (**b**) The depolarizing signal measured in the cerebellum was reduced by tetrodotoxin (TTX) treatment. Modified from Miyazawa et al. [38])

external solution, 300 μL, 20 min). Their tail tips should be cut using a surgical blade for better penetration of tubocurarine, as described previously [38, 53]. Embryos are then mounted dorsal side up in 1.8% low melting-point agarose containing 0.1 mM tubocurarine. To increase the efficiency of drug treatment, the agarose around the tail of embryos should be carefully removed using a surgical blade (Fig. 6a). Finally, the recording chamber is filled with extracellular solution containing 0.01 mM tubocurarine. In our experimental system, the spinal cord neurons located at 3–7 somatic regions are the targets of observation (Fig. 6b). The procedure of voltage imaging of the spinal cord is basically the same as that for the cerebellum.

3.5 Drug Treatments

To confirm that the fluorescence changes represent changes in neural activity, an extracellular solution containing 1 μM tetrodotoxin, a sodium channel blocker, is perfused using a peristaltic pump (Figs. 4b and 5b). Likewise, picrotoxin (100 μM), a GABA$_A$ receptor antagonist, is used to confirm that the hyperpolarization is a GABA-mediated inhibitory signal (Fig. 3).

For perfusion, a peristaltic pump and vacuum pump are used. The flow rate is set to 2 mL/min. The recording chamber is first filled with the external solution. After starting the perfusion of the drugs, the solution is aspirated with the vacuum device at almost the same speed as the perfusion in order to keep the amount of solution in the recording chamber stable. For the efficient exchange

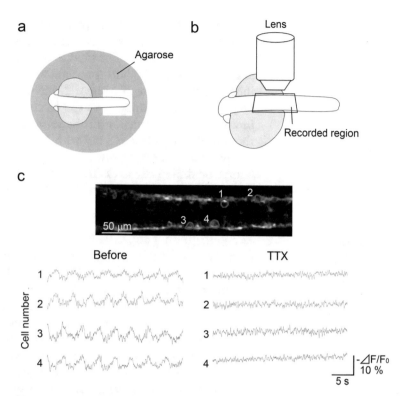

Fig. 6 Spontaneous activity of spinal cord neurons was detected at single-cell resolution via voltage imaging. (**a, b**) Schematic diagram of the spinal code imaging. Agarose around the tail of zebrafish embryo was removed (**a**). Spinal cord neurons in the 3–7 somatic regions were the targets of observation (**b**). (**c**) Monitoring of spontaneous activity of individual cells by ASAP1 imaging. (top) A fluorescence image of the ventral spinal cord of *Tg (elavl3:GAL4-VP16;UAS: ASAP1)* fish. The rostral side is to the left, and the area between 3 and 8 somites is shown. Regions of interest (ROIs) were located at individual cells (red: right side, blue: left side). (bottom) Examples of activity patterns of the 4 cells for two conditions (before and after TTX treatment). (**c** was modified from Miyazawa et al. [38])

of solutions in the chamber, the tip of the glass pipettes connected to the vacuum pump is located at the opposite side of the site of water inflow (Fig. 5b). After 5 min of perfusion, the pumps are switched off and voltage imaging is performed. To examine the effects of the drug treatment, voltage imaging is performed multiple times with an interval of approximately 5 to 10 min.

3.6 Electrical Stimulation

To characterize the responses of the voltage sensors in the zebrafish brain in vivo, we conducted electrical stimulation of the cerebellum and performed voltage imaging simultaneously (Figs. 3, 4, 5, [38, 48]). Larvae are mounted dorsal side up on 2% low melting-point agarose. For electrical stimulation, the skin above the cerebellum is carefully removed using fine forceps to expose the brain.

Electrical stimulation is applied directly to the cerebellum using a glass pipette filled with an external solution (Fig. 4b, c). We use 10 to 15 MΩ pipettes. The pipette tip is carefully positioned at the surface of the cerebellum using micromanipulators (Sutter Instrument MPC-385 or Narishige NMN-25) so that the cerebellar neurons are not damaged by the electrode itself. First, the appropriate stimulus protocol is set by varying the size and frequency of the stimuli (20 pulses of 3 mA with 1 ms duration at 33 Hz in most of the cases). Current pulses are generated by an electronic stimulator. The timing of the electrical stimulation and voltage imaging is controlled and monitored using a DAQ interface with WinWCP software.

3.7 Electrophysiological Recordings

For voltage imaging, it is important to examine the changes in membrane potential using electrophysiological recordings while performing voltage imaging. This approach provides critical information for evaluating the optical data obtained by voltage imaging, such as the sensitivity and speed of the responses. For this purpose, simultaneous recordings of the membrane potential using whole-cell patch-clamp recordings and voltage imaging are conducted to examine the membrane potential, sensitivity, and temporal resolution of responses detected using voltage imaging [18, 38].

Whole-cell patch-clamp recordings are performed as described previously at room temperature [38, 60, 61]. First, the anesthetized larva are mounted on 2% low melting-point agarose with their dorsal side up, and the skin above the cerebellum is carefully removed. Next, using wide-field microscopy, an electrode filled with an internal solution in a micromanipulator (Sutter Instruments, MPC-385) is placed above the cerebellar neuron. An internal solution consisting of (in mM): 110 K-gluconate, 15 KCl, 2 $MgCl_2$, 4 Na_2-ATP, 10 HEPES, and 10 EGTA (pH 7.2) is used. We use 10 to 15 MΩ pipettes. Electrical responses are acquired via an amplifier and DAQ interface with WinWCP software. For most cases, we analyzed the responses evoked by direct stimulation of the cerebellum. The shape of the recorded neurons can be seen by adding a fluorescent dye (Invitrogen, Alexa Fluor 594 Hydrazide) to the internal solution. Whole-cell patch-clamp recording and voltage imaging of the cerebellum are performed simultaneously.

3.8 Analysis

To extract the changes in fluorescence intensity, define regions of interest (ROIs) at desirable locations (*see* **Note 6**). In general, normalized changes in fluorescence intensity in the target cells or regions are used for quantitative analysis by calculating dF/F_0 in the ROIs ($dF/F_0 = $ ["fluorescence intensity at a certain timepoint" $- F_0]/F_0$, where F_0 is the intensity of the initial timepoints). For imaging data without stimulation, the first 15 frames of images are averaged and used as F_0. For recordings with electric stimulation, the 15 frames just before stimulation are averaged and set as F_0. These image analyses are conducted using NIS-Elements (NIKON, Calcium and FRET module) and ImageJ programs. Care should be taken for the response property of each voltage sensor, such as the direction of the changes in fluorescence intensity via depolarization. For example, ASAP1 and ArcLight show a reduction in their fluorescence intensity upon depolarization (increased intensity upon hyperpolarization). When defining ROIs in cell bodies, it is important to avoid overlap with signals from adjacent cells. If the larva/embryos move during voltage imaging, the images are aligned using ImageJ (TurboReg plugin). To improve the image quality, it is also useful to perform denoising (Nikon, NIS-Elements) and bleach correction (Fiji programs).

For quantitative analysis of the responsive area and size in the cerebellum, heatmap analysis is also useful (Fig. 3c, d, [38, 48]). First, small ROIs are placed throughout the cerebellum, and the peak amplitude of the voltage signal (dF/F) at each ROI is shown using pseudocolor. Graphing and statistical analyses are performed using Origin (Lightstone).

4   Notes

1. When conducting voltage imaging at later stages of development, the pigment cells may interfere with the detection of the voltage signals. In this case, it is preferable to use either transparent zebrafish lines that have a reduced amount of pigment cells [51] or prior treatment with phenylthiourea (PTU), which inhibits pigment formation [62].

2. Since agarose absorbs some amount of light, it is preferable to reduce the agarose amount above the samples by positioning the sample close to the surface during the mounting procedure.

3. Regarding the expression of GEVIs in zebrafish, fish with stronger GEVI fluorescence intensity are preferable for voltage imaging, especially for sensors that reduce their intensity upon depolarization [29, 30]. However, it is important to select fish with clear membrane localization of GEVIs, since some fish can show bright but ectopic signals [38]. The same is true for voltage-sensitive dyes. This point is also critical when maintaining GEVI fish lines.

4. Body movement often reduces the quality of optical detection, including voltage imaging. This is particularly true for in vivo imaging using awake animals. To prevent this, it is preferable to first put larval zebrafish in the agarose solution before transferring them to a recording chamber that contains agarose. In this way, the agarose concentration around the fish can be maintained at a sufficient level, reducing body movement.

5. A wide-field microscopy system is widely used for voltage imaging because it enables faster image acquisition than scanning microscopes. However, to maximize the quality of voltage imaging, care should be taken regarding the balance between the acquisition speed and the brightness of the images. When the acquisition speed is too fast by setting the exposure time to be short, it is difficult to obtain data with a sufficient signal-to-noise ratio (SNR) to detect changes in the membrane potential. Conversely, a longer exposure time will provide brighter images, but the temporal resolution will be lost. Thus, it is necessary to adjust the exposure time and speed of imaging. The excitation light intensity should also be set carefully, as this can significantly influence the SNR and amount of bleaching.

6. For quantitative analysis of the voltage signals, it is important to avoid mixing the signals from adjacent cells when using ROIs in individual cells. Since the cellular membrane regions are the major source of voltage signals and target of observation, ROIs should be carefully defined to cover membrane regions, especially around the cell soma. For the same reason, it is preferable to perform voltage imaging in the specimen where the membrane-localized GEVI signals are clearly observed. An alternative approach is to drive the expression of GEVIs in a mosaic manner using the injection of GEVI plasmid DNA. These help in the clearer detection of voltage signals.

Acknowledgements

This work was supported by JSPS KAKENHI (Grants-in-Aid for Scientific Research, Grant-in-Aid for JSPS Fellows), JST Tenure Track program at Saitama University, Kao Foundation grant, Narishige Foundation, and Asahi Glass Foundation.

References

1. Boyden ES et al (2005) Millisecond-timescale, genetically targeted optical control of neural activity. Nat Neurosci 8(9):1263–1268

2. Mancuso JJ et al (2011) Optogenetic probing of functional brain circuitry. Exp Physiol 96 (1):26–33

3. Yizhar O et al (2011) Optogenetics in neural systems. Neuron 71(1):9–34

4. Lin MZ, Schnitzer MJ (2016) Genetically encoded indicators of neuronal activity. Nat Neurosci 19(9):1142–1153

5. Cohen LB, Salzberg BM (1978) Optical measurement of membrane potential. Rev Physiol Biochem Pharmacol 83:35–88

6. Loew LM et al (1979) Evidence for a charge-shift electrochromic mechanism in a probe of membrane potential. Nature 281 (5731):497–499

7. Ross WN et al (1977) Changes in absorption, fluorescence, dichroism, and birefringence in stained giant axons: : optical measurement of membrane potential. J Membr Biol 33 (1–2):141–183

8. Storace D et al (2016) Toward better genetically encoded sensors of membrane potential. Trends Neurosci 39(5):277–289

9. Knopfel T, Song C (2019) Optical voltage imaging in neurons: moving from technology

development to practical tool. Nat Rev Neurosci 20(12):719–727

10. Zecevic D et al (2003) Imaging nervous system activity with voltage-sensitive dyes. Curr Protoc Neurosci. Chapter 6: p. Unit 6 17

11. Homma R et al (2009) Wide-field and two-photon imaging of brain activity with voltage- and calcium-sensitive dyes. Philos Trans R Soc Lond Ser B Biol Sci 364(1529):2453–2467

12. Thach WT (1968) Discharge of Purkinje and cerebellar nuclear neurons during rapidly alternating arm movements in the monkey. J Neurophysiol 31(5):785–797

13. Llinas R, Sugimori M (1980) Electrophysiological properties of in vitro Purkinje cell dendrites in mammalian cerebellar slices. J Physiol 305:197–213

14. Zhou WL et al (2007) Intracellular long-wavelength voltage-sensitive dyes for studying the dynamics of action potentials in axons and thin dendrites. J Neurosci Methods 164 (2):225–239

15. Wuskell JP et al (2006) Synthesis, spectra, delivery and potentiometric responses of new styryl dyes with extended spectral ranges. J Neurosci Methods 151(2):200–215

16. Fromherz P et al (2008) ANNINE-6plus, a voltage-sensitive dye with good solubility, strong membrane binding and high sensitivity. Eur Biophys J 37(4):509–514

17. Gao W et al (2006) Cerebellar cortical molecular layer inhibition is organized in parasagittal zones. J Neurosci 26(32):8377–8387

18. Tsuda S et al (2013) Probing the function of neuronal populations: combining micromirror-based optogenetic photostimulation with voltage-sensitive dye imaging. Neurosci Res 75(1):76–81

19. Roome CJ, Kuhn B (2018) Simultaneous dendritic voltage and calcium imaging and somatic recording from Purkinje neurons in awake mice. Nat Commun 9(1):3388

20. Cohen D, Yarom Y (2000) Unravelling cerebellar circuitry: an optical imaging study. Prog Brain Res 124:107–114

21. Kajiwara R, Tominaga Y, Tominaga T (2019) Network plasticity involved in the spread of neural activity within the Rhinal cortices as revealed by voltage-sensitive dye imaging in mouse brain slices. Front Cell Neurosci 13:20

22. Tominaga Y, Ichikawa M, Tominaga T (2009) Membrane potential response profiles of CA1 pyramidal cells probed with voltage-sensitive dye optical imaging in rat hippocampal slices reveal the impact of GABA(A)-mediated feedforward inhibition in signal propagation. Neurosci Res 64(2):152–161

23. Kee MZ et al (2009) Imaging activity of neuronal populations with new long-wavelength voltage-sensitive dyes. Brain Cell Biol 36(5-6):157–172

24. Peterka D et al (2011) Imaging voltage in neurons. Neuron 69(1):9–21

25. Cohen D, Yarom Y (2000) Cerebellar on-beam and lateral inhibition: two functionally distinct circuits. J Neurophysiol 83(4):1932–1940

26. Rokni D, Llinas R, Yarom Y (2007) Stars and stripes in the cerebellar cortex: a voltage sensitive dye study. Front Syst Neurosci 1:1

27. Akemann W et al (2012) Imaging neural circuit dynamics with a voltage-sensitive fluorescent protein. J Neurophysiol 108(8):2323–2337

28. Bando Y et al (2019) Comparative evaluation of genetically encoded voltage indicators. Cell Rep 26(3):802–813. e4

29. St-Pierre F et al (2014) High-fidelity optical reporting of neuronal electrical activity with an ultrafast fluorescent voltage sensor. Nat Neurosci 17(6):884–889

30. Jin L et al (2012) Single action potentials and subthreshold electrical events imaged in neurons with a fluorescent protein voltage probe. Neuron 75(5):779–785

31. Hochbaum DR et al (2014) All-optical electrophysiology in mammalian neurons using engineered microbial rhodopsins. Nat Methods 11 (8):825–833

32. Gong Y et al (2015) High-speed recording of neural spikes in awake mice and flies with a fluorescent voltage sensor. Science 350 (6266):1361–1366

33. Piatkevich KD et al (2018) A robotic multidimensional directed evolution approach applied to fluorescent voltage reporters. Nat Chem Biol 14(4):352–360

34. Abdelfattah AS et al (2019) Bright and photostable chemigenetic indicators for extended in vivo voltage imaging. Science 365 (6454):699–704

35. Nakajima R, Tsuda S, Kim J, Augustine G (2016) Optogenetics enables selective control of cellular electrical activity. In: Molecular neuroendocrinology: from genome to physiology. Wiley-Blackwell, Hoboken, New Jersey, pp 275–300

36. Storace DA, Cohen LB (2017) Measuring the olfactory bulb input-output transformation reveals a contribution to the perception of odorant concentration invariance. Nat Commun 8(1):81

37. Nakajima R et al (2016) Optogenetic monitoring of synaptic activity with genetically encoded voltage indicators. Front Synaptic Neurosci 8:22

38. Miyazawa H et al (2018) Optical interrogation of neuronal circuitry in zebrafish using genetically encoded voltage indicators. Sci Rep 8(1):6048

39. Friedrich RW, Jacobson GA, Zhu P (2010) Circuit neuroscience in zebrafish. Curr Biol 20(8):R371–R381

40. Tsuda S (2020) Optogenetics. In: Behavioral and neural genetics of zebrafish. Elsevier Academic Press, Cambridge, Massachusetts, pp 272–292

41. Del Bene F, Wyart C (2012) Optogenetics: a new enlightenment age for zebrafish neurobiology. Dev Neurobiol 72(3):404–414

42. Baier H, Scott EK (2009) Genetic and optical targeting of neural circuits and behavior–zebrafish in the spotlight. Curr Opin Neurobiol 19(5):553–560

43. Portugues R et al (2013) Optogenetics in a transparent animal: circuit function in the larval zebrafish. Curr Opin Neurobiol 23(1):119–126

44. Ahrens MB, Engert F (2015) Large-scale imaging in small brains. Curr Opin Neurobiol 32:78–86

45. Hashimoto M, Hibi M (2012) Development and evolution of cerebellar neural circuits. Develop Growth Differ 54(3):373–389

46. Bae YK et al (2009) Anatomy of zebrafish cerebellum and screen for mutations affecting its development. Dev Biol 330(2):406–426

47. Friedrich RW, Habermann CJ, Laurent G (2004) Multiplexing using synchrony in the zebrafish olfactory bulb. Nat Neurosci 7(8):862–871

48. Okumura K et al (2018) Optical measurement of neuronal activity in the developing cerebellum of zebrafish using voltage-sensitive dye imaging. Neuroreport 29(16):1349–1354

49. Inoue M et al (2015) Rational design of a high-affinity, fast, red calcium indicator R-CaMP2. Nat Methods 12(1):64–70

50. Klapoetke NC et al (2014) Independent optical excitation of distinct neural populations. Nat Methods 11(3):338–346

51. Lister JA et al (1999) Nacre encodes a zebrafish microphthalmia-related protein that regulates neural-crest-derived pigment cell fate. Development 126(17):3757–3767

52. Tominaga T, Kajiwara R, Tominaga Y (2013) VSD imaging method of ex vivo brain preparation. J Neurosci Neuroeng 2(2):9

53. Hiyoshi K et al (2021) In vivo wide-field voltage imaging in zebrafish with voltage-sensitive dye and genetically encoded voltage indicator. Dev Growth Differ 63(8):417–428

54. Asakawa K et al (2008) Genetic dissection of neural circuits by Tol2 transposon-mediated Gal4 gene and enhancer trapping in zebrafish. Proc Natl Acad Sci U S A 105(4):1255–1260

55. Scheer N, Campos-Ortega JA (1999) Use of the Gal4-UAS technique for targeted gene expression in the zebrafish. Mech Dev 80(2):153–158

56. Takeuchi M et al (2015) Establishment of Gal4 transgenic zebrafish lines for analysis of development of cerebellar neural circuitry. Dev Biol 397(1):1–17

57. Hibi M, Shimizu T (2012) Development of the cerebellum and cerebellar neural circuits. Dev Neurobiol 72(3):282–301

58. Muto A et al (2011) Genetic visualization with an improved GCaMP calcium indicator reveals spatiotemporal activation of the spinal motor neurons in zebrafish. Proc Natl Acad Sci U S A 108(13):5425–5430

59. Warp E et al (2012) Emergence of patterned activity in the developing zebrafish spinal cord. Curr Biol 22(2):93–102

60. Kimura Y et al (2013) Hindbrain V2a neurons in the excitation of spinal locomotor circuits during zebrafish swimming. Curr Biol 23(10):843–849

61. Harmon TC et al (2017) Distinct responses of Purkinje neurons and roles of simple spikes during associative motor learning in larval zebrafish. Elife 6:e22537

62. Karlsson J, von Hofsten J, Olsson PE (2001) Generating transparent zebrafish: a refined method to improve detection of gene expression during embryonic development. Mar Biotechnol (NY) 3(6):522–527

Chapter 13

Imaging Neuronal Activity in Cerebellar Cortex of Behaving Mice

Mikhail Kislin, Gerard Joey Broussard, Ben Deverett, and Samuel S.-H Wang

Abstract

In vivo imaging allows recording of population activity from many cerebellar circuit elements at once. In particular, Purkinje cells, molecular layer interneurons, and granule cells are differentially targetable for imaging studies. In this chapter, we first describe robust gene delivery strategies for these specific cerebellar cell types. We describe protocols for in vivo imaging and constraints that arise in head-fixed and freely moving behavioral paradigms. Finally, we offer practical guidelines for experimental design and data analysis, including the separation of signals from nearby cells.

Key words Purkinje cells, Two-photon microscopy, Miniscope, Virus, Calcium imaging

1 Introduction

The cerebellar cortex is a well-organized structure with characteristic cell morphology and an elaborate crystal-like organization [1, 2]. Its two input routes, the climbing fiber and mossy fiber pathways, both synapse onto Purkinje neurons (PN), the largest neurons of the mammalian brain and the sole output of the cerebellar cortex. Input to the PNs arrives by various cell types, most prominently parallel fibers arising from granule cells, which comprise the great majority of neurons in the mammalian brain and convey mossy fiber-based excitation, as well as climbing fibers arising from the inferior olive. In addition, parallel fibers and climbing fibers also excite molecular layer interneurons (MLIs), which in turn inhibit PNs. Together, granule cells, molecular layer interneurons, and Purkinje neurons provide a detailed window into cerebellar processing.

Functionally, the cerebellar cortex is organized parasagittally into microzones. Microzones are biochemically defined by distinctive antigens [3], and each microzone receives climbing fiber inputs

Roy V. Sillitoe (ed.), *Measuring Cerebellar Function*, Neuromethods, vol. 177, https://doi.org/10.1007/978-1-0716-2026-7_13,

from the same part of the inferior olive and innervates the same region of the deep cerebellar nuclei [4–6]. PN dendrites reflect this organization, as they ramify extensively with arbors spanning hundreds of microns within a thin (<10 μm) parasagittal plane. PN dendrites fire calcium-based dendritic action potentials accompanied by a "complex spike" waveform at the cell body. Each complex spike event is associated with a large dendritic calcium transient, which can be imaged to allow single-PN monitoring as well as the detection of simultaneous complex spike (CS) generation by PNs that are near one another in the mediolateral direction [7–9].

The highly modular circuitry of the cerebellum raises expectations that a realistic computational model of cerebellar circuit operations during behavior will be feasible. To date, the electrophysiological and molecular properties of most cerebellar cell types have been characterized in great detail and integrated into realistic single-cell models. However, extending these to inform a circuit-level model of cerebellar function will require research methods capable of probing cell physiology in behaving animals at the relevant spatial scales, in many cells at once. Such methods include optical imaging techniques coupled to methods for delivering activity sensors to specific cerebellar cell types. Here we review the current ability of calcium imaging techniques to serve this purpose, give specific protocols for imaging granule and Purkinje cells, and describe biological discoveries that have been enabled by the application of imaging techniques.

2   Materials and Methods

2.1   Preparing Animals for In Vivo Imaging

2.1.1   Surgery: Implantation and Craniotomy

The intact skull of adult mice is opaque, blocking optical access. This limitation can be overcome by thinning the bone, clearing the skull chemically, or permanently replacing a portion of the bone with a cranial window. Chronic cranial window preparation has been instrumental in advancing in vivo optical imaging studies in rodents [10]. This single surgical procedure with fast recovery of the experimental animal, combined with stable neuronal labeling, permits longitudinal high-resolution imaging of individual animals during development and adulthood.

Several attributes of the rodent cerebellum make it an unusually difficult site to perform window implantation. Unlike the cerebral cortex, the cerebellum is covered by thick neck muscles. Additionally, the occipital bone is not as flat as the interparietal and parietal bones that overlie the cerebrum. Several articles have refined the cranial window preparation and describe detailed methods enabling long-term in vivo imaging of cerebellum [11, 12]. Here we provide a summary protocol for these procedures optimized for cerebellum.

Anesthetize a mouse with isoflurane (5% for induction, 1.0–2.5% for maintenance in 100% oxygen at 0.5 l/min). During the procedure, monitor respiration to ensure one breath is taken roughly every second. Immobilize the head in a stereotaxic frame with coronal tiltability, and apply eye ointment to prevent ocular dehydration. Cranial window implantation usually lasts around 1 h; combining implantation with virus focal injection can prolong surgical procedures up to 4 h. Maintaining body temperature at 37.0 °C with a heating pad during surgery is crucial for small animals' survival and recovery. To reduce the surgery-induced inflammation and cerebellar edema, administer dexamethasone (2 mg/kg) by subcutaneous injection before surgery. Systemic D-mannitol (15% in saline, 1 g/kg) administration can be performed intraperitoneally 15 min before focal virus injection to widen the viral spread in tissue and improve the viability of cells near the virus injection site [13]. Prepare the surgical site by removing the hair with a scalpel and washing the site with 70% (vol/vol) isopropyl alcohol and betadine. Before removing the scalp overlaying the skull and neck muscles, apply 1% lidocaine on the incised area for local analgesia. After making an incision in the skin, detach the muscle and fascia from the interparietal bone with a sterile cotton applicator and cut them away. Avoid cutting major muscle vasculature, and control bleeding with sterile cotton-tipped applicators moistened with 0.3% hydrogen peroxide. Use tissue adhesive (Vetbond) to fix retracted neck muscles in place to maintain a low profile. Scrape off the periosteum and clean any remnants from the skull.

Under a surgical microscope, install the titanium headplate (Fig. 1i) over the parietal bone and secure it with cyanoacrylate, then trace the future craniotomy with a surgical skin marker (Fig. 1b). Thin the skull around the outside perimeter of the craniotomy with a high-speed dental drill and a conical-shaped drill bit. Stop drilling periodically to remove any dust with compressed air and sterile saline. Complete drilling of the bone flap will cause the surrounding skull to separate from the underlying meninges.

Apply copious sterile saline and then carefully lift and remove the bone fragment with forceps to create a craniotomy, keeping the dura mater intact (Fig. 1c). Cover the craniotomy with a collagen hemostatic sponge moistened with sterile saline. If needed, perform injection of virus using a borosilicate glass pipette (World Precision Instruments, 1B100F-4, 1/0.58 mm OD/ID) beveled to 30° with a ~10-μm tip opening and an automated injector system (World Precision Instruments Micro4). Add 1 μM FastGreen dye into the virus solution (5% glycerol in PBS) to visualize injection spread. When advancing the pipette into the brain tissue, use a micromanipulator to advance smoothly at approximately 10 μm per second. Position the tip of the pipette 150–350 μm below the brain surface

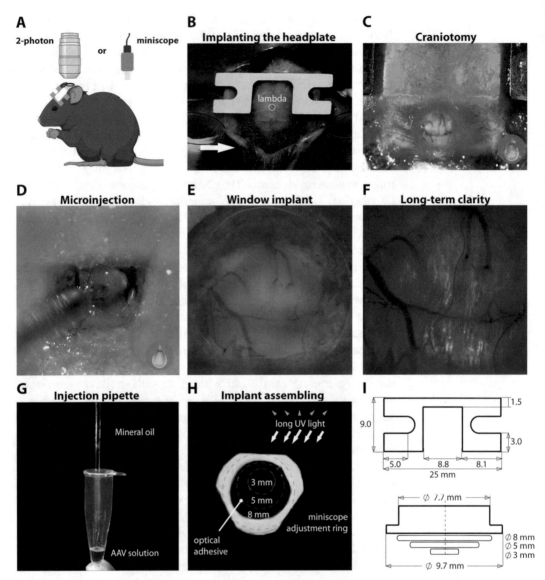

Fig. 1 Cranial window preparation for long-term in vivo calcium imaging of cerebellar neurons. (**a**) Schematic (created with BioRender.com) illustrating how the custom materials involved in the headplate and cranial window implant are combined and secured on a mouse head by dental cement for two-photon and wide-field microscopy. (**b**) Image of the mouse loaded in the stereotactic frame after the headplate has been implanted. The retracted muscle is fixed in place using Vetbond (white arrow). The headplate is centered on the skull and secured in place using a small amount of cyanoacrylate. (**c**) Image of the exposed cerebellum submerged in sterile saline (indicated by the blue water drop symbol) immediately after the craniotomy. (**d**) Image of the 250 nl AAV injection solution infusion at the rate of 3 nl/s. (**e**) Image of a 3-mm cranial window implant over the lobule IV/V of an adult mouse, taken on the day of surgery (anesthetized). The 3-mm glass coverslip locates completely inside of the edge of the craniotomy. (**f**) Green epifluorescence image illustrates window clarity and emergence of GCaMP6f expression (awake and head-restrained). (**g**) The microinjection setup fixed on a stereotactic frame allows us to safely immerse the tip of the glass pipette into the tube with the AAV injection solution. (**h**) Image of a three-layered cranial window (view of the surface that will contact the brain) assembled by using an optical adhesive and a UV light curing system. (**i**) Schematics illustrating the headplate and the three layers of coverglass (no. 1 thickness) used to construct the cranial window implant on a miniscope adjustment ring

and make a maximum of 4 injections, delivering 200–250 nl per injection at 3–5 nl/s (Fig. 1d, g). Following completion of each injection, leave the injection pipette in place for 5 min to allow for diffusion of the solution around its tip. When retracting the pipette, again use the micromanipulator to move the pipet smoothly and slowly.

Next, implant a coverglass over the craniotomy to provide clear optical access and control brain motion. Since the cerebellar surface is curved, implanting a flat coverglass might leave space between the window and the brain surface allowing bone to regrow. To prevent this, construct a stacked coverglass (Fig. 1h, i) by using two layers of different diameter coverglasses (no. 1 thickness, smallest diameter should fit to the craniotomy) glued with UV-curing optical adhesive (Norland optical adhesive, 7106, Norland Products and Light curing system, CS2010, Thorlabs) [14]. To hold the coverglass, use a glass Pasteur pipette attached to a vacuum and fixed on a micromanipulator or stereotaxic probe holder. Position the coverglass above the craniotomy, apply a drop of saline to the coverglass surface facing the brain and slowly touch the brain surface. Applying excessive pressure can impede cortical blood flow. Absorb the saline with a sterile cotton applicator, making sure that no bleeding occurs under the coverglass, and then apply cyanoacrylate to the edge of the coverglass. Apply C&B Metabond adhesive cement, allow it to fill all spaces under and around the headplate including the edges of the coverglass and also cover all areas of exposed skull. Ensure that C&B Metabond does not rise appreciably above the cranial window margins.

At the end of the surgery, discontinue anesthesia and allow the animal to wake up, keeping the mouse on the heating pad before returning to the home cage. During postsurgical care, administer an analgesic such as carprofen (5 mg/kg) for 2 days. Wait for at least 10 days after cranial window implantation before imaging. This period allows the animal to recover from anesthesia, minimizes the effects of immune response and drugs used during surgery, and ensures mechanical stability of the preparation. Inspect window clarity with epifluorescence (Fig. 1f).

2.1.2 Targeted Genetically Encodable Activity Indicator Expression in Cerebellum

Calcium imaging with genetically encodable calcium indicators (GECIs) is routinely used to measure neural activity in intact nervous systems. GECIs provide a solution to problems found with bolus loading of cell-permeant small-molecule calcium indicator dyes, namely: (1) labeling of off-target neurons and glia, (2) imaging period limited to hours, and (3) concerns about the state of astrocyte and microglia activity in the recently exposed brain. The green fluorescent protein- (GFP-) based GCaMP sensors have been iteratively engineered to enhance the signal-to-noise ratio (SNR) for detection of neural activity [15, 16]. The widely used GCaMP6f sensor now enables the detection of single action potentials under favorable conditions, and further improvements are continuing [17, 18]. GCaMPs have been used to monitor the activity of large groups of Purkinje neurons, molecular layer interneurons, and granule cells in vivo using two-photon microscopy and wide-field imaging. They can also measure activity-induced calcium changes in small subcellular compartments such as parallel fiber boutons on timescales of tens of milliseconds to months.

Cell-type specific expression of GECIs can be achieved using cell-specific gene promoters. Cell-specific promoters drive expression at higher levels in a subset of cells relative to the baseline expression across all other cells. They often play critical roles in the biological functions unique to those cells [19]. Some specific promoters do not drive protein expression strongly enough to allow functional imaging, while the strongest promoters might drive overexpression. Here we provide details of several strategies including transgenic techniques and viral infections to optimally target GECIs in selected cerebellar cortex cell types, MLIs, granule cells, and Purkinje neurons.

One powerful technology for driving cell-type specific expression is the use of genetically engineered mouse models. The Cre-loxP system is widely used to generate spatiotemporally controlled transgene expression [20]. Two elements are needed in the Cre-loxP system: (1) a Cre-driver strain in which Cre recombinase is expressed by a promoter that specifically targets the cell type of interest and (2) a Cre-dependent reporter strain with the loxP flanked (floxed) DNA. For example, development of the L7ΔAUG minigene [21] allowed generation of the first transgenic mouse strain (B6.129-Tg(Pcp2-cre)2Mpin/J, 004146, Jackson Laboratory) with Purkinje cell-specific expression of Cre recombinase [22]. However, off-target expression was reported in this strain. Later, a bacterial artificial chromosome (BAC) carrying the entire intact Purkinje cell protein 2 (Pcp2) gene was used to generate BAC/Cre transgenic mouse strain (B6.Cg-Tg(Pcp2-cre) 3555Jdhu/J, 010536, Jackson Laboratory) that have exclusive Cre expression in Purkinje and bipolar cells [23]. Developments in the generation of Cre-dependent transgenic mouse lines generated through knockin at the TIGRE locus [24, 25] include a collection of reporter lines for observing and manipulating cell-type functions. In particular, Ai93-96D (B6J.Cg-Gt(ROSA)26Sor$^{tm95.1(CAG-GCaMP6f)Hze}$/MwarJ, 028865, Jackson Laboratory) and Ai148D (TIT2L-GC6f-ICL-tTA2, 030328, Jackson Laboratory) strains provide stable transgenic expression of GCaMP6f and GCaMP6s and enable targeting of GECIs to specific cell populations in the cerebellum. Mice heterozygous for Ai148D exhibit no fluorescence, but when bred to the Pcp2-cre Cre-driver line, the resulting double transgenic animals (GCaMP6f+-tTA2+-/

Cre+-) exhibit robust expression of GCaMP6f in Purkinje neurons with no viability problems [26].

Another common strategy for expressing transgenes is the use of recombinant adeno-associated virus (AAV) and lentiviral vectors [27, 28]. Stable transgene expression is achieved by permanent random integration of lentiviral-delivered transgenes or by episomal maintenance of integrase-deficient lentivirus or AAV. An important advantage of lentiviral vectors and AAVs over other vector systems is that inflammatory and immune responses are limited [29]. Unlike AAVs, which have a diameter of approximately 20 nm, the larger particle size of lentiviruses (200 nm) limits spread through the extracellular space [30]. Hyperosmotic solution in the form of mannitol can be systemically injected before virus injection to widen extracellular spaces [31]. This strategy shows improved diffusion for AAVs within the brain parenchyma and increases the time window over which functional imaging is practical [13]. Many other factors can affect how efficiently cells of interest are targeted by AAV. Experimental parameters such as AAV titer and dosage can impact AAV efficiency, and these details are often omitted from experimental methods in the literature and can be expensive and time-consuming to determine empirically for each study. Suboptimal experimental design can drastically reduce the quality of results. Scientists using AAV for GECIs transfer must consider various questions about experimental design, including: (1) how to best deliver/administer AAV, (2) which AAV serotype to use, and (3) how to drive gene expression with gene regulatory elements. Though currently in early stages of development, efforts such as the non-profit plasmid repository Addgene's AAV Data Hub (http://datahub.addgene.org/aav/) are attempting to aggregate experimental data on optimal titers for specific viruses and injection sites. Scientists can contribute information directly to these repositories to reduce duplication of effort across the community.

AAV tropism is dictated by the interaction between cell surface proteins and glycans and AAV capsid proteins derived from both naturally occurring and synthetic AAV strains. Tropism is a critical determinant of transduction efficiency, thus serotype and route of delivery should be chosen carefully. When injected directly into the brain parenchyma, many of the naturally occurring AAV capsids have distinct local tropisms and distribution characteristics [32]. In addition to exhibiting local transduction, several serotypes exhibit transduction distal to the injection site [33], and some variants have been engineered for enhanced retrograde transport [34]. Several natural AAV capsids, like AAV9 and AAVrh10, and engineered AAV-PHP.eB cross the blood–brain barrier and systemic delivery can provide brain-wide gene delivery [35, 36]. Vectors can be also injected into the cerebral spinal fluid via the lateral ventricle, the cisterna magna, subpial or the intrathecal space. Achieving

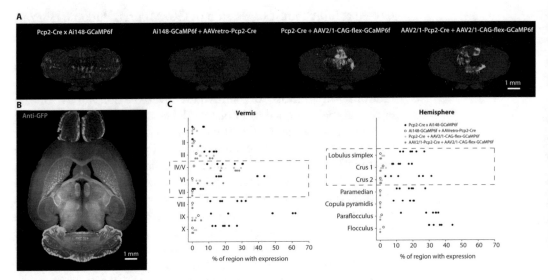

Fig. 2 Effective strategies for targeting Purkinje neurons. (**a**) Examples of immunostaining for anti-GFP presented as three-dimensional projection rendering of cerebellum light-sheet volumes overlaid over Allen Brain Atlas (gray) (created with FluoRender). Pcp2-Cre x Ai148-GCaMP6f double transgenic mouse shows near-ubiquitous labeling of PNs cross cerebellum (subpanel 1). Focal injections of rAAV2-retro carrying Cre recombinase transgene under Pcp2 promoter into the DCN in Ai148-GCaMP6f transgenic mouse results in sparse, cerebellum-wide expression (subpanel 2). Focal injections of Cre-dependent rAAV2/1 carrying GCaMP6f into a defined portion of cerebellar cortical regions in Pcp2-Cre mouse provide strong region-restricted expression of GECIs in PNs (subpanel 3). Similar expression pattern can be achieved by focal co-injections of Cre-dependent rAAV2/1 carrying GCaMP6f and rAAV2/1 with Cre recombinase transgene under Pcp2 promoter (subpanel 4). (**b**) Representative horizontal plane image of Pcp2-Cre x Ai148-GCaMP6f mouse brain stained with anti-GFP (green), autofluorescent (gray) used for anatomical reference. Note bright stripes in the molecular layer and the absence of label in the granular layer of cerebellar cortex. (**c**) Quantification of GCaMP6 expressing cells was expressed as percent of cerebellar cortical region covered after automated segmentation. Cerebellar regions accessible for optical physiology are marked with a dashed box. Coverage levels per brain show the difference in variability within and between strategies for GECIs delivery in PNs

efficiency with systemic and cerebral spinal fluid delivery requires relatively high viral doses ($>10^{10}$ viral genomes), thus requiring large volumes of high titer virus [37]. Tighter cell-type specificity can be achieved by combining AAV with the Cre-loxP system in transgenic animals. Both components of the system can be supplied by viral vector AAVs or a transgenic animal, either of which can be controlled by a cell-type specific promoter.

To characterize the PN-specific expression patterns achieved with the Pcp2-Cre x Ai148-GCaMP6f double transgenic strain (Fig. 2b) and other gene targeting strategies (Fig. 2a) we used iDISCO+ tissue clearing [38] and whole-brain immunostaining using primary chicken anti-GFP and secondary donkey anti-chicken 647 antibodies. Cleared brain samples were imaged using a light-sheet microscope (Ultramicroscope II, LaVision Biotec., Bielefeld, Germany). We counted the indicator-expressing voxels per cerebellar region and normalized on the total count of voxels in the region. Pcp2-Cre x Ai148-GCaMP6f double transgenic strain exhibited widespread and high-density expression across all cerebellar cortical regions (Fig. 2c).

We have developed multiple methods using rAAVs to deliver GECIs to cerebellar Purkinje neurons. Focal co-injection of rAAV2/1-CMV-Cre and rAAV2/1-CAG-Flex-GCaMP6f results in strong expression [17, 39]. Titer calibration is required to prevent excessive protein overexpression and, in the case of GECIs, excessive buffering of intracellular calcium signals. The use of a single vector approach with Purkinje neuron-specific Pcp2 promoter in lentiviral vectors [40] and AAVs [41] permits cell-specific expression. Originally used in transgenic mice, the Pcp2 promoter has a 4 kilobase (kb) 5′ flanking region of the pcp2 gene [42] and is not a good fit for AAV and lentiviral vectors due to their limited packaging capacity. Several studies have revealed that a shorter 0.8 kb Pcp2 promoter drives efficient and specific expression in Purkinje neurons [41, 43]. This Pcp2 promoter has a considerably weaker promoter strength compared with constitutive viral promoters such as the cytomegalovirus (CMV), murine stem cell virus (MSCV), elongation factor 1 alpha (EF1α), and CMV enhancer—chicken beta-actin (CAG) promoters. We generated rAAV2/1-Pcp2-Cre and rAAVretro-Pcp2-Cre vectors and characterized them after focal injection into Ai148D transgenic mice (Fig. 2c). Both vectors demonstrated the expected specificity of expression (Fig. 2a). By adjusting the titer of those Cre carrying AAVs one can achieve desirable sparse Golgi-like indicator expression (Fig. 6a, b). We also co-injected rAAV2/1-Pcp2-Cre and rAAV2/1-CAG-Flex-GCaMP6f and observed levels of expression similar to Pcp2-Cre x Ai148-GCaMP6f double transgenic mice. In addition, long-term, sustained expression of GCaMP6f did not result in PN nuclear invasion, a classic sign of poor cell health [44]. These AAV based strategies further expand the repertoire of viral tools that can be used for highly specific GECI delivery to PNs.

Overall, consistency and repeatability of gene delivery can be improved by following the practices described above. While powerful technologies are being developed, each has technical limits that need to be considered both when designing experiments and when interpreting results. In choosing methods for GECI expression one must consider the complex trade-offs between genetic precision, technical ease, and the availability of existing resources.

2.2 Imaging and Data Analysis

2.2.1 Imaging Cerebellar Granule Cells and Molecular Layer Interneurons In Vivo

Cerebellar granule cells form the densest neuronal layer in the brain. Granule cells are a hub of sensorimotor integration and send out one axon that branches to form a parallel fiber that runs across hundreds of Purkinje cells in a perpendicular direction to their dendritic arbors. At least three Cre-driver lines—Math1-Cre (B6.Cg-Tg(Atoh1-cre)1Bfri/J, 011104, Jackson Laboratory), GABA(A)Rα6-Cre [45], and NeuroD1-cre (B6.Cg-Tg.NeuroD1-

Cre.GN135Gsat)—have been successfully used for functional imaging of granule cells. Math1 is a transcription factor involved in the development of the cerebellum and hair cells of the cochlear and vestibular organs [46]. The Ai93/ztTA/Math1-Cre triple transgenic mice show widespread GCaMP-expression throughout most granule cells and only rarely resulted in PN expression [47]. Expression of Cre recombinase under the GABA$_A$ receptor α6 subunit promoter occurs in cells derived from a common germinal zone, the rhombic lip, and granule cell is the only cerebellar cell type with this developmental lineage [45]. To achieve GECI expression, GABA(A)Rα6-Cre driver mice were crossed to mice expressing the calcium indicator GCaMP6f in the Cre-dependent Ai95 strain [48]. NeuroD1 is a transcription factor involved in several organ developmental programs. NeuroD1-cre mice focally injected with rAAV2/1-CAG-Flex-GCaMP6f demonstrate a widespread expression in the granule layer (Fig. 3), mostly observed in granule cells and occasionally in large cells, tentatively classified as Golgi or Lugaro cells [49]. Those expression strategies were combined with large field, high-speed two-photon microscopy and computational methods, which allowed separation of signals from overlapping cells in several studies [47, 49, 50]. This approach allows monitoring of large populations of granule cells in awake behaving animals at a single granule cell resolution and enables new insights into the organization of information arriving via the mossy fiber-granule cell system.

Imaging activity of cerebellar granule cells in awake, behaving mice has revealed multiple features of encoding which were impossible to unmask with electrophysiological methods. Several recent studies have demonstrated that when mice are presented with a randomized battery of sensorimotor stimuli during free and forced locomotion, the majority (75%) of granule cells are activated in response to multiple dissimilar sensorimotor stimuli presented at the same time [48, 51]. Some granule cells may encode non-sensorimotor information, such as reward expectation [47]. Granule cells can also change their coding properties following learning [49, 50, 52]. Wagner et al. subjected mice to an operant, voluntary forelimb motor task, during which they were trained to push a lever in order to receive a sucrose-water reward with varied delay times. They showed that some granule cells responded preferentially to reward or reward omission, whereas others selectively encoded expectation of reward [47]. Furthermore, learning increased correlations among initially dissimilar neocortical layer 5 pyramidal neurons and granule cells and those neurons converged to similar, low-dimensional, task-encoding activity [52].

Giovannucci et al. and Yamada et al. used two different cerebellar forms of motor learning: classical eyeblink conditioning [49] and delay tactile startle conditioning [50]. Granule cells of naïve

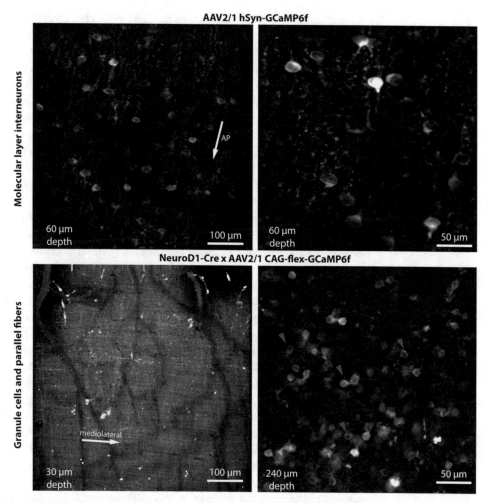

Fig. 3 Effective strategies for targeting molecular layer interneurons (MLIs) and granule cells. Top panel, mean images from representative two-photon calcium imaging movies, brightly labeled somata of MLIs, dendrites and axons are oriented along anteroposterior axis marked by the arrow labeled (AP). Lower panel, standard deviation images from representative two-photon calcium imaging movies from molecular and granular layers. The arrow marks mediolaterally aligned boutons of parallel fibers. Red arrows denote example granule cells. Numbers in the lower left corner indicate imaging depth

mice showed responses to different components of the task. The ability to monitor the same neurons over time enabled characterization of the acquisition of activity time-locked to or even preceding the learned movements [49, 50]. Population imaging studies in behaving mice have provided compelling evidence that large ensembles of granule cells are activated during sensorimotor integration and challenged the assumption that granule cells use ultra-sparse encoding of sensorimotor context.

Stellate and basket cells of the mammalian cerebellar cortex are GABAergic neurons, collectively called molecular layer interneurons [53]. The transcriptomes of PNs and MLIs are highly dynamic throughout postnatal development, making gene targeting challenging [54]. High selectivity of receptor tyrosine kinase and proto-oncogene (c-Kit) was indicated in MLIs over mature PNs [55]. Well-isolated fluorescence signals were obtained from c-kit$^{IRES-Cr}$ mice sparsely expressing rAAV2/1-CAG-flex-GCaMP6f [56]. Another method of MLI-specific expression mainly relies on the differential activity of the hSyn promoter (Fig. 3) between those neurons [13]. Indeed, Parvalbumin-Cre mice (008069, Jackson Laboratory) [57] injected with rAAV2/1-hSyn-Flex-GCaMP6f and C57BL/6 injected with rAAV2/5-hSyn-GCaMP6s or rAAV2/1-hSyn-GCaMP6f show exclusive expression in MLIs [58].

Spatial and temporal characteristics of MLI population activity have been optically examined in behaving mice during motor behavior [56, 59] and in a go/no-go task [60, 61]. During cued-licking task and rhythmic licking, groups of neighboring MLIs are engaged in highly coherent waves of synchronous neuronal activity and encode kinematic variables describing temporal aspects of rhythmic movement [56, 59]. These imaging studies also indicate that MLI activity influences motor output during continuous movement. In the odorant or audible tone cued go/no-go tasks, differential MLI activity developed during learning and this learned activity switched when the rewarded cue was reversed [61]. It was demonstrated that MLIs gate PN dendritic response during performance of learned movements [60, 62]. This evidence suggests MLIs constrain learning in the absence of error signals. Calcium activity of the MLIs presumably captures a parvalbumin-filtered combination of spontaneous electrical activity, coordinated effects of electrical synapses, potentially inhibitory effects of chemical MLI synapses, and excitatory synaptic inputs from parallel and climbing fibers.

2.2.2 Purkinje Cell-Specific Two-Photon Imaging for Elucidation of Olivocortical Modules

Calcium signaling plays a key role in all steps of the input to output translation process in Purkinje neurons which receive inputs from three major sources: excitation from parallel fibers and climbing fibers, as well as inhibition from molecular layer interneurons. Synaptic events and intrinsic firing of Purkinje neurons are the main mechanisms responsible for activation of voltage gated calcium channels and calcium influx into the cytosol. G-protein coupled receptors (e.g. metabotropic glutamate receptor 1) also contribute to rise of cytosolic calcium by activation of signal transduction pathways linked to calcium release from endoplasmic reticulum via IP3 and ryanodine receptors [63–65]. Buffering and extrusion of free calcium from the cytosol are mainly mediated by calcium-binding proteins (e.g. calbindin, parvalbumin) and calcium pumps (e.g. sarcoplasmic/endoplasmic reticulum or plasma membrane calcium ATPase, sodium/calcium exchanger).

PNs exhibit a rich repertoire of dendritic excitability, most notably dendritic calcium spikes triggered by either climbing fibers or parallel fibers and associated with substantial elevation of calcium in the dendrite [66–70]. Activation by parallel fibers triggers local dendritic calcium increases confined to small regions of the spiny branchlets [71], while climbing fiber calcium transients propagate across the entire dendritic tree [67, 72, 73]. Quantitative characterization of the magnitude and spatial spread of the calcium signals and their precise relationship with electrical activity has been performed in vitro and in vivo [64, 70, 72–74]. The calcium transients coincide precisely with the onset of the CSs measured electrophysiologically, have a rise time of tens of milliseconds, and decay with a half-time of 150–200 ms as measured by GCaMP6f. The CS firing rate is on average 0.5–2.0 Hz with inter-complex-spike interval ranging from 30 ms to 3.5 s, with response peaks being followed by a period of reduced intertrial firing. Several in vivo studies showed that the calcium transients are triggered primarily by climbing fiber input and not by parallel fibers or spontaneous calcium spikes [7, 72].

Anatomical organization of olivo-cerebellar climbing fibers and electrical coupling between neurons in the inferior olive [75] suggest that neighboring PNs located within defined parasagittal areas should tend to show synchronous complex spike activity. Studies utilizing high spatial resolution two-photon microscopy have revealed spatially fine parasagittal regions with PNs showing high CS synchrony [7, 76]. These imaging studies have shown that microzones are activated spontaneously or by sensorimotor inputs, indicating that microzones represent functional units in the cerebellar cortex. In contrast to individual Purkinje cells, ensembles of simultaneously firing Purkinje cells provide a reliable encoding of the sensory input [7].

Hereafter, we will use the term "calcium event" to refer to these climbing fiber-triggered calcium transients. Fast-responding GECIs, like GCaMP6f, have been shown to robustly report dendritic calcium events in PNs with excellent resolution [17, 39]. GCaMP6f kinetics (Fig. 4e) can closely track climbing fiber-induced calcium changes in PNs; however, in experimental conditions where complex spikes doublets occur with smaller than 150 ms intervals [77, 78], electrical recordings more clearly show whether one or two events have occurred. GECIs continue to improve and recent enhanced GCaMP sensors [18] which are tailored to detect neuronal activity in specific applications might further improve tracking of calcium events in PNs.

Several recent studies have combined optical recording methods with animal behavioral paradigms to study how PN activity relates to sensorimotor and cognitive information processing during behavior. These studies have demonstrated that population

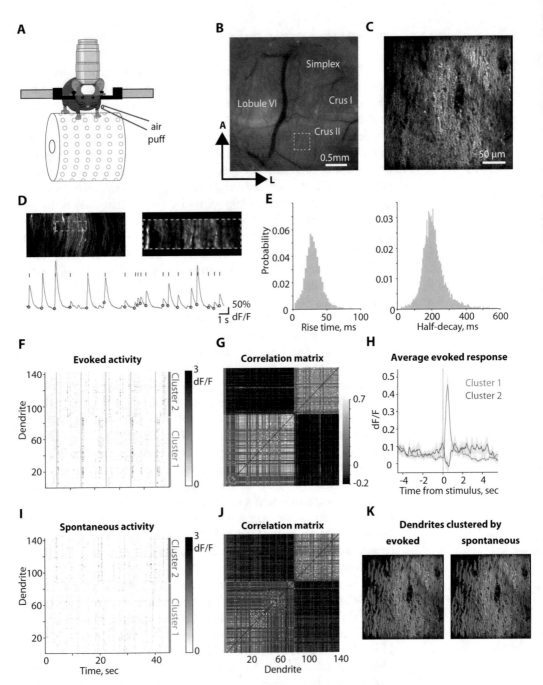

Fig. 4 In vivo imaging of Purkinje dendrites with two-photon microscopy during sensory stimulation. (**a**) Schematic of head-fixed animal on running wheel during two-photon imaging. (**b**) Wide-field image of the analyzed ROI in Crus II with anatomical orientation as indicated by arrows. (**c**) Session average of two-photon data. (**d**) Average two-photon zoomed-in images of Purkinje dendrites expressing GCaMP6f (top view) and gray traces represent calcium-mediated fluorescence change over time. The tick marks represent detected peaks and cycles—starts of dendritic transients. (**e**) Distributions of rise (10–90%) and half-decay times of GCaMP6f (103 dendrites from $n = 4$). (**f, g**) Sample evoked activity from all detected PN dendrites clustered by correlation with heatmap of pairwise correlations. (**h**) Trial-averaged evoked activity from the clusters presented in (**f**) and (**g**). (**i, j**) Sample spontaneous activity from all detected PC dendrites clustered by correlation with heatmap of pairwise correlations. (**k**) Spatial organization of the PC dendrite clusters during evoked (left) and spontaneous (activity). The clusters are identical in each case

activity of PNs encodes information related to the stimuli presented to animals, the choices animals make [26], and the representation of reward-related information [9, 26, 79–81]. Heffley et al. reported that microzones in lobule simplex exhibited enhanced activity during correctly executed movements that subsequently yielded reward which was absent in error trials. Omission of expected rewards and unexpected rewards elicited signals similar to those during successful trials [9]. Kostadinov et al. used a different cued timed forelimb operant behavior and found that some microzones became active during reward, while others were suppressed, with larger changes evoked by unexpected rewards. Optical monitoring of the same population of PNs over training revealed that while animals learned the reward-predicting cue the reward-delivery-evoked calcium events were gradually suppressed [79]. Expanding these findings, Tsutsumi et al. explored neuronal heterogeneity between aldolase C (zebrin) zone in auditory go/no-go task and identified a comprehensive functional map of crus II [81]. Population imaging studies could unmask how temporally orchestrated activation of spatially organized modules by relevant sensory cues, motor responses, and reward outcomes could be a basis for implementing sophisticated behavior.

The physiology and morphology of Purkinje neurons impose performance criteria on recording population activity optically. Separation of individual Purkinje dendrites requires high spatial resolution—<1 μm in the xy plane and <10 μm along the z-axis. To detect PN synchrony changes in intracellular calcium should be recorded at a minimum temporal resolution of 20 ms, faster than the shortest interspike intervals and comparable to physiological events such as rebound in target neurons of the cerebellar nuclei. To capture millisecond synchrony that may underlie synchronous multi-dendrite signals, two-photon microscopy can be used in line-scan mode at 2 kHz (0.5 ms resolution).

Two-photon (2P) microscopy is an indispensable tool in in vivo neurophysiology because it can be used to image cells expressing activity indicators deep in tissue with minimal photon contamination from out-of-focus cells [82, 83]. The effective axial extent of the excitation volume of a 2P microscope can be adjusted by several mechanisms, including extent of back aperture fill where slightly underfilling increases the axial point spread function (PSF) [84]. In the case of Purkinje cell dendrites, which extend along the axial plane when imaging normal to a folium, the effect of increased axial PSF can be harnessed to reduce the effects of z-axis motion artifact.

Using this principle, we imaged calcium signals triggered by activation of climbing fiber inputs to PNs expressing GCaMP6f using the Pcp2-cre x Ai148 strategy. Mice were head-fixed on top of a freely rotating cylindrical treadmill and allowed to locomote in place either spontaneously or while we delivered 40-ms air-puffs to the whiskers (Fig. 4a–c). During periods lacking whisker stimulation, we acquired high time resolution (60 Hz) movies of spontaneous calcium activity (Fig. 4d, e). In crus II (Fig. 4b), the sensory stimulation reliably triggered dendritic spiking across the PNs ensemble that was time-locked to the presentation of individual stimuli in a subpopulation of imaged PNs (Fig. 4f–h). Correlation-based clustering of PN activity during periods without sensory stimulation (Fig. 4i, j) showed neuronal clusters that aligned with clusters derived from analysis of stimulus-evoked activity (Fig. 4k). The cross-correlation coefficient reflects the degree of the CS synchrony. These data are consistent with the observation [7, 76] that intrinsic connectivity within the inferior olive gives rise to synchronized evoked firing of PN dendrites.

Calcium signals in the somata of PNs also convey information about PN activity. Unlike the dendritic calcium signal, which mostly conveys information about the complex spiking activity of PNs, somatic calcium conveys information about PN simple spikes. It has been shown that both in slice and in vivo, two-photon fluorescence signals from the somata of GCaMP-expressing PNs correlate well with simple-spike firing rate [85]. Owing to increased calcium buffering within PN somata and to the high baseline firing rate of PNs (approximately 40–50 Hz), these signals are slower and exhibit lower signal-to-noise ratios, limiting their utility in capturing precise temporal firing patterns. Nevertheless, PN somatic calcium fluorescence signals have been shown to track behaviorally relevant sensorimotor variables during rodent decision-making [26].

2.2.3 Targeting Approaches to Overcome Constraints of Wide-Field Imaging of Purkinje Cells

Miniaturized fluorescent microscopes ("miniscopes") are lightweight imaging devices that allow optical recordings from neurons in freely moving animals over the course of weeks [86]. Initial miniscope systems were optical fiber-based [87], later, fluorescence excitation and detection were combined onboard the microscope housing [88]. The availability of such miniscopes through commercial vendors and open-source miniscope projects has already led to a multitude of studies. GRIN (gradient refractive index) lenses allow endoscopic two-photon microscopy, and a snap-in fluorescence microscope (model S, Doric Lenses Inc) enables recording Purkinje dendrites through a glass window. Miniaturized head-mounted fluorescence microscopes have been used for cellular-resolution imaging of several brain regions in freely moving animals [89–91]. It has proven possible to extract multiple signals from an imaging video using analytical techniques, including a combination of principal and independent component analysis (PCA/ICA) and more recently constrained nonnegative matrix factorization (CNMF-E) with an added term to model a time-varying fluorescence background signal [92, 93]. Here we test the limits of miniscope recordings to extract distinguishable Purkinje cell signals in wide-field mode, using two-photon microscopy of the same region to provide a benchmark.

We conducted in vivo miniaturized and two-photon microscopy of the same cerebellar region in Pcp2-cre x Ai148D mice with dense GCaMP6f expression (Fig. 5a). For both imaging techniques, mice were head-restrained and allowed to freely locomote on a spherical treadmill for 1 h during imaging. We observed different distributions of signal-to-noise ratio mainly explained by photon sensitivity of the detectors: a CMOS sensor in the case of miniscope and a photomultiplier tube for two-photon microscopy. Brain motion artifacts are common in both head-fixed and unrestrained mice. Imaging data from both techniques were therefore first motion corrected using algorithmic approaches [94]. To compensate for the differences in optical sectioning of both microscopes, we aligned a time-averaged single field of view recorded with the miniaturized microscope to a two-photon microscope stack, collapsed along the axial axis spanning a depth of 150 μm (Fig. 5b) using elastix [95]. Next, regions of interest (ROI) were selected manually based upon morphology and calcium activity. The manually selected dendritic ROIs were used as seeds for constrained nonnegative matrix factorization (CNMF and CNMF-E), which models the calcium dynamics by decomposing the recorded fluorescence into background, spatial footprints, and temporal components [93, 96].

Extracted spatial footprints from two-photon data showed expected PN dendritic morphological properties [1], while spatial footprints from the miniscope extend beyond the anticipated boundaries, especially along the minor axis (Fig. 5c–e). We computed the Pearson correlation coefficient for extracted temporal components and confirmed that nearby spatial footprints had highly correlated activity. With two-photon microscopy the strongest dendrite–dendrite correlations were detected at distances of tens of micrometers. Correlations dropped gradually in the parasagittal direction while declining steeply in the mediolateral direction (Fig. 5f), consistent with previous published observations.

However, miniscope recordings showed higher footprint–footprint time cross-correlations which dropped gradually in both mediolateral and parasagittal directions (Fig. 5f). Correlation-based clustering of PN activity during periods of spontaneous behavior resulted in smoothed borders between clusters and few pairs of uncorrelated footprints (Fig. 5g). We further observed lower calcium event rates for the miniscope. Together, these data indicate that wide-field imaging using a miniscope in densely labeled Purkinje neurons results in unavoidable mixing of fluorescence signals from adjacent structures. Such recordings should be

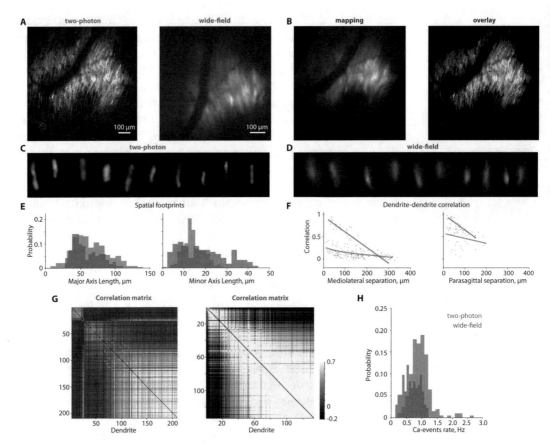

Fig. 5 Lack of single-dendrite resolution under dense expression with miniature wide-field microscopy. (a) Representative two-photon and wide-field images of Purkinje dendrites from Pcp2-Cre x Ai148-GCaMP6f mouse acquired from the same cerebellar region. The time-averaged projection across imaging frames is shown. (b) To map between two-photon and wide-field space, the wide-field image was transformed into two-photon space. (c, d) Spatial footprints of Purkinje cell dendrites imaged with two-photon and wide-field were extracted using constrained nonnegative matrix factorization CNMF and CNMF-E, respectively. (e) Distributions of major and minor axes of extracted spatial footprints. Note distribution of minor axis length is broader for wide-field imaging. (f) Cross-correlations of event times in pairs of dendrites as a function of mediolateral or parasagittal separation. (g) Heatmap of the pairwise comparisons of the correlation between event times of the dendrites. Note that the population of Purkinje dendrites in wide-field imaging show higher correlation and less clear spatial clusters. (h) Calcium event rate across two-photon (green; $n = 221$ dendrites) and wide-field (red; $n = 134$ dendrites) dendrite imaging session

analyzed with great caution [9] and should avoid the assumption that signals from individual dendrites can be resolved.

To minimize recording fluorescence from out-of-focus Purkinje dendrites, we used focal injection of low titer rAAVretro-Pcp2-Cre into Ai148D transgenic mice. Sparse labeling of PNs was achieved by reduction of the titer of the Cre-expressing viral vector (Fig. 6a, b). As a result of the sparse labeling with GCaMP6f, background modeling and temporal component extraction with CNMF-E were similar to data acquired with the two-photon

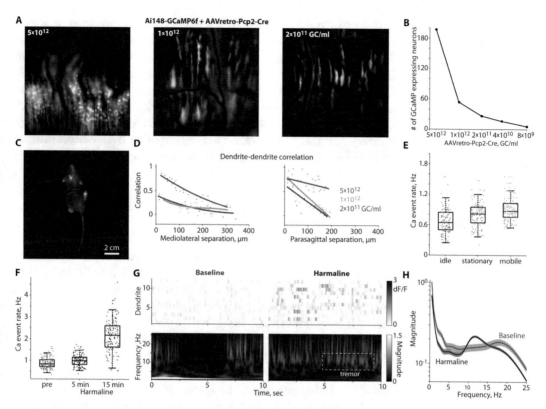

Fig. 6 Sparse expression enables extraction of Purkinje dendrite calcium activity with miniaturized wide-field fluorescence microscopy. (**a, b**) A broad range of GCaMP transduction patterns can be achieved by adjusting genetic copies of rAAVretro-Pcp2-Cre when injected into the DCN of Ai148-GCaMP6f transgenic mouse. Representative wide-field images of Purkinje dendrites acquired with miniscope and number of dendrites detected in field of view (**b**) from mice harvested 3 weeks after craniotomy and viral injection. (**c**) Bottom view of a mouse running with a miniscope mounted over lobule IV/V of cerebellum. (**d**) Cross-correlations of event times in pairs of dendrites as a function of mediolateral or parasagittal separation for different expression density. (**e**) Calcium event rate for individual dendrites plotted for periods of idle, stationary, and mobile mouse behavior. (**g**) Sample dendritic activity and sonogram view (continuous power spectrum) of mouse centroid trace at baseline and at 15 min after harmaline (10 mg/kg, i.p. administration), peak around 10–15 Hz represents the frequency of harmaline-induced tremor. (**f**) Calcium event rate prior and after harmaline administration, at 15-min mice develop pronounced tremor. (**h**) Power spectral density of centroid under baseline and harmaline (note the logarithmic scale on the *y*-axes, solid line—mean, shaded region—95% CI)

microscope. The steep drop in correlation along the mediolateral axis was similar to the higher resolution two-photon data (Fig. 6d). To further test the reliability of Purkinje dendrite imaging with a miniscope, we examined event detection and stability after systemic injection of the tremorgenic drug harmaline. Harmaline induces a severe 10–15 Hz tremor (Fig. 6h) by increasing rhythmic firing rate (Fig. 6f) and synchrony of neurons in the inferior olive [97]. The average calcium event rate for each Purkinje dendrite was computed from 5 min samples of activity taken 30 min prior to the drug injection then 5 and 15 min after harmaline injection. During the first 5 min, calcium event rate remained at the pre-injection level (0.99 ± 0.18 Hz compared to 0.84 ± 0.19 Hz, mean $\pm$ SD), while 15 min after injection the harmaline-induced tremor became prominent and the rate increased significantly to 2.19 ± 0.44 Hz (Fig. 6f). We found that we could still correct strong tremor-associated brain motion and isolate activity from dendrites with calcium event rate as high as 4.6 Hz in mice expressing GCaMP6f sparsely in Purkinje neurons. We conclude that under conditions of wide-field microscopy, single dendrites can be resolved optically if probe expression is sufficiently sparse.

3   Future Directions

Investigations of cerebellar cortical function have been ongoing for more than 50 years. While monitoring the activity of hundreds of neurons chronically in the cerebellum was a distant vision just a few decades ago, it is now routinely performed in many laboratories through optical methods. Despite major advances, calcium imaging is still limited by the biophysical properties of existing GECIs, including affinity, signal-to-noise ratio, rise and decay kinetics, and dynamic range. With the rapid improvement of genetically encoded indicators and imaging techniques, recordings at >1 kHz frame rates will greatly improve effective temporal sampling of cerebellar activity. This will help to explore how the distributed mossy fiber-parallel fiber system functionally and adaptively interacts with individual olivocorticonuclear modules. Future developments will likely improve imaging recordings in freely moving mice and help in uncovering neural activity patterns occurring during natural behaviors with fully intact vestibular input. The cerebellar cortex, in which anatomical circuits have been defined particularly well, is a perfect candidate for circuit-level functional analyses of how specific animal behaviors are generated and controlled.

Acknowledgments

We thank Andrea Giovannucci and Pengcheng Zhou for discussion and advice on image analysis and Laura Lynch and Zahra Dhaner-awala for technical assistance. This work was supported by grants to S.S.-H.W. from the National Institutes of Health (grant nos. NIH R01 NS045193, R01 MH115750, and U19 NS104648), and a fellowship to G.J.B. (grant no. F32 MH120887).

References

1. Palay SL, Chan-Palay V (1974) Cerebellar cortex: cytology and organization. Springer Science & Business Media, Berlin

2. Groenewegen HJ, Voogd J, Freedman SL (1979) The parasagittal zonation within the olivocerebellar projection. II. Climbing fiber distribution in the intermediate and hemispheric parts of cat cerebellum. J Comp Neurol 183:551–601. https://doi.org/10.1002/cne.901830307

3. Apps R, Hawkes R, Aoki S et al (2018) Cerebellar modules and their role as operational cerebellar processing units. Cerebellum 17:654–682. https://doi.org/10.1007/s12311-018-0952-3

4. Voogd J, Ruigrok TJH (2004) The organization of the corticonuclear and olivocerebellar climbing fiber projections to the rat cerebellar vermis: the congruence of projection zones and the zebrin pattern. J Neurocytol 33:5–21. https://doi.org/10.1023/B:NEUR.0000029645.72074.2b

5. Sugihara I, Shinoda Y (2007) Molecular, topographic, and functional Organization of the Cerebellar Nuclei: analysis by three-dimensional mapping of the Olivonuclear projection and aldolase C labeling. J Neurosci 27:9696–9710

6. Apps R, Hawkes R (2009) Cerebellar cortical organization: a one-map hypothesis. Nat Rev Neurosci 10:670–681. https://doi.org/10.1038/nrn2698

7. Ozden I, Sullivan MR, Lee HM, Wang SS-H (2009) Reliable coding emerges from coactivation of climbing fibers in microbands of cerebellar Purkinje neurons. J Neurosci 29:10463–10473. https://doi.org/10.1523/JNEUROSCI.0967-09.2009

8. De Gruijl JR, Hoogland TM, De Zeeuw CI (2014) Behavioral correlates of complex spike synchrony in cerebellar microzones. J Neurosci 34:8937–8947. https://doi.org/10.1523/JNEUROSCI.5064-13.2014

9. Heffley W, Song EY, Xu Z et al (2018) Coordinated cerebellar climbing fiber activity signals learned sensorimotor predictions. Nat Neurosci 21:1431–1441. https://doi.org/10.1038/s41593-018-0228-8

10. Holtmaat A, Bonhoeffer T, Chow DK et al (2009) Long-term, high-resolution imaging in the mouse neocortex through a chronic cranial window. Nat Protoc 4:1128–1144. https://doi.org/10.1038/nprot.2009.89

11. Nishiyama N, Colonna J, Shen E et al (2014) Long-term in vivo time-lapse imaging of synapse development and plasticity in the cerebellum. J Neurophysiol 111:208–216. https://doi.org/10.1152/jn.00588.2013

12. Roome CJ, Kuhn B (2014) Chronic cranial window with access port for repeated cellular manipulations, drug application, and electrophysiology. Front Cell Neurosci 8:379

13. Kuhn B, Ozden I, Lampi Y et al (2012) An amplified promoter system for targeted expression of calcium indicator proteins in the cerebellar cortex. Front Neural Circuits 6:49. https://doi.org/10.3389/fncir.2012.00049

14. Goldey GJ, Roumis DK, Glickfeld LL et al (2014) Removable cranial windows for long-term imaging in awake mice. Nat Protoc 9:2515–2538. https://doi.org/10.1038/nprot.2014.165

15. Broussard GJ, Liang R, Tian L (2014) Monitoring activity in neural circuits with genetically encoded indicators. Front Mol Neurosci 7:97. https://doi.org/10.3389/fnmol.2014.00097

16. Inoue M (2020) Genetically encoded calcium indicators to probe complex brain circuit dynamics in vivo. Neurosci Res 169:2–8

17. Badura A, Sun XR, Giovannucci A et al (2014) Fast calcium sensor proteins for monitoring neural activity. Neurophotonics 1:025008. https://doi.org/10.1117/1.NPh.1.2.025008

18. Dana H, Sun Y, Mohar B et al (2019) High-performance calcium sensors for imaging activity in neuronal populations and microcompartments. Nat Methods 16:649–657. https://doi.org/10.1038/s41592-019-0435-6

19. Luo L, Callaway EM, Svoboda K (2018) Genetic dissection of neural circuits: a decade of Progress. Neuron 98:865. https://doi.org/10.1016/j.neuron.2018.05.004

20. Huang ZJ, Josh Huang Z, Zeng H (2013) Genetic approaches to neural circuits in the mouse. Annu Rev Neurosci 36:183–215

21. Smeyne RJ, Chu T, Lewin A et al (1995) Local control of granule cell generation by cerebellar Purkinje cells. Mol Cell Neurosci 6:230–251. https://doi.org/10.1006/mcne.1995.1019

22. Barski JJ, Dethleffsen K, Meyer M (2000) Cre recombinase expression in cerebellar Purkinje cells. Genesis 28:93–98

23. Zhang X-M, Ng AH-L, Tanner JA et al (2004) Highly restricted expression of Cre recombinase in cerebellar Purkinje cells. Genesis 40:45–51. https://doi.org/10.1002/gene.20062

24. Madisen L, Garner AR, Shimaoka D et al (2015) Transgenic mice for intersectional targeting of neural sensors and effectors with high

specificity and performance. Neuron 85: 942–958. https://doi.org/10.1016/j.neuron.2015.02.022

25. Daigle TL, Madisen L, Hage TA et al (2018) A suite of transgenic driver and reporter mouse lines with enhanced brain-cell-type targeting and functionality. Cell 174:465–480.e22. https://doi.org/10.1016/j.cell.2018.06.035

26. Deverett B, Koay SA, Oostland M, Wang SS-H (2018) Cerebellar involvement in an evidence-accumulation decision-making task. Elife 7: e36781

27. Naldini L, Blömer U, Gallay P et al (1996) In vivo gene delivery and stable transduction of nondividing cells by a lentiviral vector. Science 272:263–267. https://doi.org/10.1126/science.272.5259.263

28. Tenenbaum L, Chtarto A, Lehtonen E (2004) Recombinant AAV-mediated gene delivery to the central nervous system. J Gene Med 6: S212–S222

29. Nayak S, Herzog RW (2010) Progress and prospects: immune responses to viral vectors. Gene Ther 17:295–304. https://doi.org/10.1038/gt.2009.148

30. Cetin A, Komai S, Eliava M et al (2006) Stereotaxic gene delivery in the rodent brain. Nat Protoc 1:3166–3173. https://doi.org/10.1038/nprot.2006.450

31. Burger C, Nguyen FN, Deng J, Mandel RJ (2005) Systemic mannitol-induced hyperosmolality amplifies rAAV2-mediated striatal transduction to a greater extent than local co-infusion. Mol Ther 11:327–331. https://doi.org/10.1016/j.ymthe.2004.08.031

32. Drouin LM, Agbandje-McKenna M (2013) Adeno-associated virus structural biology as a tool in vector development. Future Virol 8: 1183–1199. https://doi.org/10.2217/fvl.13.112

33. Zingg B, Chou X-L, Zhang Z-G et al (2017) AAV-mediated anterograde Transsynaptic tagging: mapping Corticocollicular input-defined neural pathways for defense behaviors. Neuron 93:33–47. https://doi.org/10.1016/j.neuron.2016.11.045

34. Tervo DGR, Hwang B-Y, Viswanathan S et al (2016) A designer AAV variant permits efficient retrograde access to projection neurons. Neuron 92:372–382. https://doi.org/10.1016/j.neuron.2016.09.021

35. Deverman BE, Pravdo PL, Simpson BP et al (2016) Cre-dependent selection yields AAV variants for widespread gene transfer to the adult brain. Nat Biotechnol 34:204–209. https://doi.org/10.1038/nbt.3440

36. Maturana CJ, Verpeut JL, Pisano TJ et al (2020) Small Alphaherpesvirus latency-associated promoters drive efficient and long-term transgene expression in the central nervous system. Mol Ther Methods Clin Dev 17: 843–857

37. Kim J-Y, Grunke SD, Levites Y et al (2014) Intracerebroventricular viral injection of the neonatal mouse brain for persistent and widespread neuronal transduction. J Vis Exp 91: 51863. https://doi.org/10.3791/51863

38. Renier N, Adams EL, Kirst C et al (2016) Mapping of brain activity by automated volume analysis of immediate early genes. Cell 165: 1789–1802. https://doi.org/10.1016/j.cell.2016.05.007

39. Najafi F, Giovannucci A, Wang SS-H, Medina JF (2014) Sensory-driven enhancement of calcium signals in individual Purkinje cell dendrites of awake mice. Cell Rep 6:792–798. https://doi.org/10.1016/j.celrep.2014.02.001

40. Tsubota T, Ohashi Y, Tamura K et al (2011) Optogenetic manipulation of cerebellar Purkinje cell activity in vivo. PLoS One 6: e22400. https://doi.org/10.1371/journal.pone.0022400

41. El-Shamayleh Y, Kojima Y, Soetedjo R, Horwitz GD (2017) Selective Optogenetic control of Purkinje cells in monkey cerebellum. Neuron 95:51–62.e4. https://doi.org/10.1016/j.neuron.2017.06.002

42. Oberdick J, Smeyne RJ, Mann JR et al (1990) A promoter that drives transgene expression in cerebellar Purkinje and retinal bipolar neurons. Science 248:223–226. https://doi.org/10.1126/science.2109351

43. Nitta K, Matsuzaki Y, Konno A, Hirai H (2017) Minimal Purkinje cell-specific PCP2/L7 promoter virally available for rodents and non-human primates. Mol Ther Methods Clin Dev 6:159–170. https://doi.org/10.1016/j.omtm.2017.07.006

44. Tian L, Hires SA, Mao T et al (2009) Imaging neural activity in worms, flies and mice with improved GCaMP calcium indicators. Nat Methods 6:875–881. https://doi.org/10.1038/nmeth.1398

45. Fünfschilling U, Reichardt LF (2002) Cre-mediated recombination in rhombic lip derivatives. Genesis 33:160–169. https://doi.org/10.1002/gene.10104

46. Chow LML, Tian Y, Weber T et al (2006) Inducible Cre recombinase activity in mouse cerebellar granule cell precursors and inner ear hair cells. Dev Dyn 235:2991–2998

47. Wagner MJ, Kim TH, Savall J et al (2017) Cerebellar granule cells encode the expectation of reward. Nature 544:96–100. https://doi. org/10.1038/nature21726

48. Markwalter KH, Yang Y, Holy TE, Bonni A (2019) Sensorimotor coding of Vermal granule neurons in the developing mammalian cerebellum. J Neurosci 39:6626–6643. https://doi. org/10.1523/JNEUROSCI.0086-19.2019

49. Giovannucci A, Badura A, Deverett B et al (2017) Cerebellar granule cells acquire a widespread predictive feedback signal during motor learning. Nat Neurosci 20:727–734. https:// doi.org/10.1038/nn.4531

50. Yamada T, Yang Y, Valnegri P et al (2019) Sensory experience remodels genome architecture in neural circuit to drive motor learning. Nature 569:708–713. https://doi.org/10. 1038/s41586-019-1190-7

51. Ozden I, Dombeck DA, Hoogland TM et al (2012) Widespread state-dependent shifts in cerebellar activity in locomoting mice. PLoS One 7:e42650. https://doi.org/10.1371/ journal.pone.0042650

52. Wagner MJ, Kim TH, Kadmon J et al (2019) Shared cortex-cerebellum dynamics in the execution and learning of a motor task. Cell 177: 669–682.e24. https://doi.org/10.1016/j. cell.2019.02.019

53. Jörntell H, Bengtsson F, Schonewille M, De Zeeuw CI (2010) Cerebellar molecular layer interneurons—computational properties and roles in learning. Trends Neurosci 33:524–532

54. Paul A, Cai Y, Atwal GS, Huang ZJ (2012) Developmental coordination of gene expression between synaptic partners during GABAergic circuit assembly in cerebellar cortex. Front Neural Circuits 6:37. https://doi. org/10.3389/fncir.2012.00037

55. Amat SB, Rowan MJM, Gaffield MA et al (2017) Using c-kit to genetically target cerebellar molecular layer interneurons in adult mice. PLoS One 12:e0179347. https://doi. org/10.1371/journal.pone.0179347

56. Gaffield MA, Christie JM (2017) Movement rate is encoded and influenced by widespread, coherent activity of cerebellar molecular layer interneurons. J Neurosci 37:4751–4765. https://doi.org/10.1523/JNEUROSCI. 0534-17.2017

57. Hippenmeyer S, Vrieseling E, Sigrist M et al (2005) A developmental switch in the response of DRG neurons to ETS transcription factor signaling. PLoS Biol 3:e159. https://doi.org/ 10.1371/journal.pbio.0030159

58. Astorga G, Bao J, Marty A et al (2015) An excitatory GABA loop operating in vivo. Front Cell Neurosci 9:275. https://doi.org/ 10.3389/fncel.2015.00275

59. Astorga G, Li D, Therreau L et al (2017) Concerted interneuron activity in the cerebellar molecular layer during rhythmic Oromotor behaviors. J Neurosci 37:11455–11468. https://doi.org/10.1523/JNEUROSCI. 1091-17.2017

60. Gaffield MA, Rowan MJM, Amat SB et al (2018) Inhibition gates supralinear Ca^{2+} signaling in Purkinje cell dendrites during practiced movements. Elife 7:e36246. https://doi.org/10.7554/eLife.36246

61. Ma M, Futia GL, Ozbay BN et al (2020) Molecular layer interneurons in the cerebellum encode for valence in associative learning. Nat Commun 11(1):4217

62. Rowan MJM, Bonnan A, Zhang K et al (2018) Graded control of climbing-fiber-mediated plasticity and learning by inhibition in the cerebellum. Neuron 99:999–1015.e6. https:// doi.org/10.1016/j.neuron.2018.07.024

63. Sarkisov DV, Wang SS-H (2008) Order-dependent coincidence detection in cerebellar Purkinje neurons at the inositol trisphosphate receptor. J Neurosci 28:133–142. https://doi. org/10.1523/JNEUROSCI.1729-07.2008

64. Wang SS-H, Denk W, Häusser M (2000) Coincidence detection in single dendritic spines mediated by calcium release. Nat Neurosci 3: 1266–1273. https://doi.org/10.1038/81792

65. Llano I, González J, Caputo C et al (2000) Presynaptic calcium stores underlie large-amplitude miniature IPSCs and spontaneous calcium transients. Nat Neurosci 3: 1256–1265. https://doi.org/10.1038/81781

66. Llinás R, Sugimori M (1980) Electrophysiological properties of in vitro Purkinje cell somata in mammalian cerebellar slices. J Physiol 305:171–195. https://doi.org/10. 1113/jphysiol.1980.sp013357

67. Tank DW, Sugimori M, Connor JA, Llinás RR (1988) Spatially resolved calcium dynamics of mammalian Purkinje cells in cerebellar slice. Science 242:773–777. https://doi.org/10. 1126/science.2847315

68. Konnerth A, Dreessen J, Augustine GJ (1992) Brief dendritic calcium signals initiate long-lasting synaptic depression in cerebellar Purkinje cells. Proc Natl Acad Sci U S A 89: 7051–7055. https://doi.org/10.1073/pnas. 89.15.7051

69. Lev-Ram V, Miyakawa H, Lasser-Ross N, Ross WN (1992) Calcium transients in cerebellar Purkinje neurons evoked by intracellular stimulation. J Neurophysiol 68:1167–1177. https://doi.org/10.1152/jn.1992.68.4.1167

70. Miyakawa H, Lev-Ram V, Lasser-Ross N, Ross WN (1992) Calcium transients evoked by climbing fiber and parallel fiber synaptic inputs in Guinea pig cerebellar Purkinje neurons. J Neurophysiol 68:1178–1189. https://doi.org/10.1152/jn.1992.68.4.1178

71. Eilers J, Augustine GJ, Konnerth A (1995) Subthreshold synaptic Ca2+ signalling in fine dendrites and spines of cerebellar Purkinje neurons. Nature 373:155–158. https://doi.org/10.1038/373155a0

72. Kitamura K, Häusser M (2011) Dendritic calcium signaling triggered by spontaneous and sensory-evoked climbing fiber input to cerebellar Purkinje cells in vivo. J Neurosci 31:10847–10858. https://doi.org/10.1523/JNEUROSCI.2525-10.2011

73. Roome CJ, Kuhn B (2018) Simultaneous dendritic voltage and calcium imaging and somatic recording from Purkinje neurons in awake mice. Nat Commun 9:3388. https://doi.org/10.1038/s41467-018-05900-3

74. Maeda H, Ellis-Davies GC, Ito K et al (1999) Supralinear Ca^{2+} signaling by cooperative and mobile Ca^{2+} buffering in Purkinje neurons. Neuron 24:989–1002. https://doi.org/10.1016/s0896-6273(00)81045-4

75. Llinas R, Baker R, Sotelo C (1974) Electrotonic coupling between neurons in cat inferior olive. J Neurophysiol 37:560–571. https://doi.org/10.1152/jn.1974.37.3.560

76. Schultz SR, Kitamura K, Post-Uiterweer A et al (2009) Spatial pattern coding of sensory information by climbing fiber-evoked calcium signals in networks of neighboring cerebellar Purkinje cells. J Neurosci 29:8005–8015. https://doi.org/10.1523/JNEUROSCI.4919-08.2009

77. Ten Brinke MM, Boele HJ, De Zeeuw CI (2019) Conditioned climbing fiber responses in cerebellar cortex and nuclei. Neurosci Lett 688:26–36. https://doi.org/10.1016/j.neulet.2018.04.035

78. Titley HK, Kislin M, Simmons DH et al (2019) Complex spike clusters and false-positive rejection in a cerebellar supervised learning rule. J Physiol 597:4387–4406. https://doi.org/10.1113/JP278502

79. Kostadinov D, Beau M, Blanco-Pozo M, Häusser M (2020) Publisher correction: predictive and reactive reward signals conveyed by climbing fiber inputs to cerebellar Purkinje cells. Nat Neurosci 23:468. https://doi.org/10.1038/s41593-020-0594-x

80. Heffley W, Hull C (2019) Classical conditioning drives learned reward prediction signals in climbing fibers across the lateral cerebellum. eLife 8:e46764

81. Tsutsumi S, Hidaka N, Isomura Y et al (2019) Modular organization of cerebellar climbing fiber inputs during goal-directed behavior. eLife 8:e47021

82. Denk W, Strickler JH, Webb WW (1990) Two-photon laser scanning fluorescence microscopy. Science 248:73–76. https://doi.org/10.1126/science.2321027

83. Stosiek C, Garaschuk O, Holthoff K, Konnerth A (2003) In vivo two-photon calcium imaging of neuronal networks. Proc Natl Acad Sci 100:7319–7324

84. Zipfel WR, Williams RM, Webb WW (2003) Nonlinear magic: multiphoton microscopy in the biosciences. Nat Biotechnol 21:1369–1377. https://doi.org/10.1038/nbt899

85. Ramirez JE, Stell BM (2016) Calcium imaging reveals coordinated simple spike pauses in populations of cerebellar Purkinje cells. Cell Rep 17(12):3125–3132. https://doi.org/10.1016/j.celrep.2016.11.075

86. Malvaut S, Constantinescu V-S, Dehez H et al (2020) Deciphering brain function by miniaturized fluorescence microscopy in freely behaving animals. Front Neurosci 14:819. https://doi.org/10.3389/fnins.2020.00819

87. Helmchen F, Fee MS, Tank DW, Denk W (2001) A miniature head-mounted two-photon microscope. High-resolution brain imaging in freely moving animals. Neuron 31:903–912. https://doi.org/10.1016/s0896-6273(01)00421-4

88. Ghosh KK, Burns LD, Cocker ED et al (2011) Miniaturized integration of a fluorescence microscope. Nat Methods 8:871–878. https://doi.org/10.1038/nmeth.1694

89. Shemesh OA, Linghu C, Piatkevich KD et al (2020) Precision calcium imaging of dense neural populations via a cell-body-targeted calcium indicator. Neuron 107:470–486.e11. https://doi.org/10.1016/j.neuron.2020.05.029

90. Shuman T, Aharoni D, Cai DJ et al (2020) Breakdown of spatial coding and interneuron synchronization in epileptic mice. Nat Neurosci 23:229–238. https://doi.org/10.1038/s41593-019-0559-0

91. Siciliano CA, Noamany H, Chang C-J et al (2019) A cortical-brainstem circuit predicts and governs compulsive alcohol drinking. Science 366:1008–1012. https://doi.org/10.1126/science.aay1186

92. Mukamel EA, Nimmerjahn A, Schnitzer MJ (2009) Automated analysis of cellular signals from large-scale calcium imaging data. Neuron 63:747–760. https://doi.org/10.1016/j.neuron.2009.08.009

93. Zhou P, Resendez SL, Rodriguez-Romaguera J et al (2018) Efficient and accurate extraction of in vivo calcium signals from microendoscopic video data. elife 7:e28728. https://doi.org/10.7554/eLife.28728

94. Pnevmatikakis EA, Giovannucci A (2017) NoRMCorre: an online algorithm for piecewise rigid motion correction of calcium imaging data. J Neurosci Methods 291:83–94. https://doi.org/10.1016/j.jneumeth.2017.07.031

95. Klein S, Staring M, Murphy K et al (2010) Elastix: a toolbox for intensity-based medical image registration. IEEE Trans Med Imaging 29:196–205. https://doi.org/10.1109/TMI.2009.2035616

96. Giovannucci A, Friedrich J, Gunn P et al (2019) CaImAn an open source tool for scalable calcium imaging data analysis. Elife 8:e38173

97. de Montigny C, Lamarre Y (1973) Rhythmic activity induced by harmaline in the olivo-cerebello-bulbar system of the cat. Brain Res 53:81–95. https://doi.org/10.1016/0006-8993(73)90768-3

Chapter 14

Measuring Cerebellar Processing and Sensorimotor Functions in Non-Human Primates

Nico A. Flierman, Eric Avila, Chris I. De Zeeuw, and Aleksandra Badura

Abstract

The cerebellum is well known for integrating and processing sensory inputs to guide motor functions. Lesions to this brain region result in loss of movement precision, problems with balance, and difficulty in acquiring new motor skills. Moreover, in recent years it has been shown that the cerebellum is also involved in many cognitive functions. In order to study the cerebellar neural mechanisms that underlie these processes we need to measure the cerebellar activity accompanying these complex behaviors. We can achieve this by performing electrophysiological recordings from the neural tissue to obtain signals from cerebellar neurons in awake, behaving animals. Even though many experimental questions can be addressed in rodents, some can only be answered by using non-human primates (NHPs). Particularly, the low acuity of the visual system and less developed neocortex make rodents undesirable models when higher order sensorimotor functions are being probed. For analogous reasons NHPs are indispensable when testing cortical prosthesis. In cerebellar research, NHPs have most commonly been used to study the role of Purkinje cell processing in integrating different forms of sensory information to acquire, plan, and coordinate eye and limb movements. Our goal in this chapter is to provide the reader with guidelines on how to perform measurements of cerebellar function in NHPs, specifically using single-unit extracellular recordings. We highlight the advantages and limitations of this approach focusing on the surgical and technical aspects of these experiments, and we describe standard and novel approaches to quantify the behavior of NHPs during experimental manipulations.

Key words Non-human primates, Cerebellum, Customized implants, Single-unit recordings, Electrophysiology, Extracellular recordings

1 Introduction

Neuroscience research employs a variety of species to study brain mechanisms underlying behaviors. Over the last few decades, the field has made incredible progress in studying even seemingly complex functions in worms [1] and flies [2]. When it comes to the mammalian brain, the most commonly used species are mice and rats. In fact, in 2015 over 30% of all neuroscience articles used rodents as the model organism [3]. While rodents in general, and mice in particular, offer many advantages, such as the wide

Roy V. Sillitoe (ed.), *Measuring Cerebellar Function*, Neuromethods, vol. 177, https://doi.org/10.1007/978-1-0716-2026-7_14,
© Springer Science+Business Media, LLC, part of Springer Nature 2022

availability of genetic tools and high-resolution reference atlases [4], many scientists argue that diversifying animal models is highly advantageous [3, 5, 6]. Non-human primates (NHPs, predominantly rhesus macaques, i.e., *macaca mulatta*) represent the model organisms closest to humans and are therefore indispensable for many lines of fundamental and applied research [7–9]. NHP research is of particular importance for studying the structure and function of the neocortex as well as the way its activity drives eye movements and dexterity [10]. A relatively prominent lateralization and interhemispheric connectivity are also distinctive features of the primate brain [11, 12], although some evidence of lateralization has recently been shown in mice as well [13, 14]. The opinion piece by Roelfsema and Treue compiles an extensive list of research topics where input from NHP studies is required to further our knowledge on brain functions in health and disease states [8].

In cerebellar research, NHPs have long been used to study oculomotor adaptation and visual functions [15–22]. Notably, the eyes of some rodents, e.g., gerbils, possess features that resemble the primate macula [23], and research in mice can answer many fundamental questions about vision [24]. However, NHPs are uniquely fitted for this line of research because, unlike rodents, they have a fovea and use saccadic eye movements to track objects in their field of view, making them a perfect model organism for the human visual system [25, 26]. Further, studies on the development of the upper limb neuroprosthetics are limited in rodents and have been almost exclusively conducted in NHPs [27, 28]. In fact, many fundamental cerebellar mechanisms that govern grasping and pointing motions have been uncovered, thanks to NHP research [29–31]. Sensorimotor integration of vestibular inputs is another area where studies done in NHPs aid our understanding of how the cerebellum controls balance, coordination, and self-motion [32–34].

In the last 10 years the insights gained from NHP research showed that the cerebellum is crucial for many functions that are highly relevant for understanding human behavior. For example, it is involved in integrating sensory information to guide motor functions in simple and complex tasks, learning new motor sequences [35–37], and processing of reward signals during ongoing acquisition of motor tasks [38]. Thus, while research on complex behaviors and cognitive processes can, to a certain degree, be also conducted in rodents [39, 40], the superb level of NHPs' trainability and the anatomical similarities within the primate lineage [41] offer indispensable advantages so as to further our understanding of the human brain.

The above mentioned studies are not exhaustive and are merely representative examples of how cerebellar research benefits from NHP studies. This is in part due to the anatomical similarities between the human and NHP cerebellum. Even though the NHP cerebellum has only 33% of the surface area of the neocortex, as compared to 78% in humans [42], it is still most similar to the human one with the same neuronal cell types and conserved micro-circuitry [43, 44]. Particularly, the expansion of the dentate nucleus and the surface of the hemispheres, which expanded in parallel with the neocortex during evolution, are shared among primates [45, 46]. The expansion of the cortices in primates results in extensive cerebello-cerebral connections, which have been studied using long-distance viral tracing predominantly in NHPs [47]. These studies indicated that the cerebellum is indeed not only a central hub in the motor-control system but is also a part of the non-motor circuitry [48].

Evident from the examples listed above, the ability to obtain neuronal signals from the NHP cerebellum and directly link it to the observed behaviors provides major insights to our understanding of higher cerebellar functions. However, neuronal recordings in NHPs in general, and from the cerebellum in particular, present many technical challenges. First, the surface of the cerebellum in NHPs is not readily accessible, which hampers the use of multielec-trode arrays, commonly used for the neocortical recordings in NHPs [49, 50] or glass electrodes, used in rodents [51]. In fact, the majority of cerebellar recordings from macaque monkeys have been obtained using tungsten electrodes, which, on the one hand, offer excellent signal isolation, but, on the other hand, can only record a single neuron at the time. Notably, several groups are actively working on incorporating novel multichannel recording methods into cerebellar research in NHPs, with the first successful experiments performed in marmosets [52]. Second, due to the location of the NHP cerebellum, electrodes have to travel through the neocortex to reach the cerebellar cortex or the cerebellar nuclei. Most commonly, electrodes are inserted in guide tubes passing through the primary visual cortex. This is an important point of consideration, since repeated measurements can damage this structure. This is particularly important for visual tasks, and therefore careful calibration of visual field acuity is strongly advised prior to performing any experiments. Third, because the surface of the cerebellar cortex is highly folded and visual inspection of the insertion point on the electrode is not possible, it is possible to miscalculate the final position of the electrode. This can be prevented by obtaining an MRI image confirmation of the electrode placement.

In this chapter we will describe step-by-step procedures that enable successful cerebellar recordings in NHPs performing complex tasks. We will highlight techniques that allow for customization of the materials to an individual NHP, which in turn increases the comfort of the animal as well as the durability of the implants. We will also outline training methods and tools that allow quantification of behavioral output, with specific focus on the eye movements. Finally, we will illustrate how to perform electrophysiological recordings from the cerebellum and how its activity can be modulated pharmacologically.

2   Materials: Implants Designs and Manufacturing Process

The importance of NHPs well-being used in research extends to all devices and implants used on them. Therefore, novel technological advancements such as CNC machines, 3D printers, biocompatible materials, etc. are highly recommended to use in favor of NHP health to increase the implants' longevity and provide better data acquisition. Creating personalized, skull-formed, cranial implants for each animal has become the state-of-the-art standard and a critical and necessary step in the experimental planning process. Commercially available products do not account for individual differences between NHPs, which in turn may have a negative effect on the implant stability, bone health, and precision of the neural target location. Customized implants based on detailed anatomical scans reduce the surgery duration, lower the chance of postoperative complications (e.g., granulation and bacterial growth), and improve the rate of osseointegration. Implementing such customized solutions is in line with the recommendation of the European Union Scientific Committee, which recommends adopting "the highest standards of NHP housing and husbandry and (...) scientific procedures" [53].

2.1   CT/MRI Image Processing

In order to design customizable implants for chronic cerebellar recordings, 3D models of the skull and the cerebellum need to be generated. Obtaining proper imaging of a skull and brain areas to record from is critical to generate a suitable model, which in turn facilitates implant design, surgical planning, and neurophysiological recordings. Computerized tomography (CT) is fast, widely available, and particularly well suited for obtaining high-resolution images of bone. It has exquisite contrast resolution and the ability to scan with metal objects. Depending on the scanner, ideal minimum spatial resolution should be of at least 0.5 mm. The imaging is done under anesthesia (*see* Subheading 4.1.1). After obtaining the CT images, freely available software can be used to segment (i.e., isolate the structures of interest) and to render a 3D image of a skull. Note that the process of segmentation and reconstruction of the surface of a skull is the same for both image sources (CT and MRI); however, CT provides superior results for bone segmentation.

Magnetic resonance imaging (MRI) is an imaging technique that uses a magnetic field and radio waves to generate anatomical images. MRI is especially useful for soft tissues; hence, it is recommended to obtain head images using both technologies. If both are

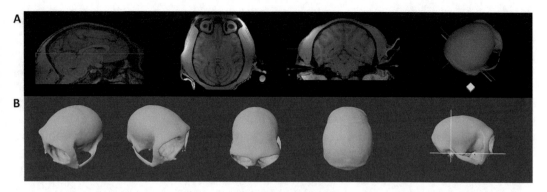

Fig. 1 Segmentation and 3D reconstruction. (**a**) Segmentation of skull (magenta) and brain (red) from an MRI of one monkey using ITK-Snap. (**b**) Digital skull reconstruction by ITK-Snap (exported as .stl) as observed in any 3D computer graphics software. Yellow lines in the rightmost image show the stereotaxic planes

available, this step provides an extraordinary resource for three main reasons:

1. The CT provides a high-resolution image of the skull to design a customized form-fitted implant;

2. The MRI provides detailed soft tissue images to localize and map the areas to record for the subsequent electrode implantation;

3. Obtaining anatomical high-quality images from both of these sources allows co-registration using landmarks. Our recommendation is using an MRI-compatible stereotaxic device. In case it is not available, the second-best option is to use MRI-compatible ear bars and additional MRI-compatible markers [54, 55] with visible interaural markers that will provide helpful alignment markers (Fig. 1a, right column). If none of these is available, one of the freely available software provides the best user-friendly approach to re-align the MRI and CT images.

Here we provide information to process the CT and MRI from two open-source software systems developed to segment structures in 3D from medical images. They both offer similar features and results: ITK-Snap (http://www.itksnap.org/ [56]) and 3D slicer (https://www.slicer.org/ [57]). Both offer comprehensive tutorials online so we will only cover the main steps.

2.1.1 Import and Re-alignment

3D Slicer: "Load Data" or "Add Data" tool to import image series, and adjust contrast and brightness. Save data as a medical record bundle (.mrb).

ITK-Snap: "Open main image" to import image series, and adjust contrast and brightness. Save the main image as a .nifti file (Neuroimaging Informatics Technology Initiative).

If no MRI-compatible stereotaxic frame was used and the image is rotated it needs to be re-aligned before proceeding to the next steps. This will allow alignment of the MRI images to a stereotaxic coordinate system. Having the MRI stereotaxically aligned will help to use an NHP reference brain atlas during surgery and electrophysiological recordings. This is particularly important for electrode position planning.

These are the steps of the alignment using the 3D Slicer interactive tool:

1. Orientation labels: Make sure that the labels for orientation are correct (left-right, anterior-posterior). Go to the "Transforms" module and create a "New Linear Transform" and use the "Rotation" menu to rotate the MRI to its proper orientation. Start by doing it ±90°. Once completed, go to the bottom section "Apply transform," select the MRI you are working on, and click on the right arrow, so the name of the MRI moves to the right. Check whether the labels are correct in the 3D view of the MRI. Lock the changes by clicking the "Harden transform" in the "Apply transform" menu section.

2. Fiducial markers: Fiducial markers can be created in the "Markups" module to be able to align the MRI in a stereotaxic plane. Create a new Markups fiducial list and position 4 fiducials, 2 in the ear canals in each side (external *meatus acusticus*) where the bony part begins, and 2 in the lower ridge of the orbital bone. Lock the fiducials. If a frameless stereotaxic device was used, the fiducial markers can be placed on the tip of the ear bars. Make sure that the 2D projection is toggled on (Advanced menu of the "Markups" module). This will allow visibility for all fiducial markers in all views.

3. MRI rotation: This step allows to rotate and align the ear and eye fiducials in the same plane. This is achieved in the "Transforms" module. Under "Transformable," select the MRI and the recently created fiducials and click on the right green arrow. Then, use the "Rotation" sliders to align the fiducial markers to the same plane. You can use the crosshairs as a visual help tool. Once completed, "Harden" the transform.

4. Reset to stereotaxic origin: This function sets all the MRI slices to stereotactic coordinates. Create a new fiducial marker at the stereotactic origin (midline at the intersection of the ear plane and the eye plane). Place the fiducial marker at the center of that intersection. The location of this fiducial marker (R, A, and S) will be used to translate the MRI to the stereotaxic coordinates. Open the "Transforms" module, create a new transform, move from "Transformable" all previous transformations, and input in the "Translation" section, with the opposite sign (!), the stereotaxic origin fiducial marker location (i.e., if positive,

input negative in R in LR, A in PA, and S in IS), and harden transform. Note, if done correctly, when checking the fiducial marker location again, this should now be at the origin (0, 0, 0).

2.1.2 Segmentation

Segmentation (i.e., isolating the regions of interest) is achieved by labeling the skull and the target areas for recordings (Fig. 1).

ITK-Snap: After adjusting the contrast and brightness, this process can be completed by first using automatic segmentation and then adjusting with manual segmentation. Adjusting contrast and brightness will significantly help the algorithm to consider the intensities of the image. Progress and current state of the segmentation can be observed in the lower left window by clicking the "update" button. After the automatic segmentation is done, manually scroll through the cuts to assess the result. If there are any gaps, fill in manually. For CT this is rarely the case, but more commonly found for MRI. Depending on the brain area, the target areas for electrophysiology can be automatically segmented. However, the cerebellum often needs to be manually segmented.

3D Slicer: For CT, select the CT and go to the "Volumes" module. Select the CT-brain view under display. Then, go to the "Segment Editor" module and click on "+ add" to add a segmentation. Under the "Effects" menu, select "Threshold" and adjust the range slider until the colored part only includes the skull. Click on "Apply." Due to the low contrast between tissues in the MRI, segmenting the skull will require automatic and manual segmentation. Select the MRI and go to the "Segment Editor" module and click on "+ add" to add a segmentation. Under the "Effects" menu, select "Threshold" and adjust the range slider until the colored part only includes the outside of the skull, subcutaneous tissue and skin. Click on "Apply." Using the sagittal or coronal cut, select "Islands" under the "Effects" menu, select "Keep selected island" and click anywhere on the colored segmentation in the MRI. Now, click on "Margin" in the "Effects" menu, and then "Apply." Now, create a second segmentation ("+Add"), select the "Scissors" tool, and cut in the middle of the previous segmentation to leave the top of the skull free. This does not need to be exact, and do it all around the skull. Once this is completed, the selected region will be highlighted in a separate color. Select again "Margin," and select the option "Shrink," then click "Apply." Click on "Islands" to remove the outer segmentation and leave the inner selection. Click once more on "Margin," and this time select the option "Grow," and then "Apply." The outside surface of the second segmentation will conform to the surface of the skull. The inner segmented area can now be erased. Another way to segment out the skull is to use the "Threshold" effect. This will isolate most of the skull and the remaining segments need to be adjusted manually.

2.1.3 Export of the 3D
structures (File format: .stl
or .obj)

Once the previous steps are completed, the skull and brain structures can be exported to multiple file formats. The most commonly used are: .stl and .obj.

ITK-Snap: Menu Segmentation>Export as Surface Mesh enables exporting all labels created as separate files. Once you hit "Next," the file format and the directory can be selected.

3D Slicer: Go to the "Segmentations" module. Under the "Export/Import" models and labelmaps" select "Export," "Models" and "Export to new model hierarchy."

Go to Menu File>Save will open a window with every element created, the file format and the directory to be saved.

To ease comparison and take advantage of the two imaging methods, the segmented and aligned CT can be co-registered with the MRI to identify the brain regions of interest [58]. To achieve this, both datasets can be loaded into 3D Slicer and, using the same landmark registration procedure described above, stereotaxically aligned to the same landmarks. The accuracy of the co-registration will directly depend on the user's ability to place the markers in the corresponding landmarks. The alignment can be achieved using bony structures (stereotaxic planes) and brain structures as a reference frame [59]. All horizontal slices can be aligned to the horizontal plane passing through the interaural line and the infraorbital ridge (stereotaxic plane); all coronal slices can be aligned to the vertical plane passing through the interaural line (ear-bar zero, stereotaxic plane). Brain structures commonly used to align the same planes are the anterior and posterior commissure, as well as the genu and splenium of the corpus callosum [60, 61]. These landmarks can be used to index any position in a standard macaque brain atlas [62–64]. After successful registration the location of the recording chamber can be determined on the basis of the MRI images and macaque atlases.

2.2 Adapting Implant Design to CT/MRI Images

There are numerous computer-aided design (CAD) software solutions to create and/or modify designs. They can be used to create headposts and chambers for the electrophysiological recordings of the cerebellum in NHPs. For this purpose, the surface mesh or object containing the NHP head images is used to shape the base of the implant to the skull. We will briefly discuss how to shape the bottom of any design to the shape of the skull in SolidWorks (Dassault Systemes) and Inventor (Autodesk). Both software packages have the same approach and tools.

After designing the headpost or recording chamber (discussed below) and importing the 3D image of the reconstructed skull, described in Subheading 2.1, the base of the design (body and legs) is shaped to perfectly fit the skull. This is achieved by moving the implant to the intended position on the skull and intersecting one another. For cerebellar recordings, the most common position of the headposts is fronto-parietal. However, given the advancement of technology and possibilities to record simultaneously from multiple brain regions, the headpost position and leg shape can vary accordingly. The chamber is generally placed over the occipital bone and the exact position of the chamber is based on the intended recording location(s). The location is planned on a monkey-by-monkey basis and numbers will change based on intended recording location and chamber size. The bottom base of the design needs to be thick enough as the contour will fit directly on the shape of the skull.

SolidWorks: Select the implant, go to Insert>Features>Combine, and under "Operation" select "Subtract," under "Main Body" select your implant and under "Bodies to Subtract" select the skull. The main image will highlight the area that is going to be subtracted, so this is a good time to adjust the area of the implant that will be deleted. Click "Ok" (the green check mark on the left menu) and the workspace will show the subtracted body. The subtracted body can be added at any time again. If you do not want for it to be deleted, you can use the tool "Indent" instead. To properly move the two bodies, go to Insert>Features>Move/Copy, select the body to move, and then click on "Translate/Rotate" at the bottom of the left menu. You can flip the alignment and manually rotate them and click on "Add." Select the two body faces that are intersecting and need to be added, and subsequently click on "Ok." Now go to the menu Insert>Features>Indent. In the same way as before, select your Target Body (implant), the Tool Body region (skull) and select the option "Cut" underneath this last menu.

Inventor: The approach and tools are the same. Once the implant and skull are in the workspace, click on "Combine" indicated on the left side menu. Select the implant as the "Base" and the skull as the "Tool Body"; click on "Ok." This is a good moment to save the implant again.

2.3 Headpost

Commercial headposts have been used for a long time in primate neurobiology. However, the problem with these implants is that the "feet" of the headpost that adhere to the skull have to be bent to fit the skull during the surgery. This is a delicate and time-consuming process and even when performed by an expert can still result in implants having imperfect contact with the skull. Moreover, even if good alignment is achieved the base of the implant remains flat. The gaps between the skull and the implant make osseointegration harder and thus the strength of the implant worsens over time and could result in an infection (*see* **Note 1**).

Custom implants: Standard headposts generally have 4 main feet, which are placed in a way so that they can support forces from all angles. Customized headposts generally also have 3–4 feet, which are designed in a way to efficiently redistribute the force that the headpost is exerting onto the skull from all directions

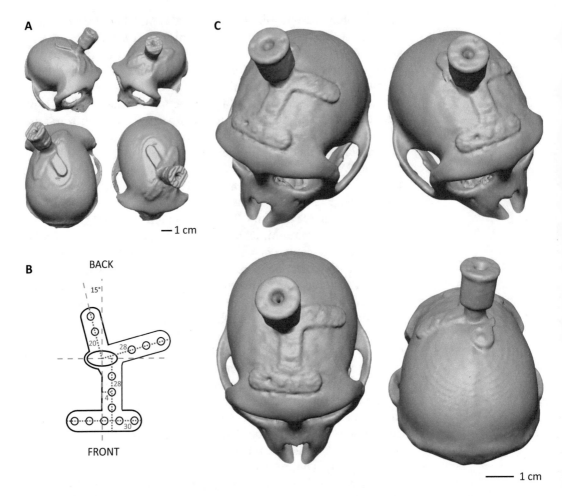

Fig. 2 Headpost design. (**a**) A CT scan of an implanted standard headpost with 4 feet. (**b**) Design of a customized headpost that accommodates recording chambers or multielectrode array implants over the cerebellum and frontal eye field. (**c**) A CT scan of the implanted headpost shown in (**b**)

(Fig. 2). Ideally, space and skull-shape permitting, each foot should have at least 2–3 screws; too few screws can cause major problems when one of them comes out. The general design of a headpost is the same across all animals. Particularly, the top part of the post that is fitted in the head fixation frame forms the standard part to which the recording equipment can be attached. The customized part is the exact shape of the headpost feet and their bottom side, which adhere to the skull. The goal is to make it fit perfectly to the NHP skull anatomy (*see* Subheading 2.2 for details on how this is achieved).

Advances in 3D printing technology allow fast and affordable creation of individualized implants from a large range of materials. Most commonly the design is printed in grade 5 titanium, with a resolution of 50 μm. When designing implants that are MRI compatible, polyetheretherketone (PEEK) is currently the material of choice, which can be attached to the skull with ceramic screws (for details on design implants in PEEK, *see* [65, 66]). Custom headposts are designed based on the individual NHP's skull shape, reconstructed using CT and MRI scans (*see* Subheading 2.1). For headposts CT images are usually sufficient.

Considerations for placement: When planning the location of the customized headposts the front legs of the headposts should be positioned close to but not less than ~2 mm from the anterior-most part of the supraorbital ridge (also known as the eyebrow bone) [67].

2.4 Recording chamber

Recording chambers for NHPs can also be purchased commercially. However, commercial chambers do not connect perfectly to the contour of the skull. This results in manual adjustment of legs and filling of gaps with dental acrylic (Jet Acrylic, Lang, Illinois, USA) during surgery. These fixes are susceptible to bacterial growth and/or could break over time [68]. Moreover, heat generation during curing of dental cement could damage osseointegration of titanium implants [69]. Thus, a chamber that connects closely to the skull is preferred. Similar to the headposts, the chambers are most commonly made from grade 5 titanium (unless MRI-compatible materials are required). The walls of the chamber are usually 1 mm thick. For the outside of the chamber the 3D printer resolution of 50 μm is smooth enough; however, this inside is electropolished to a roughness of Ra < 0.05 μm to aid cleaning and reduce the risk of infection.

Custom chambers are designed based on individual NHP's skull shape reconstructed using CT and MRI scans (*see* Subheading 2.1). While the CT is superior in determining the exact shape of the skull, the MRI is used to plan the location of the implants relative to the brain areas of interest. Therefore, it is advisable to use both techniques together for the chamber design. The size and shape of the chamber should allow the experimenters to reach the brain areas of interest and fit the contour of the skull. Small round recording chambers can be attached by 6 feet that are 1 mm thick and have one screw each. Such chambers are sufficient to record from one side of the cerebellum when implanted over one of the occipital bones. Yet, larger chambers that go past the midline are preferable for cerebellar recordings as they enable electrode insertion in the entire cerebellum (Fig. 3). They are secured with 1 mm thick feet with 2 or 3 screws. The top of the chamber should remain level to allow attachment of the electrode drivers. Here, we present a chamber that allows for recording in any region of the cerebellum (20 × 40 mm), located based on stereotaxic coordinates at anterior–posterior (AP) -11 mm (front edge), AP -28 mm (hind edge), right medio-lateral (ML) 18 mm, and left ML 22 mm. It should be noted that the smaller chambers cannot be subjected to much

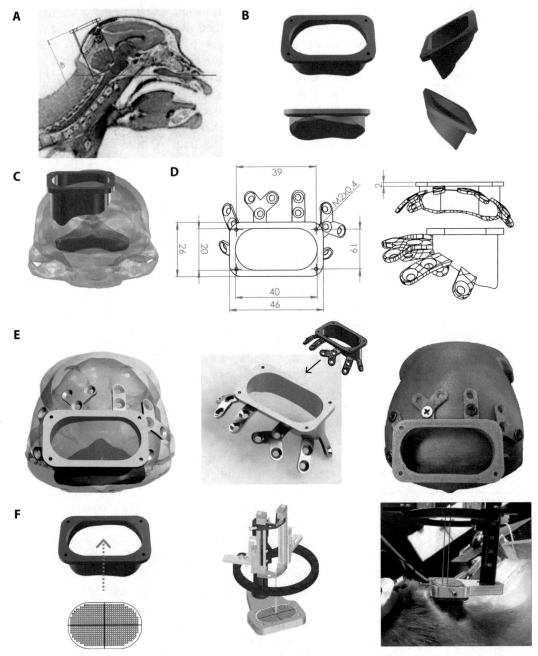

Fig. 3 Customized chamber for cerebellar recordings. (**a**) Chamber angle and depth plan from MRI sagittal cut. Based on the MRI, we calculated the chamber size and angle to be able to reach all vermal lobules and left lateral cerebellum. This can be easily expanded to cover the whole right cerebellum. (**b**) Custom cerebellar chamber rendered after shaping the bottom to the skull. (**c**) Posterior view of the chamber fitted on the MRI reconstructed skull (gray) and cerebellar (red) target locations. (**d**) Engineering drawing of the chamber with the legs to be screwed onto the skull. The legs were also skull-fitted. (**e**) **Left** Posterior view of the chamber with the legs fitted on the skull on top of the occipital bone. **Middle** After having completed the CAD chamber, the chamber was printed in 3D printed in Grade 5 Titanium alloy. **Right** Plastic 3D printed skull and chamber that facilitated surgery preparation. (**f**) **Left** Custom two-sided grid (to cover all the available recording space in the chamber) to guide electrode recordings. Every hole is 1 mm × 1 mm apart. **Middle** Custom grid mounted

force, neither are they compatible with multiple feet, which may cause scalp retraction from the chamber wall [70].

2.5   Coatings

Similar to medical implants used in humans, the bottom of chamber and headpost implants is coated with hydroxyapatite (HA) which is a synthetic calcium phosphate. This technique is commonly used to enhance the stability of dental implants. Coating different types of metals and their alloys, such as stainless steel, titanium, and magnesium, requires specific surface pre-treatment. Acid etching is most commonly used as pre-treatment for implants. For titanium, the surface pre-treatments improve the thickness of the TiO_2 passive layer, improving adhesion and bonding of the HA coating [71]. Since these pre-processing steps require a high level of expertise, they are commonly outsourced (e.g., Medicoat AG, Mägenwil, Switzerland, which offer coating and pre-treatment for around €200 per implant). Notably, the effectiveness of these coatings is still debated [66, 70, 72–74]. In vitro studies of HA coatings show that they can improve osseointegration and have a significant effect on the bone regeneration process through enhancement of cellular adhesion, proliferation, and differentiation thus helping stability and longevity of the implant [75]. In vivo studies do not always report all details of techniques used to enhance the functionalization of the coatings; factors that should be considered include: microroughness and thickness of coating, calcium phosphate solubility, and nanotopography. These details might contribute to the mixed reporting of effectiveness, (for review, *see* [76]). Furthermore, there is little data on the use of these coatings specifically in NHPs.

3   Methods: CT and MRI Scans

A computed tomography (CT) scan should be obtained to allow 3D models of the NHP's skull to be constructed. CT images have excellent contrast for bone and are thus very suitable for reconstruction of the skull. Magnetic resonance imaging (MRI) is especially useful for visualizing soft tissues and is therefore indispensable in designing the recording chambers. MRI is also very helpful in determining the exact electrode position post-chamber implantation (Fig. 4).

This step is particularly useful for cerebellar recordings since the electrode has to travel a long distance through the visual cortex and the tentorium to reach the cerebellum, enhancing the chances

Fig. 3 (continued) on the FlexMT microdrive (AlphaOmega). The microdrive screwed onto the four corners of the chamber that also served to screw the lid to cover the chamber when not in use. **Right** Photograph of an actual cerebellar recording in a monkey implanted with the chamber using two electrodes

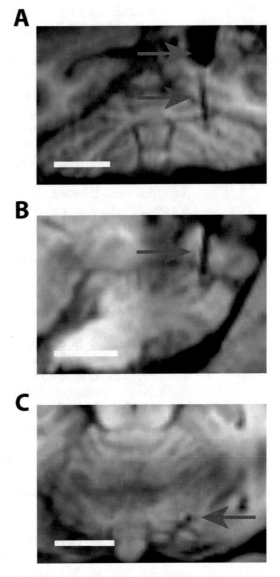

Fig. 4 Confirmation of the electrode placement. (**a**) Coronal MR image of cerebellum with artefact from tungsten electrode indicated by the red arrow. Purple arrow indicates artifact from titanium guide tube in visual cortex. (**b**) Sagittal section of the same image stack as (**a**). (**c**) Transverse section of the same image stack as (**a** and **b**). Scale bars indicate 1 cm in all panels

for potential errors in reaching the target. The CT/MRI scans often have to be repeated several times throughout the duration of the experiments (*see* **Note 2**). For this procedure the animal is lightly anesthetized while in the cage (Subheading 4.1.1). When the animal is sedated, it is transported in a cage to the scanner location.

4   Methods: Surgical Procedures

4.1   Anesthesia

Repeated anesthesia can cause discomfort. Therefore, minor procedures, such as cleaning the chamber or wound treatment post-op, are performed without anesthesia (*see* Subheading 4.4).

4.1.1   Anesthesia for Minor Procedures

Minor procedures that do not take longer than 90 min and do not require intubation, have a lighter anesthesia protocol. Types of procedures that require this form of anesthesia are: thinning of the dura, removal of stitches, small adjustments to implants, such as replacing a screw, and removal of the dental tartar. Throughout the procedure the animal's temperature is regulated through a heating mat. The animal is weighed prior to the procedure to ensure the correct dose is used. Anesthesia is given through medetomidine (0.08 mg/kg body weight) or medetomidine mixed with ketamine (7 mg/kg body weight), administered intramuscularly (i.m.). After every 20–30 min another 4 mg/kg ketamine is given to maintain the proper level of anesthesia. At the end of the procedure atipamezole 0.5 mg/kg i.m. is given as an anesthesia antidote to wake the animal up. NHPs are anesthetized in their home cage, and then transferred to the CT/MRI scanner. After the procedure they are returned to their home cage, where they are monitored during recovery from anesthesia.

4.1.2   Anesthesia for Major Procedures

To sedate an NHP a combination of ketamine and medetomidine (7 mg/kg, 0.08 mg/kg, respectively, i.m.) is administered while the animal is still in the cage. The animal is secured in the stereotaxic apparatus with ear bars that are prepared with the topical analgesic xylocaine. Eye ointment is put on the cornea to prevent dehydration. The animal is intubated and fully anesthetized with 0.8–1.5% isoflurane combined with 20% O_2 and 80% air and 0.5 mg/kg intravenous (i.v.) administration of midazolam. For general analgesia the animal receives 0.005 mg/kg of fentanyl i.v.. To prevent dehydration the animal is put on an i.v. drip of Ringer-glucose of 15 ml/kg/h.

During the surgery the animal is constantly monitored through ECG, heart frequency, external blood pressure, SpO_2, CO_2, body temperature, and breathing frequency. If the skull is opened for the implantation of a recording chamber, the animal receives 0.25 mg/kg dexamethasone i.v. 30 min before the opening of the skull, which is repeated every 30 min when the skull is open. If brain edema occurs it is treated with maximally 50 ml of mannitol in bolus through the i.v. drip. If the heart rate drops below 75 bpm 0.2 ml of atropine is administered i.v. and the anesthesia is adjusted. 10 min before the surgery is finished the animal receives i.v. 0.003 mg/kg buprenorphine, which partially antagonizes the fentanyl. The respirator is set to manual breathing and the animal is released from the stereotaxic apparatus. Lastly, when isoflurane is stopped and the animal starts to wake up, the i.v. drips are removed. The animal is then transported back to the cage and put under a heat lamp. If there was no intracranial procedure the animal receives pain relief with finadyne, (1–2 mg/kg) once every 2 days for 6 days total. If an intracranial procedure was performed the animal will also receive antibiotics on consultation of the veterinary for 10 days along with 5 days of dexamethasone in decreasing concentration (from 0.7 mg/kg to 0.1 mg/kg, i.m.).

After an intracranial procedure the animal receives 0.003 mg/kg i.m. buprenorphine as pain relief for 2 days; thereafter finadyne, 1–2 mg/kg i.m., once every 2 days for 6 days total. Alternatively, other anti-inflammatory drugs can be given orally. If an animal was socially housed before the procedure, it will be single housed for up to 9 days or until deemed fit again to be housed socially.

4.2 Headpost Implantation

Before surgery, the headpost, screws, and tools are sterilized in an autoclave. Headpost implantation is performed under general anesthesia and analgesia. The animal is head-fixed with ear bars into the stereotaxic apparatus and the scalp is shaved and prepped with alternating betadine and alcohol sponges. An incision is made in the scalp and the skin is gently pulled aside at the location where the headpost will be placed. The fit of the headpost is tested and if necessary, some bone is removed with a dental drill to ensure perfect contact. If the fit is good, the screw holes are drawn on the skull through the holes of the headpost. The holes are pre-drilled with a hand or mechanical drill. Subsequently, the headpost is placed on the skull and fastened into place with a torque measuring screwdriver until the 1.4 Nm tension is reached (alternatively: two-finger tightness). The skin is closed over the feet to fit as closely as possible around the headpost and sutured. After the headpost surgery the animal is trained at the task of interest (1–6 months). Only when the task performance is sufficient the recording chamber is implanted.

4.3 Chamber Placement and Craniotomy

Before surgery, the chamber, Ti screws, and all tools are sterilized in an autoclave. Craniotomy and chamber placement are carried out under general anesthesia and analgesia, and the animal also receives drugs that reduce intracranial pressure during the operation (e.g., mannitol, dexamethasone; for dosage see above). The duration of the procedure is approximately 1–2 h. The animal is head-fixed with ear bars in a stereotaxic device and the scalp is shaved and disinfected with alternating scrubs of betadine and alcohol. An incision is made in the scalp and the skin is retracted. Based on the stereotactic coordinates, the chamber is placed on top of the skull and the location of the screw holes is marked on the skull with a marker pen. When the recording chamber is placed on the marked location the chamber should follow the contours of the skull perfectly. Before opening the skull, screw holes of appropriate diameter are made with a hand or motor drill. Before the chamber is screwed in place, the craniotomy is made with a trephin or motor drill in the marked location, leaving the dura intact. Then the chamber is placed over the craniotomy and fastened into place with a torque measuring screwdriver until the 1.4 Nm tension is reached. To make the inside of the chamber water tight, dental cement should be applied at the intersection between the skull and the chamber. The skin is sutured up tightly surrounding the chamber, the animal is taken off the anesthesia and transported back to its cage.

4.4 Postoperative Care

After any type of anesthesia, the animals are left in the front part of the cage where they cannot climb and fall due to after-effects of the anesthesia. After major procedures the animals are treated with an infrared warmth lamp overnight. The NHPs are left in the cage to recover from the chamber implantation for one to two weeks before recordings are made, or until the implant wound completely heals. However, during that time, the chamber must be cleaned every 2–3 days to prevent infection. Throughout the duration of the experiments, the chamber is cleaned every day before and after the recording session. This procedure takes place while the animal is sitting in the primate chair. The chamber lid is removed and the interior of the chamber is flushed with anti-bacterial solutions such as chlorhexidine. Finally, the chamber is flushed with sterile saline. The total duration of the cleaning procedure is around 5 min and it causes minimal discomfort. The chamber should be filled with several milliliters of sterile saline at all times to prevent the dura mater from drying up or scabbing.

5 Methods: Behavior

5.1 Eye Tracking

To study cerebellar activity in relation to sensorimotor behavior it is essential to monitor one or more behavioral parameters. Almost all primate neurophysiology studies record eye movements. In the context of cerebellar recordings, eye movements need to be recorded as precisely as possible, since the cerebellum is involved in the execution of all types of eye movements [77, 78]. In order to parse saccadic eye movements into their minute components or study their kinematic properties, the data must be acquired at a high spatial and temporal resolution (e.g., the amplitude of glissades is usually $\pm 0.2°$, red segment of the trace in Fig. 5a, b, and duration of a saccade is between 30 and 70 ms). Historically, eye movements are tracked using sclera implanted search coils [79]. Search coils have excellent spatial and temporal resolution and do not suffer from some of the artifacts such as pupil wobble, which can be observed in pupil trackers [80, 81]. Therefore, coils are particularly useful for studies of eye movement kinematics.

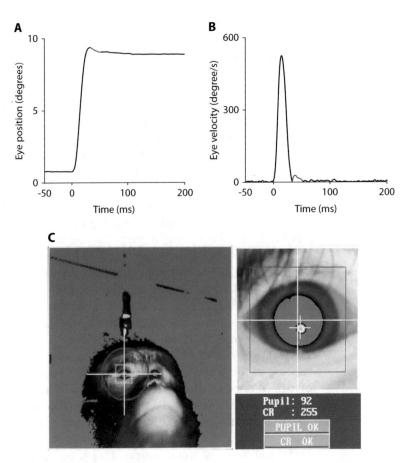

Fig. 5 Eye tracking. (**a**) Saccadic eye movement recorded with infrared video eye tracker. Glissades are marked in red. (**b**) Velocity trace of the same saccadic eye movement as in (**a**). (**c**) Tracking of the left pupil with eyelink 1000+ system. White cross in the top view displays the center of the pupil. In the bottom view the large white cross represents the center of the pupil and the small white cross through the light blue dot represents the corneal reflection

Conversely, the coil system is a lot more labor intensive, invasive, and can lead to altered saccade kinematics [80, 82]. More commonly used methods of tracking the eye movements rely on non-invasive infrared trackers [83, 84].

One well-known infrared eye movement tracker is the Eyelink 1000 plus system. It allows measurements of the position of the pupil at 1000 Hz at a precision of 0.01° (SR Research, www.sr-research.com; Fig. 5c). Given the high temporal resolution of the visual system, tasks that rely on visual cues require a monitor with a high refresh rate. Usually, the animal is placed at a viewing distance of approximately 52 cm from a monitor, e.g., a CRT monitor (100 Hz, 1152 × 864 pixels; [22]. At the beginning of every experiment the eye tracker is calibrated with the standard 5-point Eyelink calibration at 10° eccentricity. Saccades are detected online by the eye tracker when eye velocity exceeds 30°/s. Data is stored offline in the form of *x–y* coordinates and can be further analyzed offline using, e.g., MATLAB (MathWorks Inc.).

5.2 Orofacial Behavior

Recent research has shown that the cerebellum is actively involved in many cognitive tasks [38, 47, 85]. Therefore, it is increasingly important to monitor the movement and position of many body parts to distinguish the neural signals triggered by motor activity from those related to sensory information, reward signals, or attention. Particularly when it comes to reward signals it is necessary to record covariates such as tongue movements and mouth openings, as NHPs are most commonly rewarded by a drop of juice delivered via a lick spout. In some laboratories the lick spout is equipped with an infrared [86], or a capacitive touch sensor [38] that can detect the licks. However, modern tracking technologies offer an opportunity to directly measure all facial movements (Fig. 6). To track facial behavior an industrial camera can be used (e.g., The Imaging Source, GmbH), triggered by a TTL pulse coming from the Eye-link (the example data shown in Fig. 6 was recorded at 40 Hz and

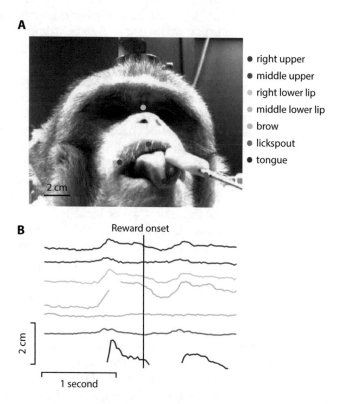

Fig. 6 Tracking of orofacial behavior. (**a**) Different points in the frame are annotated for tracking of mouth movements and tongue protrusion. The brow and lick spout can be used as reference. (**b**) Output of DeepLabCut analysis showing mouth opening and tongue protrusion before reward onset. Tracking of the points shown in (**a**). If the point is occluded the line is not visible

640 × 480 pixels). It is important to only capture as many frames and pixels as are of interest, since otherwise the computational power needed quickly increases. Data from the camera is fed into a trained neural network to extract facial markers with high precision. For this purpose, we currently use DeepLabCut, an open source and an easy to use machine learning algorithm [87]. We have recently developed a new version of the tracking software capable of extracting orofacial movements from NHPs that also takes into account previous frames [88]. Most of these programs have their own manual and are easy to set up even for those not computationally inclined.

5.3 Hand and Whole-Body Tracking

As described above, video tracking of different body parts or the whole body is a complementary, non-intrusive method to quantify behavior. The advent of continuous, unrestrained, more naturalistic tasks [89] puts 3D markerless tracking at the forefront of behavioral estimation [87, 88, 90, 91]. Hand and body tracking can be achieved just by using a single camera for tracking hand movements using a joystick [92] to 62 or more synchronized cameras [90]. Hand and whole-body tracking videos can be recorded using an industrial monochrome CCD camera (DMK 23U445, The Imaging Source LLC, North Carolina, USA) capturing a 1280 × 960 video at 30 frames/s (1.2 Megapixels) or a color camera (BFS-U3-23S3C-C) capturing 1920 × 1200 video at 163 FPS (2.3 MP), among others. Careful considerations must be taken in selecting lenses when cameras are placed at short distances and low light conditions. The start and end of the video recording is synchronized with other behavioral data using a trigger pulse sent by the behavioral PC controller. The storing of the video can be done using common video formats (such as .avi) or saving every frame and reconstructing them offline. To extract pose estimation or the trajectory of hand movements, identifiable features in the hand (i.e.., the wrist and fingers) are labeled to be tracked in a random subset of frames. Next, the researcher has to train a deep neural network model, for example, using the DeepLabCut [87] or OpenMonkey Studio [90], and extract the time course of the spatial locations of the marked features.

5.4 Training of the Behavioral Tasks

Training non-human primates on complex tasks can be quite challenging. Therefore, a project-based training program with clear objectives and training using positive reinforcement (PRT) will give the best results. Training is usually split into small segments in a build-up fashion, using the last learned behavior to continue to the next goal. PRT technique is used to obtain the desired behavior and helps to establish a positive relationship with the NHPs. When an NHP presents an undesired behavior the task is temporarily paused so the animal cannot achieve the reward. Most countries nowadays only allow PRT methods, i.e., punishment is forbidden.

Thus, undesired behavior can only be reduced by a time out. When a small segment of the task is ingrained, the task is made more complex by adding the next step and reinforcing the correct behavior again.

In head-fixed, visually guided behavioral tasks animals are seated in a primate chair. This is a chair on wheels that fits inside the setup, so the only physical act that is required of the animal is climbing in the chair. For this to happen, efficient cooperative training must be accomplished. This can sometimes be a stressful process for the animal, so selecting the optimal training approach must be planned in advance and dedicated to the type of chair and task to be accomplished. In addition, developing a positive interaction with the animal before any training begins will facilitate training and cooperation [93, 94]. Each training will depend on the type of chair selected for the experiment and the required behavior. Typical NHP chairs, where the monkey is guided to the chair, either using a pole-collar technique or chain, both require desensitization to the chair and to the element attached to his neck. Desensitization to these elements can be done early on in the process, while the animal is acclimating to their new space. For instance, guiding a monkey to a chair using a chain can take 1–1.5 weeks, while pole and collar can take up to 5–6 weeks due to the pole being a more stressful object for the monkey. Training duration times can be variable and depends on each animal [94]. Training can be performed twice a day for a duration of 15–20 min, although some reports suggest that once a day training may increase success [95]. The training includes desensitization to the chain or pole and collar, guidance to chair, and finally, closing the chair. Training a monkey to freely cooperate in climbing into the chair, without a guiding object or method of head or neck restraining, could take between 6 and 8 weeks. Providing the animal with behavioral choices using PRT, the trainer facilitates motivating effects that positively impact the behavior of NHPs during experiments [96–98].

To start the experiments, first the animals are habituated to the setup, by exposing them to the environment for increasing amounts of time while building a positive association by giving fruit juice or treats in the form of (dried) fruit. When habituated to the setup, the same procedure is repeated but with head fixation through the headpost in a stepwise procedure, where on the first day the animal is head-fixed for 15 min, then 25–30 min, and so on, adding 10–15 min to the fixation time with each step. Subsequently, the animal is trained to fixate their eyes at a dot at the center of the screen for increasing amounts of time, which is monitored by an eye tracker. From this point the animals have to input their own behavior instead of passive habituation. To improve performance the water intake is controlled prior to training. During the training or experimental sessions, the animals can earn the water in the form of fruit juice. This is a form of reinforcement where the behavior is strengthened by removing a passively applied aversive stimulus, i.e., thirst. Juice is more suitable than food rewards, since it can be stretched out over thousands of trials by giving small drops of juice for every correct trail. Also, there is a smaller chance of disturbance of electrophysiological measurements from oro-facial movements, such as chewing. Notably, some laboratories use fruit puree or food pellets instead of the juice and do not control the water or food intake prior to training/experiments [99, 100].

For traditional tasks, when an animal is consistently fixating, real task elements can be added. For instance, a saccade target or a distractor. Interestingly, several laboratories have demonstrated that fixation in the center position is not always necessary [101] and in fact the natural eye movements can reveal the NHP's internal states. More naturalistic tasks also lack a fixation point [102]. Regardless of the presence or absence of the central fixation point, the amount of task elements or distractors is very slowly built up, only if performance is constantly above chance level for multiple days. Other tools include increasing or decreasing the brightness of a stimulus that should be attended to or ignored. When a stimulus causes a forced choice later on in the task, this element should be introduced in the simplest way possible, generally in the central fixation position. When fully integrated, the element can be presented for shorter periods of time or gradually moved to an eccentric position on the screen. A possible flowchart of the procedures carried on for cerebellar neural measurements in a behaving NHPs is shown in Fig. 7. The levels of discomfort given here are estimates from a license approved by the Dutch Ethical Animal Committee.

Besides the classical type of setups involving a primate chair, an eye tracker, and a screen, a world of behavioral manipulations exists. Many cerebellar tasks either directly probe the oculomotor system or rely on the eye movements as a behavioral readout. Numerous different behavioral manipulations have been described. These include the use of touchscreens [103], an arm manipulandum, which measures and alters the force exerted [104], primate chairs that can introduce tilt, translation, and roll to affect vestibular input [105, 106]. Together, they provide a complete toolbox to study cerebellar behaviors.

A common strategy in NHP laboratories is to use the animals for several experimental protocols. Usually starting at easier tasks and slowly progressing to the most complex ones. In the context of cerebellar research that could mean studying processing of reflexive responses (e.g., reflexive eye movements) before proceeding to volitional behaviors.

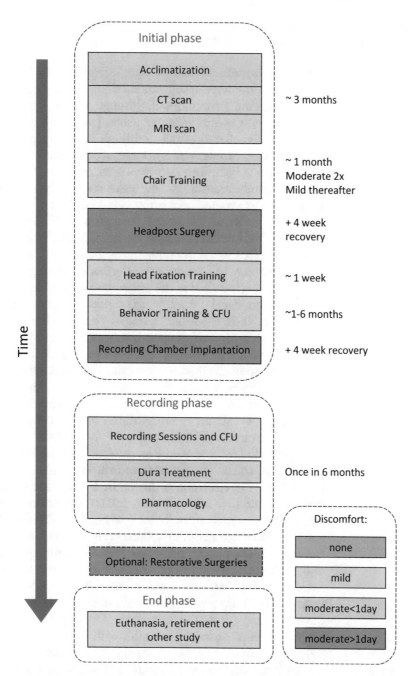

Fig. 7 Flowchart primate experiments. Initial phase: after arrival primates go through an extensive preparation and habituation phase (time estimates to the left). Recording phase: data collection according to experimental protocol. Endphase: euthanasia, retirement to a sanctuary or allocation to another study. Colors represent levels of discomfort. Abbreviation: controlled fluid uptake (CFU)

6   Methods: Cerebellar Electrophysiology

To date, cerebellar neuronal signals from NHPs have predominantly been obtained using single cell, extracellular electrophysiological recording techniques. Although we recognize that novel methods, such as multichannel arrays [52] are being introduced to NHP cerebellar research (although currently mainly in small primates such as marmosets). Here, we will focus on the single-unit recording technique, which is most commonly used.

Single-unit recordings are most commonly obtained using tungsten glass-coated electrodes (1–2 MΩ, e.g., AlphaOmega Engineering, Nazareth, Israel) through a sterile 23-gauge guide tube, which is inserted through the dura. A motorized microdriver (AlphaOmega Engineering, Nazareth, Israel) with a 1-mm spaced grid can be used to introduce the electrode and map the recording sites. The electrode inside the guide tube is carefully moved across the dura into visual cortex, which causes a brief moment of mild discomfort. It is therefore advisable to administer a few drops of fruit juice during this step to distract the animal and encourage him to sit still. Alternatively, a few drops of Lidocaine 2% can be placed on top of the dura a few minutes before inserting the guide tube to provide local analgesia. Subsequently, the electrode is moved into the cerebellar tissue in micrometer increments through the guide tube. Thus, the tentorium can be passed with just the tungsten electrode (250 μm width 60° beveled tip), not requiring the guide tube to be moved past the dura and thereby minimizing the damage to the visual cortex and the cerebellum. When in the cerebellum, the electrode is moved in 1–5 μm increments in the gray matter and 10–50 μm increments in the white matter, until the activity of a single cell is found (*see* **Note 3**). The AlphaOmega software allows for real-time spike sorting, rasters, and peristimulus histograms. The rasters and histograms can be aligned to any of the digital inputs of the signal processing box and are thus very useful tools for discriminating if a unit is task related (*see* **Note 4**). Extracellular recordings are digitized and sampled at 44 kHz and subsequently stored during the experiment. Under good conditions cells can easily be recorded for 20 min, with outliers of over 45 min, yielding hundreds of trials depending on the task.

Purkinje cells can be distinguished by the presence of spontaneous firing of simple spikes (SS) and complex spikes (CS) (Fig. 8a), as well as the presence of a concurrent short pause in SS firing following a CS (i.e., climbing fiber pause; *see* [107], as distinguished online using, for example, Multi-Spike Detector (MSD, AlphaOmega Engineering). Raw traces are offline spike sorted again for higher precision discrimination between SS, CS, and noise. Spike events are detected by a manually set positive and negative amplitude threshold for tracers and their derivatives.

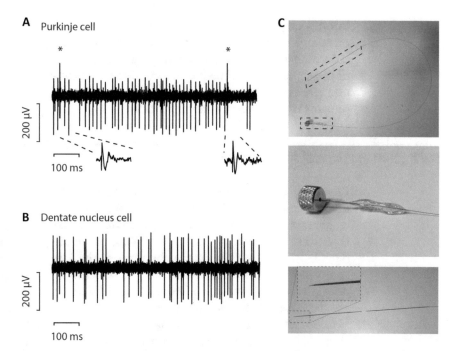

Fig. 8 Cerebellar single-unit recordings. (**a**) example trace of Purkinje cell recording exhibiting simple spikes and complex spikes (denoted with *). (**b**) Example trace of dentate nucleus cell. (**c**) Injectrode used for pharmacological interventions. Top panel shows the entire injection line within the top dashed box a tungsten electrode attached to the capillary tube and in the bottom dashed box a removable needle from Hamilton syringe attached to the capillary. Middle panel shows magnification of the removable needle. Bottom panel shows magnification of the electrode and a further magnified inset of the tip

This threshold is set for every recording session individually based on the signal quality. Threshold crossings are aligned to the peak of the derivative, which is more consistent than the peak of the raw signal, since SS have a larger negative amplitude and CS have a larger positive amplitude when the tip of the electrode is close to the cell body of the Purkinje cell. Spike waveforms are cut out around the peak from −0.5 to +6 ms for Purkinje cells, because CS have a long duration, and −0.5 to +2 for cerebellar nuclei cells, which do not have CSs (Fig. 8b).

Waveforms belonging to putative Purkinje cells are analyzed through principal component analysis (PCA) to distinguish the SS and CS waveforms. The first three components are used to select exemplary SS and CS to train a probabilistic neural network in a supervised machine learning fashion. This trained neural network is used to determine the class of the rest of the cutout spike waveforms (*see* **Notes 4** and **5**).

If pharmacological agents are used, these will be applied via a combined recording electrode/pipette (Fig. 8c), as described in the next section. The animals will perform the same tasks as outlined above. The duration of daily sessions is usually limited to 3 to 4 h, but can be prolonged depending on the type of experiment (pharmacological interventions with concurrent electrophysiological recordings can last up to 8 h with preparations). The total duration of the experimental time is quite variable across laboratories but on average data is collected for several years, although the chamber remains usable after this time-period.

7   Methods: Pharmacological Interventions

Reversible pharmacological agents can be applied locally through the implanted recording chamber, to the area of the cerebellum where task related neurons are recorded from. If the site of application is in the cortex, a sharp glass pipette or combined electrode/pipette will be lowered across the dura in the same manner as a recording electrode. In case of deeper structures, a glass pipette does not work. Instead, an injectrode can be made by attaching a capillary tube (150 μm outer diameter, 40 μm inner diameter (Molex Polymicro) to a tungsten electrode (AlphaOmega) (Fig. 8c). This injectrode is connected to the needle of a 10 μl syringe (Hamilton, model 701), which is used to inject the pharmacological agents. This injectrode can be lowered to the target in the same way as the electrode. The pharmacological agent will either be slowly injected using pressure or iontophoresis. The approach used will depend upon the properties of the drug to be applied and the desired volume of the effect; iontophoresis produces a more local effect, whereas pressure injections can affect larger volumes.

To estimate to what extent a cerebellar region that is being recorded from might be causally related to a particular behavior, inactivation studies can be informative [19, 38, 108]. Most of the published studies use muscimol, a potent, selective agonist of the GABAA receptors, which inhibits activity in Purkinje cells, i.e., the sole output neurons of the cerebellar cortex [109]. Most commonly between 2 and 10 μl of 5 μg/ml muscimol is injected directly into the cerebellum [19, 110]. Muscimol is dissolved in sterile saline and injected with a syringe pump with an average speed of 1 μl per 10 min. The injection is preceded by the same injection with only saline to ensure any effect was not based on volumetric tissue displacement. The behavior of interest is recorded before and directly after the injection. If the injection is correctly targeted, a behavioral effect should be visible after 15 min, and the effect of the drug could increase up to 60 min [108, 111]. Depending on the type of drug, suppression of activity can be up to 24 h; therefore, no more than 3 sessions per week can be performed to ensure complete washout before starting a session.

8 Notes

1. The edge of the skin near the implant will often retract somewhat over time. This could uncover the feet of the implant and
   make the areas with imperfect contact between implant and
   skull a place of infection that is very difficult to keep clean.
   Therefore, a customized post, which adheres tightly to the
   skull, is favorable. In general, skin retraction is a known effect
   of cranial implants, especially the ones that use acrylic and
   titanium. This happens because skin cannot adhere to these
   materials well. In addition, if the implant causes some skin
   tension, this effect will be amplified. When this happens, it is
   a common practice to cover the exposed area with acrylic or any
   other type of cement like Metabond. There are a few things
   that can be done to try to prevent this, depending on the type
   of implant: (1) Leaving space between the "legs" of the implant
   so the skin can bond to the skull; (2) positioning the screws
   inside the implant when possible; and (3) improving the surgical technique. Specifically, the tissue should be properly positioned ("wrapped") around the implant to decrease skin
   tension; this gives a chance for the skin to adhere to the
   underlying tissue. This can be achieved by opening the skin
   layer by layer and attaching it back in the same fashion.

2. The precise number of MRI scans depends on several variables:

   (a) The reliability of the localization of the brain structures
       involved; in case of cerebellar recordings, it is advisable to
       get at least one scan with the electrode present conform-
       ing to the targeted location.

   (b) The growth rate and age of the NHP; since the experi-
       ments usually take several years, the skull can change the
       shape considerably in that period. This is particularly
       important if the implant has to be adjusted at some
       point. Scans are performed under ketamine anesthesia
       (*see* Subheading 4.1.1) with an electrode and titanium
       guide tube (typically a 28–30G needle) in place when
       determining the recording location. Scanning with elec-
       trodes in the brain is a well-known technique [112], and
       there is no real risk on implant heating for titanium
       implants [113]; when using limited, yet sufficient, time
       to visualize electrode position, artifacts remain usually
       local, i.e., close to the implant/guide tube itself (*see*
       Fig. 4 for details). It should be noted though that the
       artifact can be substantially larger than the electrode,
       which is 250 μm in diameter.

3. When the sound of a neuronal unit can be seen or heard
   ramping up quickly there is some movement in the tissue, this

can be accounted for by retracting the electrode 50–100 μm or more if necessary to prevent penetrating the cell.

4. It should be noted that single-unit recordings from Purkinje cells or interneurons in the cerebellar cortex of NHPs can differ from those in other animals such as rodents or fish due to the different sizes of the molecular, granular, and Purkinje cell layers involved. For example, even though the amplitudes of the spikes of Purkinje cells can be substantially higher in NHPs, with tungsten electrodes it is often relatively difficult to record complex spikes and simple spikes of a single Purkinje cell at the same time in NHPs. Since CSs are predominantly dendritic and SSs somatic, satisfactory isolation of both types of spike at the same time requires positioning of the electrode close to both compartments.

5. When performing electrophysiological recordings, it is essential to analyze the data as soon as possible to see whether the recorded cell shows any task related activity. If this is not sufficiently clear on the online rasters, it should be done offline immediately after the recording.

9   Conclusions

Measurements of single neuron activity in awake and behaving NHPs offer a unique insight into the brain of a species most closely resembling that of a human. Considering that electrophysiological measurements in humans in vivo are predominantly performed intraoperatively and never in healthy individuals, NHP research is the only alternative in studying the primate brain. Techniques presented in this chapter allow researchers to answer relevant questions about the role of the cerebellum in planning, learning, and coordination of sensorimotor tasks. Historically, NHP cerebellar recordings have been instrumental in defining many characteristics of such tasks; for example: encoding of saccade kinematics, movement predictions, and error-based adjustment of future performance [78, 106, 114, 115]. Although the field of movement control and learning is far from solved and many experiments are still focused on these functions, there is an observable shift in the field toward more cognitive functions of the cerebellum. In recent years, the techniques we describe in this chapter have provided insights into the role of the cerebellum in behaviors classically attributed to cerebral cortex and basal ganglia, such as reward processing, attention, working memory, and reinforcement learning [38, 108]. Developments in other fields, such as machine learning, have been incorporated into the field of NHP research of the cerebellum, further fuelling this shift. For instance, the orofacial behavior tracking shown in Fig. 7 is applied to distinguish motor-related neuronal activity (such as licking) from reward signals. Notably, the use of camera methods for behavior tracking omits the necessity to put markers on the animals.

Similarly, techniques previously only available in rodent research are slowly starting to become accessible to primate neuroscientists. Several studies using optogenetics in cortical neurons have been published in recent years [116, 117]. Calcium imaging and chemogenetic manipulations have also made their first appearances in the literature together with the use of high-density silicon probes and arrays [118–120]. The latter have successfully been applied in the cerebellar research [52]. Advances in the 3D printing technology, described in detail in this chapter, have made it possible to fully customize the headposts and recording chambers, which offers much improvement with respect to the older, non-customized models. The use of these custom-made implants leads to fewer rejections, shorter recovery times, and less infections, which all contribute to the well-being of the animal and success in research efforts.

In conclusion, the techniques described here provide a guide to study motor and cognitive behavior in combination with cellular activity in NHPs, specifically in rhesus macaques.

References

1. Stern S, Kirst C, Bargmann CI (2017) Neuromodulatory control of long-term Behavioural patterns and individuality across development. Cell 171:1649–1662.e10. https://doi.org/10.1016/j.cell.2017.10.041

2. Guo A, Zhang K, Ren QZ, Su HF, Chen NN (2016) Functional connectivity mapping of decision-making in drosophila melanogaster. In: Advances in cognitive Neurodynamics (V). Springer, Singapore, pp 35–40

3. Keifer J, Summers CH (2016) Putting the "biology" Back into "neurobiology": the strength of diversity in animal model Systems for Neuroscience Research. Front Syst Neurosci 10:69. https://doi.org/10.3389/fnsys.2016.00069

4. Wang Q, Ding S-L, Li Y, Royall J, Feng D, Lesnar P, Graddis N, Naeemi M, Facer B, Ho A, Dolbeare T, Blanchard B, Dee N, Wakeman W, Hirokawa KE, Szafer A, Sunkin SM, Oh SW, Bernard A, Phillips JW, Hawrylycz M, Koch C, Zeng H, Harris JA, Ng L (2020) The Allen mouse brain common coordinate framework: a 3D reference atlas. Cell 181:936–953.e20

5. Hale ME (2019) Toward diversification of species models in neuroscience. Biotechnol Bioprocess Eng 93:166–168

6. Maximino C, do Carmo Silva RX, de Nazaré Santos da Silva S, do Socorro Dos Santos Rodrigues L, Barbosa H, de Carvalho TS, Leão LKDR, Lima MG, Oliveira KRM, Herculano AM (2015) Non-mammalian models in behavioural neuroscience: consequences for biological psychiatry. Front Behav Neurosci 9:233. https://doi.org/10.3389/fnbeh.2015.00233

7. Mitchell AS, Thiele A, Petkov CI, Roberts A, Robbins TW, Schultz W, Lemon R (2018) Continued need for non-human primate neuroscience research. Curr Biol 28:R1186

8. Roelfsema PR, Treue S (2014) Basic neuroscience research with nonhuman primates: a small but indispensable component of biomedical research. Neuron 82:1200–1204

9. Jennings CG, Landman R, Zhou Y, Sharma J, Hyman J, Movshon JA, Qiu Z, Roberts AC, Roe AW, Wang X, Zhou H, Wang L, Zhang F, Desimone R, Feng G (2016) Opportunities and challenges in modeling human brain disorders in transgenic primates. Nat Neurosci 19:1123–1130

10. Hutchison RM, Everling S (2012) Monkey in the middle: why non-human primates are needed to bridge the gap in resting-state investigations. Front Neuroanat 6:29

11. Procyk E, Wilson CRE, Stoll FM, Faraut MCM, Petrides M, Amiez C (2016) Midcingulate motor map and feedback detection: converging data from humans and monkeys. Cereb Cortex 26:467–476

12. Passingham R (2009) How good is the macaque monkey model of the human brain? Curr Opin Neurobiol 19:6–11

13. Stoodley CJ, D'Mello AM, Ellegood J, Jakkamsetti V, Liu P, Nebel MB, Gibson JM, Kelly E, Meng F, Cano CA, Pascual JM, Mostofsky SH, Lerch JP, Tsai PT (2017) Altered cerebellar connectivity in autism and cerebellar-mediated rescue of autism-related behaviours in mice. Nat Neurosci 20:1744–1751

14. Kelly E, Meng F, Fujita H, Morgado F, Kazemi Y, Rice LC, Ren C, Escamilla CO, Gibson JM, Sajadi S, Pendry RJ, Tan T, Ellegood J, Albert Basson M, Blakely RD, Dindot SV, Golzio C, Hahn MK, Katsanis N, Robins DM, Silverman JL, Singh KK, Wevrick R, Taylor MJ, Hammill C, Anagnostou E, Pfeiffer BE, Stoodley CJ, Lerch JP, du Lac S, Tsai PT (2020) Regulation of autism-relevant behaviours by cerebellar-prefrontal cortical circuits. Nat Neurosci 23(9):1102–1110. https://doi.org/10.1038/s41593-020-0665-z

15. Optican LM, Robinson DA (1980) Cerebellar-dependent adaptive control of primate saccadic system. J Neurophysiol 44:1058–1076

16. Ethier V, Zee DS, Shadmehr R (2008) Changes in control of saccades during gain adaptation. J Neurosci 28:13929–13937

17. Thier P, Dicke PW, Haas R, Thielert C-D, Catz N (2002) The role of the oculomotor vermis in the control of saccadic eye movements. Ann N Y Acad Sci 978:50–62

18. Golla H, Tziridis K, Haarmeier T, Catz N, Barash S, Thier P (2008) Reduced saccadic resilience and impaired saccadic adaptation due to cerebellar disease. Eur J Neurosci 27:132–144

19. Kunimatsu J, Suzuki TW, Tanaka M (2016) Implications of lateral cerebellum in proactive control of saccades. J Neurosci 36:7066–7074

20. Medina JF, Lisberger SG (2009) Encoding and decoding of learned smooth-pursuit eye movements in the floccular complex of the monkey cerebellum. J Neurophysiol 102:2039–2054

21. Fuchs AF, Brettler S, Ling L (2010) Head-free gaze shifts provide further insights into the role of the medial cerebellum in the control of primate saccadic eye movements. J Neurophysiol 103:2158–2173

22. Flierman NA, Ignashchenkova A, Negrello M, Thier P, De Zeeuw CI, Badura A (2019) Glissades are altered by lesions to the oculomotor vermis but not by saccadic adaptation. Front Behav Neurosci 13:194

23. Huber G, Heynen S, Imsand C, vom Hagen F, Muehlfriedel R, Tanimoto N, Feng Y, Hammes H-P, Grimm C, Peichl L, Seeliger MW, Beck SC (2010) Novel rodent models for macular research. PLoS One 5:e13403

24. Huberman AD, Niell CM (2011) What can mice tell us about how vision works? Trends Neurosci 34:464–473

25. Zeki SM (1978) The cortical projections of foveal striate cortex in the rhesus monkey. J Physiol 277:227–244

26. Chen C-Y, Hoffmann K-P, Distler C, Hafed ZM (2019) The foveal visual representation of the primate superior colliculus. Curr Biol 29:2109–2119.e7

27. Aggarwal V, Tenore F, Acharya S, Schieber MH, Thakor NV (2009) Cortical decoding of individual finger and wrist kinematics for an upper-limb neuroprosthesis. Conf Proc IEEE Eng Med Biol Soc 2009:4535–4538

28. Adewole DO, Serruya MD, Harris JP, Burrell JC, Petrov D, Chen HI, Wolf JA, Cullen DK (2016) The evolution of Neuroprosthetic interfaces. Crit Rev Biomed Eng 44:123–152

29. Ojakangas CL, Ebner TJ (1992) Purkinje cell complex and simple spike changes during a voluntary arm movement learning task in the monkey. J Neurophysiol 68:2222–2236

30. Wang J-J, Kim JH, Ebner TJ (1987) Climbing fiber afferent modulation during a visually guided, multi-joint arm movement in the monkey. Brain Res 410:323–329

31. Albert ST, Hadjiosif AM, Jang J, Zimnik AJ, Soteropoulos DS, Baker SN, Churchland MM, Krakauer JW, Shadmehr R (2020) Postural control of arm and fingers through integration of movement commands. elife 9:e52507. https://doi.org/10.7554/eLife.52507

32. Cullen KE, Brooks JX (2015) Neural correlates of sensory prediction errors in monkeys: evidence for internal models of voluntary self-motion in the cerebellum. Cerebellum 14:31–34

33. Angelaki DE, Yakusheva TA, Green AM, Dickman JD, Blazquez PM (2010)

Computation of egomotion in the macaque cerebellar vermis. Cerebellum 9:174–182

34. Laurens J, Angelaki DE (2020) Simple spike dynamics of Purkinje cells in the macaque vestibulo-cerebellum during passive whole-body self-motion. Proc Natl Acad Sci U S A 117:3232–3238

35. Herzfeld DJ, Shadmehr R (2014) Cerebellum estimates the sensory state of the body. Trends Cogn Sci 18:66–67

36. Shadmehr R (2017) Learning to predict and control the physics of our movements. J Neurosci 37:1663–1671

37. Popa LS, Streng ML, Hewitt AL, Ebner TJ (2016) The errors of our ways: understanding error representations in cerebellar-dependent motor learning. Cerebellum 15:93–103

38. Sendhilnathan N, Semework M, Goldberg ME, Ipata AE (2020) Neural correlates of reinforcement learning in mid-lateral cerebellum. Neuron 106:188–198.e5

39. Gao Z, Davis C, Thomas AM, Economo MN, Abrego AM, Svoboda K, De Zeeuw CI, Li N (2018) A cortico-cerebellar loop for motor planning. Nature 563:113–116

40. Chabrol FP, Blot A, Mrsic-Flogel TD (2019) Cerebellar contribution to preparatory activity in motor neocortex. Neuron 103: 506–519.e4

41. Van Essen DC, Donahue CJ, Glasser MF (2018) Development and evolution of cerebral and cerebellar cortex. Brain Behav Evol 91:158–169

42. Sereno MI, Diedrichsen J, Tachrount M, Testa-Silva G, d'Arceuil H, De Zeeuw C (2020) The human cerebellum has almost 80% of the surface area of the neocortex. Proc Natl Acad Sci U S A 117:19538–19543

43. Clark DA, Mitra PP, Wang SS (2001) Scalable architecture in mammalian brains. Nature 411:189–193

44. Gerrits NM, Voogd J (1989) The topographical Organization of Climbing and Mossy Fiber Afferents in the Flocculus and the ventral Paraflocculus in Rabbit, Cat and Monkey. In: Strata P (ed) The olivocerebellar system in motor control. vol 17. Springer, Berlin Heidelberg New York, pp 26–29

45. Balsters JH, Cussans E, Diedrichsen J, Phillips KA, Preuss TM, Rilling JK, Ramnani N (2010) Evolution of the cerebellar cortex: the selective expansion of prefrontal-projecting cerebellar lobules. NeuroImage 49:2045–2052

46. Barton RA, Venditti C (2014) Rapid evolution of the cerebellum in humans and other great apes. Curr Biol 24:2440–2444

47. Caligiore D, Pezzulo G, Baldassarre G, Bostan AC, Strick PL, Doya K, Helmich RC, Dirkx M, Houk J, Jorntell H, Lago-Rodriguez A, Galea JM, Miall RC, Popa T, Kishore A, Verschure PF, Zucca R, Herreros I (2017) Consensus paper: towards a systems-level view of cerebellar function: the interplay between cerebellum, basal ganglia, and cortex. Cerebellum 16:203–229

48. Strick PL, Dum RP, Fiez JA (2009) Cerebellum and nonmotor function. Annu Rev Neurosci 32:413–434

49. Spira ME, Hai A (2013) Multi-electrode array technologies for neuroscience and cardiology. Nat Nanotechnol 8:83–94

50. Rajan AT, Boback JL, Dammann JF, Tenore FV, Wester BA, Otto KJ, Gaunt RA, Bensmaia SJ (2015) The effects of chronic intracortical microstimulation on neural tissue and fine motor behaviour. J Neural Eng 12:066018

51. Badura A, Schonewille M, Voges K, Galliano E, Renier N, Gao Z, Witter L, Hoebeek FE, Chedotal A, De Zeeuw CI (2013) Climbing fiber input shapes reciprocity of Purkinje cell firing. Neuron 78:700–713

52. Sedaghat-Nejad E, Herzfeld DJ, Hage P, Karbasi K, Palin T, Wang X, Shadmehr R (2019) Behavioural training of marmosets and electrophysiological recording from the cerebellum. J Neurophysiol 122:1502–1517

53. SCHEER (Scientific Committee on Health, Environmental and Emerging Risks), Final Opinion on 'The need for non-human primates in biomedical research, production and testing of products and devices (update 2017)', 18 May 2017.

54. Laurens J, Kim B, Dickman JD, Angelaki DE (2016) Gravity orientation tuning in macaque anterior thalamus. Nat Neurosci 19 (12):1566–1568. https://doi.org/10.1038/nn.4423. Epub 2016 Oct 24

55. Avila E, Lakshminarasimhan KJ, DeAngelis GC, Angelaki DE (2019) Visual and Vestibular Selectivity for Self-Motion in Macaque Posterior Parietal Area 7a. Cereb Cortex. 29 (9):3932–3947. https://doi.org/10.1093/cercor/bhy272

56. Yushkevich PA, Piven J, Hazlett HC, Smith RG, Ho S, Gee JC, Gerig G (2006) User-guided 3D active contour segmentation of anatomical structures: significantly improved efficiency and reliability. NeuroImage 31: 1116–1128

57. Kikinis R, Pieper SD, Vosburgh KG (2014) 3D slicer: a platform for subject-specific image analysis, visualization, and clinical support. In: Jolesz F (ed) Intraoperative Imaging and

Image-Guided Therapy. Springer, New York, NY, pp 277–289

58. Tek P, Chiganos TC, Mohammed JS, Eddington DT, Fall CP, Ifft P, Rousche PJ (2008) Rapid prototyping for neuroscience and neural engineering. J Neurosci Methods 172:263–269

59. Avila E, Lakshminarasimhan KJ, DeAngelis GC, Angelaki DE (2019) Visual and vestibular selectivity for self-motion in macaque posterior parietal area 7a. Cereb Cortex 29:3932–3947

60. Dubowitz DJ, Scadeng M (2011) Suppl 2: A Frameless Stereotaxic MRI Technique for Macaque Neuroscience Studies. Open Neuroimaging J 5:198

61. Frey S, Pandya DN, Chakravarty MM, Bailey L, Petrides M, Collins DL (2011) An MRI based average macaque monkey stereotaxic atlas and space (MNI monkey space). NeuroImage 55:1435–1442

62. Paxinos G, Huang X-F, Toga AW (2000) The rhesus monkey brain in stereotaxic coordinates. Academic Press, Cambridge, Massachusetts

63. Saleem KS, Logothetis NK (2012) A combined MRI and histology atlas of the rhesus monkey brain in stereotaxic coordinates. Academic Press, Cambridge, Massachusetts

64. Reveley C, Gruslys A, Ye FQ, Glen D, Samaha J, Russ BE, Saad Z, Seth AK, Leopold DA, Saleem KS (2017) Three-dimensional digital template atlas of the macaque brain. Cereb Cortex 27:4463–4477

65. Mulliken GH, Bichot NP, Ghadooshahy A, Sharma J, Kornblith S, Philcock M, Desimone R (2015) Custom-fit radiolucent cranial implants for neurophysiological recording and stimulation. J Neurosci Methods 241:146–154

66. Ortiz-Rios M, Haag M, Balezeau F, Frey S, Thiele A, Murphy K, Schmid MC (2018) Improved methods for MRI-compatible implants in nonhuman primates. J Neurosci Methods 308:377–389

67. Chen X, Possel JK, Wacongne C, van Ham AF, Klink PC, Roelfsema PR (2017) 3D printing and modelling of customized implants and surgical guides for non-human primates. J Neurosci Methods 286:38–55

68. McAndrew RM, Lingo VanGilder JL, Naufel SN, Helms Tillery SI (2012) Individualized recording chambers for non-human primate neurophysiology. J Neurosci Methods 207:86–90

69. Ormianer Z, Laufer BZ, Nissan J, Gross M (2000) An investigation of heat transfer to the implant-bone interface related to exothermic heat generation during setting of autopolymerizing acrylic resins applied directly to an implant abutment. Int J Oral Maxillofac Implants 15:837–842

70. Adams DL, Economides JR, Jocson CM, Parker JM, Horton JC (2011) A watertight acrylic-free titanium recording chamber for electrophysiology in behaving monkeys. J Neurophysiol 106:1581–1590

71. Huynh V, Ngo NK, Golden TD (2019) Surface activation and pretreatments for biocompatible metals and alloys used in biomedical applications. Int J Biomater 2019:3806504

72. Welch JM, Lu J, Rodriguiz RM, Trotta NC, Peca J, Ding JD, Feliciano C, Chen M, Adams JP, Luo J, Dudek SM, Weinberg RJ, Calakos N, Wetsel WC, Feng G (2007) Cortico-striatal synaptic defects and OCD-like behaviours in Sapap3-mutant mice. Nature 448:894–900

73. Lanz F, Lanz X, Scherly A, Moret V, Gaillard A, Gruner P, Hoogewoud HM, Belhaj-Saif A, Loquet G, Rouiller EM (2013) Refined methodology for implantation of a head fixation device and chronic recording chambers in non-human primates. J Neurosci Methods 219:262–270

74. Overton JA, Cooke DF, Goldring AB, Lucero SA, Weatherford C, Recanzone GH (2017) Improved methods for acrylic-free implants in nonhuman primates for neuroscience research. J Neurophysiol 118:3252–3270

75. Surmenev RA, Surmeneva MA, Ivanova AA (2014) Significance of calcium phosphate coatings for the enhancement of new bone osteogenesis--a review. Acta Biomater 10:557–579

76. Bral A, Mommaerts MY (2016) In vivo biofunctionalization of titanium patient-specific implants with nano hydroxyapatite and other nano calcium phosphate coatings: a systematic review. J Craniomaxillofac Surg 44:400–412

77. Popa LS, Streng ML, Ebner TJ (2017) Long-term predictive and feedback encoding of motor signals in the simple spike discharge of Purkinje cells. eNeuro 4:ENEURO.0036-17.2017. https://doi.org/10.1523/ENEURO.0036-17.2017

78. Hong S, Negrello M, Junker M, Smilgin A, Thier P, De Schutter E (2016) Multiplexed coding by cerebellar Purkinje neurons. elife 5:1–19

79. Robinson DA (1963) A method of measuring eye movement using a scleral search coil in a magnetic field. IEEE Trans Biomed Eng 10:137–145

80. Kimmel DL, Mammo D, Newsome WT (2012) Tracking the eye non-invasively: simultaneous comparison of the scleral search coil and optical tracking techniques in the macaque monkey. Front Behav Neurosci 6:49

81. Nyström M, Hooge I, Holmqvist K (2013) Post-saccadic oscillations in eye movement data recorded with pupil-based eye trackers reflect motion of the pupil inside the iris. Vis Res 92:59–66

82. Frens MA, van der Geest JN (2002) Scleral search coils influence saccade dynamics. J Neurophysiol 88:692–698

83. Machado CJ, Nelson EE (2011) Eye-tracking with nonhuman primates is now more accessible than ever before. Am J Primatol 73: 562–569

84. Ryan AM, Freeman SM, Murai T, Lau AR, Palumbo MC, Hogrefe CE, Bales KL, Bauman MD (2019) Non-invasive eye tracking methods for New World and Old World monkeys. Front Behav Neurosci 13:39

85. D'Angelo E, Casali S (2012) Seeking a unified framework for cerebellar function and dysfunction: from circuit operations to cognition. Front Neural Circuits 6:116

86. Tsutsui K-I, Hosokawa T, Yamada M, Iijima T (2016) Representation of functional category in the monkey prefrontal cortex and its rule-dependent use for Behavioural selection. J Neurosci 36:3038–3048

87. Mathis A, Mamidanna P, Cury KM, Abe T, Murthy VN, Mathis MW, Bethge M (2018) DeepLabCut: markerless pose estimation of user-defined body parts with deep learning. Nat Neurosci 21:1281–1289

88. Liu X, Yu SY, Flierman NA, Loyola S, Kamermans M, Hoogland TM, De Zeeuw CI. OptiFlex: Multi-Frame Animal Pose Estimation Combining Deep Learning With Optical Flow. Front Cell Neurosci. 2021 May 28;15:621252. https://doi.org/10.3389/fncel.2021.621252

89. Huk A, Bonnen K, He BJ (2018) Beyond trial-based paradigms: continuous behaviour, ongoing neural activity, and natural stimuli. J Neurosci 38:7551–7558

90. Bala PC, Eisenreich BR, Yoo SBM, Hayden BY, Park HS, Zimmermann J. Automated markerless pose estimation in freely moving macaques with OpenMonkeyStudio. Nat Commun. 2020 Sep 11;11(1):4560. https://doi.org/10.1038/s41467-020-18441-5.

91. Berger M, Agha NS, Gail A (2020) Wireless recording from unrestrained monkeys reveals motor goal encoding beyond immediate

reach in frontoparietal cortex. Elife 9: e51322. https://doi.org/10.7554/eLife. 51322

92. Heimbauer LA, Conway CM, Christiansen MH, Beran MJ, Owren MJ (2012) A serial reaction time (SRT) task with symmetrical joystick responding for nonhuman primates. Behav Res Methods 44:733–741

93. Bliss-Moreau E, Theil JH, Moadab G (2013) Efficient cooperative restraint training with rhesus macaques. J Appl Anim Welf Sci 16: 98–117

94. McMillan JL, Perlman JE, Galvan A, Wichmann T, Bloomsmith MA (2014) Refining the pole-and-collar method of restraint: emphasizing the use of positive training techniques with rhesus macaques (Macaca mulatta). J Am Assoc Lab Anim Sci 53:61–68

95. Fernström A-L, Fredlund H, Spångberg M, Westlund K (2009) Positive reinforcement training in rhesus macaques-training progress as a result of training frequency. Am J Primatol 71:373–379

96. Reinhardt V (2003) Working with rather than against macaques during blood collection. J Appl Anim Welf Sci 6:189–197

97. Reinhardt V (2004) Common husbandry-related variables in biomedical research with animals. Lab Anim 38:213–235

98. Laule GE, Bloomsmith MA, Schapiro SJ (2003) The use of positive reinforcement training techniques to enhance the care, management, and welfare of primates in the laboratory. J Appl Anim Welf Sci 6:163–173

99. Westlund K (2015) Training laboratory primates--benefits and techniques. Primate Biol 2:119

100. Prescott MJ, Brown VJ, Flecknell PA, Gaffan D, Garrod K, Lemon RN, Parker AJ, Ryder K, Schultz W, Scott L, Watson J, Whitfield L (2010) Refinement of the use of food and fluid control as motivational tools for macaques used in behavioural neuroscience research: report of a working group of the NC3Rs. J Neurosci Methods 193:167–188

101. Lakshminarasimhan KJ, Avila E, Neyhart E, DeAngelis GC, Pitkow X, Angelaki DE (2020) Tracking the Mind's eye: primate gaze behaviour during virtual Visuomotor navigation reflects belief dynamics. Neuron 106:662–674.e5

102. Knöll J, Pillow JW, Huk AC (2018) Lawful tracking of visual motion in humans, macaques, and marmosets in a naturalistic, continuous, and untrained behavioural context. Proc Natl Acad Sci U S A 115: E10486–E10494

103. Norris SA, Hathaway EN, Taylor JA, Thach WT (2011) Cerebellar inactivation impairs memory of learned prism gaze-reach calibrations. J Neurophysiol 105:2248–2259

104. Miall RC, Weir DJ, Stein JF (1987) Visuomotor tracking during reversible inactivation of the cerebellum. Exp Brain Res 65:455–464

105. Laurens J, Meng H, Angelaki DE (2013) Neural representation of orientation relative to gravity in the macaque cerebellum. Neuron 80:1508–1518

106. Streng ML, Popa LS, Ebner TJ (2018) Modulation of sensory prediction error in Purkinje cells during visual feedback manipulations. Nat Commun 9:1099

107. De Zeeuw CI, Hoebeek FE, Bosman LWJ, Schonewille M, Witter L, Koekkoek SK (2011) Spatiotemporal firing patterns in the cerebellum. Nat Rev Neurosci 12:327–344

108. Kunimatsu J, Suzuki TW, Ohmae S, Tanaka M (2018) Different contributions of preparatory activity in the basal ganglia and cerebellum for self-timing. Elife 7:e35676. https://doi.org/10.7554/eLife.35676

109. Caesar K, Thomsen K, Lauritzen M (2003) Dissociation of spikes, synaptic activity, and activity-dependent increments in rat cerebellar blood flow by tonic synaptic inhibition. Proc Natl Acad Sci U S A 100:16000–16005

110. Monzée J, Drew T, Smith AM (2004) Effects of muscimol inactivation of the cerebellar nuclei on precision grip. J Neurophysiol 91: 1240–1249

111. Kunimatsu J, Tanaka M (2010) Roles of the primate motor thalamus in the generation of antisaccades. J Neurosci 30:5108–5117

112. Logothetis NK, Pauls J, Augath M, Trinath T, Oeltermann A (2001) Neurophysiological investigation of the basis of the fMRI signal. Nature 412:150–157

113. Buchli R, Boesiger P, Meier D (1988) Heating effects of metallic implants by MRI examinations. Magn Reson Med 7:255–261

114. Herzfeld DJ, Kojima Y, Soetedjo R, Shadmehr R (2015) Encoding of action by the Purkinje cells of the cerebellum. Nature 526: 439–442

115. Junker M, Endres D, Sun ZP, Dicke PW, Giese M, Thier P (2018) Learning from the past: a reverberation of past errors in the cerebellar climbing fiber signal. PLoS Biol 16: e2004344

116. Jazayeri M, Lindbloom-Brown Z, Horwitz GD (2012) Saccadic eye movements evoked by optogenetic activation of primate V1. Nat Neurosci 15:1368–1370

117. Chernov MM, Friedman RM, Chen G, Stoner GR, Roe AW (2018) Functionally specific optogenetic modulation in primate visual cortex. Proc Natl Acad Sci U S A 115: 10505–10510

118. Seidemann E, Chen Y, Bai Y, Chen SC, Mehta P, Kajs BL, Geisler WS, Zemelman BV (2016) Calcium imaging with genetically encoded indicators in behaving primates. Elife 5:e16178. https://doi.org/10.7554/eLife.16178

119. Mitz AR, Bartolo R, Saunders RC, Browning PG, Talbot T, Averbeck BB (2017) High channel count single-unit recordings from nonhuman primate frontal cortex. J Neurosci Methods 289:39–47

120. Deffains M, Nguyen TH, Orignac H, Biendon N, Dovero S, Bezard E, Boraud T (2020) In vivo electrophysiological validation of DREADD-based modulation of pallidal neurons in the non-human primate. Eur J Neurosci 53(7):2192–2204. https://doi.org/10.1111/ejn.14746

Chapter 15

Optogenetics in Complex Model Systems (Non-Human Primate)

Robijanto Soetedjo and Yoshiko Kojima

Abstract

Primates use rapid eye movements, called saccades, to scan their surroundings. Most of the well-studied neuronal elements that generate saccades are located in the brainstem, but recently our labs and others showed that the midline cerebellum is required for the production of accurate and stereotypical saccades. This oculomotor cerebellum receives mossy fiber inputs from saccade-related areas in the brainstem and sends its outputs to the brainstem saccade burst generator. How Purkinje cell simple spike activity is formed, and how the activity affects the movement in real-time are not well understood. In this chapter, we describe techniques to address these questions. Using optogenetics we manipulate the simple spike activity of cerebellar Purkinje cells while the saccade is ongoing. We also express optogenetic opsin to inhibit a subset of mossy fiber inputs to the cerebellum and examine the effects of the inhibition on simple spike activity.

Key words Optogenetics, Cerebellum, Vermis, Saccade, Oculomotor, Brainstem

1 Introduction

Saccades are rapid, stereotypical, and accurate eye movements, which we use to scan our surroundings [1, 2]. We make about three saccades every second. Critical neuronal substrates for generating saccades are located in the brainstem (Fig. 1). The intermediate layer of the superior colliculus (SC) [3], which receives projections from oculomotor areas of the cerebral cortex, provides the main pathway to drive burst neurons in both the pontine and mesencephalic reticular formations (the brainstem burst generator) [4]. In turn, burst neurons in the reticular formations drive the motoneurons of the extraocular muscles. The burst of the motoneurons quickly rotates the eyes to bring the object of interest to the fovea. A gaze holding signal which is thought to originate from a mathematical integration of the burst activity keeps the eyes from sliding back toward the center position [5–7].

Single unit recording has been successful in understanding the brainstem burst generator and motoneurons in great detail [8–

Roy V. Sillitoe (ed.), *Measuring Cerebellar Function*, Neuromethods, vol. 177, https://doi.org/10.1007/978-1-0716-2026-7_15,
© Springer Science+Business Media, LLC, part of Springer Nature 2022

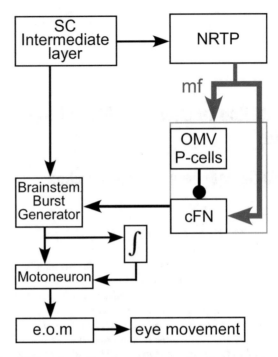

Fig. 1 Simplified schematic of neural circuits for saccade generation. Abbreviations: SC, superior colliculus; NRTP, nucleus reticularis tegmenti pontis; mf, mossy fibers; OMV, oculomotor vermis (midline, lobules VIc and VII); P-cells, cerebellar cortex Purkinje cells; cFN, caudal fastigial nucleus; e.o.m, extraocular muscles; $\int$, mathematical integration. ●, OMV P-cells inhibition on cFN neurons

15]. Individual neuron in the brainstem burst generator exhibits activity that is highly correlated with the metrics, kinematics, and the timing of the saccades. These data allow for mathematical models [16, 17] of brainstem saccade generator to be developed and guide further studies. Electrical stimulation [11, 12, 18], pharmacological inactivation [10, 19, 20], and lesion studies [5] further refined our understanding of the circuitry of the brainstem saccade generator.

Research starting in the mid-1970s revealed that the integrity of the midline cerebellum also is required for the production of accurate saccades [21, 22]. This cerebellar area, vermis lobules VIc and VII (the oculomotor vermis, OMV) and the caudal fastigial nucleus (cFN) to which OMV's Purkinje cells (P-cells) project, receives input from the SC through the nucleus reticularis tegmenti pontis (NRTP) [23–26]. In turn, the cFN projects monosynaptically to various neuron types in the brainstem burst generator [23, 27]. Therefore, it constitutes a parallel pathway by which the cerebellum can influence the premotor signals that produce saccades. Permanent lesions of the OMV in rhesus monkeys abolish the accuracy and stereotypical nature of saccades [22, 28–30]. Also, pharmacological inactivation of either the cFN or OMV causes dysmetric and highly variable saccades [31–33] and electrical stimulation of the cFN [34] or OMV [35, 36] during the course of saccades makes them dysmetric. Thus, the cerebellar output signal to the burst generator is crucial for keeping saccades accurate. In addition, many studies indicate that the pathway via the cerebellum is responsible for motor learning of saccades [37].

In contrast to motoneurons and neurons in the brainstem burst generator, neurons in the cFN and the OMV P-cell's simple spike (SS) exhibit a highly variable temporal pattern of saccade-related activity which may include small or large bursts, multiple bursts, burst-pause, or pause-burst discharge patterns [38, 39]. Moreover, individual neuron's activity is weakly correlated with the metrics and kinematics of the saccades. Although these data appear to contradict the evidences that support the roles of midline cerebellum in keeping saccades accurate, we found that when we grouped P-cells according to their complex spike preferred directional tuning and then averaged the activities across cells and trials, the correlation between saccade kinematics and the average SS activity substantially improved [40]. Similarly, averaging cFN activity across many neurons improves its correlation with saccade kinematics [41].

The findings from population analysis and other experiments led to the suggestion that the population activities of neuron groups in the contra- and ipsilateral OMV and cFN affect both the acceleration and deceleration phases of saccades [31, 42–44]. However, how cerebellar output affects the brainstem burst generator moment-by-moment during a saccade is not well understood. Electrical stimulation and pharmacological inactivation experiments are not specific enough to answer this question due to their non-specificity of the affected neurons and poor time resolution for the inactivation. Conversely, specific expression of channel rhodopsin (ChR2) in OMV P-cells would allow for brief inhibition of cFN activity at any time while a saccade is unfolding. In this chapter, we present an experiment that perturbs P-cell SS activity in real-time to better understand the functional significances of the cerebellar moment-by-moment control of the burst generator.

Another mystery of the oculomotor vermis is the origin of highly varied saccade-related discharge patterns of SS across the P-cells. However, these diverse saccade-related discharge patterns must, in large part, result from the rather stereotypical NRTP saccade-related bursts [45]. Mathematical modeling study suggested that different subsets of the large number of granule cells may process the mossy fiber signals differently to provide a large variety of signal patterns in the parallel fibers to allow for efficient learning [46]. Given that a single P-cell dendritic tree makes synaptic contacts with different efficacies with >170,000 parallel fibers

[47, 48], the highly varied P-cell's saccade-related SS discharges may occur as predicted by the model. Therefore, eliminating some mossy fiber inputs would change the contribution of the parallel fibers to P-cell SS activity and change its discharge pattern. In this chapter, we also describe a technique that exploits the expression optogenetic opsin at axon terminals to silence a subset of mossy fiber terminals in the OMV during saccades and observe how a P-cell's SS discharge pattern changes.

2 Material and Methods

2.1 Animal Preparations

We use rhesus monkey (*Macaca mulatta*) in our studies. Once an animal is well trained for coming out of its home cage and sitting in an acrylic primate chair (homemade, Washington National Primate Research Center [WaNPRC] Bioengineering Core), we perform an aseptic surgery to implant a preformed 3-turn scleral search coil (38 AWG Teflon insulated stranded stainless steel wire, Cooner Wire Co., Chatsworth, CA, USA) on one eye under the conjunctiva, as described in these two publications [49, 50]. During the same surgery, we also implant 3 dental cement head stabilization lugs, one at the front which also functions as the eye coil connector, and two are implanted behind the ears. Post-surgical treatments are performed by the Veterinarian staff of WaNPRC.

Once the animal recovers from the surgery, it is placed in an experimental booth ($60'' \times 58'' \times 91''$) and positioned so that the eye with the coil is in the center of two orthogonal pairs of square Helmholtz coils ($54'' \times 54''$) which are driven to produce alternating magnetic fields at two different frequencies in the range of 60–130 kHz. The two pairs of coils induce electromotive force voltages at two different frequencies that are proportional to the rotation angle of the coiled eye [51]. A phase sensitive detector [52, 53] demodulates the eye coil voltages to produce horizontal and vertical voltage outputs proportional to the animal gaze position. Our scleral search coil system was manufactured by C-N-C Engineering, Seattle, WA, USA. Because the company is not in business anymore, for a new recording setup, the WaNPRC Bioengineering Core manufactures both the power driver for the Helmholtz coils and the phase detector using modern parts.

The animal is trained to track jumping laser target-spot ($0.25°$ diameter) projected on a $2' \times 4'$ screen [19]. A dollop of applesauce reward is given every time its gaze is within a preset window around the target-spot. The applesauce delivery system is a variable speed peristaltic drive (Masterflex 7553-60, Cole Palmer, Vernon Hills, IL, USA) modified to allow control by 5 V pulses.

Once the animal is well trained, in a second aseptic surgery, we implant a 20 mm ID recording chamber made of 1 mm thick stainless steel cylinder placed over a trephined opening in the skull and held in place with dental cement and titanium screws. The chamber is perpendicular to the horizontal stereotaxic plane, placed in the midline with the center at 14.5 mm posterior to the inter-aural axis to allow access to the OMV [54].

All experiments were performed in accordance with the *Guide for the Care and Use of Laboratory Animals* [55] and exceeded the minimal requirements recommended by the Institute of Labora-tory Animal Resources and the Association for Assessment and Accreditation of Laboratory Animal Care International. All of the procedures were evaluated and approved by the local Animal Care and Use Committee of the University of Washington.

2.2 Single Unit Recording

2.2.1 Materials

1. Alpha-Omega glass tungsten microelectrode.

2. Iron solution.

3. 21G hypodermic stainless steel cannula.

4. Homemade cannula holder with hydraulic Microdrive.

5. Homemade analog amplifiers and filters.

    Extracellular single unit activity is recorded using a 250 μm diameter glass tungsten microelectrode (Alpha-Omega, Nazar-eth, Israel). The tip has 60° taper angle and is coated with iron (Dalic 2060 iron solution or equivalent, Sifco, Independence OH, USA) to bring the impedance to ~100 kΩ at 1 kHz (IMP-1 Impedance tester, BAK electronics, Umatilla, FL, USA). The microelectrode is inserted into a 21G hypodermic needle tube that serves as a guide cannula during an electrode penetration. The guide cannula holder is part of a homemade hydraulic microdrive system designed and built by the WaNPRC Bioengineering Core. The Bioengineering Core also designed and built the head stage amplifier, post-amplifier, and signal filter units.

2.3 Injectrode

2.3.1 Materials

1. 35G and 30G stainless steel hypodermic tubes.

2. Medical grade epoxylite (#6001M, and its solvent #6001S).

3. Polyimide tubing.

4. PTFE (Teflon) tubing.

5. Stainless steel solder.

    To construct the injection tube (injectrode, Fig. 2), we insert a 20 mm, 35-gauge stainless steel hypodermic tube (K-Tube Technologies, Poway, CA. USA) into a thin wall 30-gauge hypodermic tube stock for a distance of ~5 mm and solder the junction between the two. About 2 cm of this assembly is dipped into epoxylite several times. Once the epox-ylite dries, we insert the exposed part of the tube assembly into a polyimide sleeve and use epoxylite to seal the interface between the polyimide and the dried epoxylite. We then

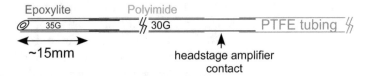

Fig. 2 Construction of an injectrode. Diameters are in American Wire Gauge standard

grind the tip of the insulated 35-gauge tube to a ~45° bevel. We then dip the tip into thinner epoxylite to insulate the bevel while keeping the hole open by pumping air through the tube. Under a 30× dissection microscope, we then scrape off some of the epoxylite at the tip of the bevel to expose the metal, which has an impedance of ~20 to 50 kΩ at 1 kHz. The other end of the injectrode is connected to both a Teflon tubing (ID of 0.009″, Zeus Industrial Products, Inc.; Orangeburg, SC), which is calibrated to move 40 nL of liquid for 1 mm movement of the meniscus and the terminal of the recording amplifier. The end of the Teflon tubing is connected to a pico pump (WPI, Sarasota, FL, USA) via a 30G needle (with the sharp tip cut) and a luer-lock 3-way stopcock. We fill the injectrode and Teflon tubing with 40–50 μL viral vector solution. The exposed tip of this injectrode records multiunit and background activity, which can be used to confirm the injection location. We mount the injectrode inside the same 21G guide cannula and microdrive system used for single unit recording.

2.4 Data Acquisition and Behavioral Control System

2.4.1 Materials

1. Power 1401 digitizer/controller unit.

   We sample eye and target position signals at 1 kHz and the unit signal at 50 kHz using a Power 1401 digitizer/controller unit (Cambridge Electronic Designs, UK). The microprocessor inside the Power 1401 allows us to run homemade sequencer programs that control the galvanometers, turn the laser target-spot ON/OFF, deliver food reward, and deliver stimulation pulses. In addition to the sequencer programs, a Microsoft windows computer runs a homemade Spike2 (Cambridge Electronic Designs, UK) script to control the parameters of the sequencer programs, save the recorded data, and run on-line analysis scripts to analyze action potentials and saccades.

2.5 Light Delivery to Target Area

2.5.1 Materials

1. 24G and 17G stainless steel hypodermic tubes.

2. Aluminum stock.

3. Delrin.

4. Silver brazing alloy wire.

5. Alpha-Omega EPS micropositioner.

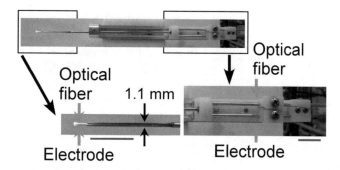

Fig. 3 Construction of double-barrel guide cannula and its holder. Left inset, double-barrel guide cannula tip with both optical fiber and microelectrode extended. Right inset, cannula holder and independent optical fiber and microelectrode receptacles. Red bars, 10 mm

6. FT200EMT optical fiber with FC/PC connectors and FC/PC adapters.

7. Alpha-Omega glass tungsten microelectrode.

Recording the light responses of neurons near the tip of the optical fiber is necessary for evaluating the success of the expression of the optogenetic opsin electrophysiologically. The Bioengineering Core designed a custom holder that drives a microelectrode and an optical fiber independently (Fig. 3). It uses a double-barrel stainless steel cannula (1.1×0.54 mm) to guide a microeletrode and an optical fiber (200 μm, FT200EMT, Thorlabs, New Jersey, USA) into a target area in the brain. The cannula is made of two 24G stainless steel hypodermic tubes, and a thin wall 17G stainless steel tube covers and reinforces the two 24G tubes. Silver brazing is used to join the tubes. The electrode-fiber distance when both are extended 10 mm out is about 0.4 mm.

The holder of the double-barrel cannula is mostly made of Delrin, and it has two receptacles that clamp the optical fiber and microelectrode separately. An Alpha-Omega (Nazareth, Israel) EPS micropositioner drives the two receptacles to move independently along two metal guides (right inset). The micropositioner is controlled by a Microsoft windows computer. The Delrin structure slides on an aluminum base so that the depth of the cannula penetration can be adjusted. During an experiment, the aluminum holder is attached to the recording chamber cannula positioner.

To reduce tissue damage and increase the light spread, we grind the animal end of the optical fiber into a cone with an apical angle of 20°. During an experiment, we maintained the tip of the optical fiber 0.5–1 mm above the tip of the micro-electrode. Before reaching the area, we turn on the laser at the intended optical power (30–60 mW) several times at different depths to identify light-induced artifacts generated by photo-voltaic (Becquerel) effect [56]. With our setup, we rarely observed the photovoltaic artifacts. We measure optical power using a Thorlabs PM100D meter equipped with a standard photodiode power sensor module (S121C). We use FC/PC connector adapters to couple optical fiber from the source to the animal.

2.6 Laser Light Source

2.6.1 Materials

1. High-power laser diodes.

2. Homemade laser diode driver.

3. 589 nm DPSS laser source.

    The mass production of laser illuminated computer projectors gives rise to the wide availability of inexpensive high-power (>700 mW) laser diodes. Some of the part numbers are Osram PLTB450B (450 nm), Nichia NUGM03 (525 nm), and Sharp GH0637AA2G (638 nm). The high-power laser diodes allow us to tolerate lossy, but inexpensive lower-precision optical parts used to couple the laser diode output into an optical fiber tip. Our homemade 450 nm laser source is capable to produce >100 mW of optical power at the optical fiber tip.

    We designed and built our own 450 nm laser light source for activating ChR2. Because laser diode light intensity is proportional to the applied current, the laser diode driver is based on a voltage-to-current amplifier (Fig. 4a, a simplified schematic diagram) [57, 58]. Briefly, when a 0.5 V is presented at the positive input of a differential amplifier (A, representing an operational-amplifier driver circuit), the transistor will start to conduct, and current will flow through the laser diode, transistor, and 1 Ω resistor. Due to Ohm's law, the voltage across the resistor is linearly proportional to the amount of current flow. The negative input of the differential amplifier compares the feedback voltage developed across the 1 Ω resistor with the 0.5 V at the positive input, and the amplifier circuit automatically corrects any discrepancy between the two inputs by adjusting the amount of drive to the transistor. This automatic feedback control keeps the amount of current that flows through the laser diode to a constant 0.5 A. Voltage input to this type of laser diode driver allows for intensity control by the Power 1401 digital-to-analog converter (DAC) output.

    For activating ArchT opsin, we used a commercially built 589 nm diode-pumped solid-state laser (DPSSL) source (MGL-F-589 nm to 50 mW, Opto Engine LLC, Midvale, UT, USA). This laser has a rise-time of >500 ms. Fig. 4b compares the light intensity time course produced by both the homemade 450 nm (blue traces) and 589 nm DPSS (orange traces) laser sources. The 450 nm intensity has a rise-time of <250 μs and an overshoot of 7.2%. In contrast, the

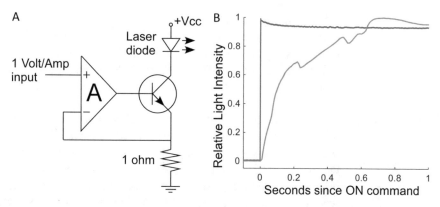

Fig. 4 (**a**) A simplified circuit of a laser diode driver. *A* a simplified operational-amplifier driver circuit, *+Vcc* positive power supply. (**b**) Comparison of light intensity time courses of 589 nm DPSS laser (orange) and 450 nm laser (blue)

589 nm DPSSL intensity increases much slower with a rise-time of >500 ms and an overshoot of 5.0%. Therefore, this DPSSL may not be suitable for experiments that require short light pulses, unless the DPSSL is turned ON constantly and a shutter mechanism is used to generate the light pulses.

2.7 Experiment Specific Methods

We describe two experiments that involve specific expression of ChR2 in P-cells and expression of ArchT in mossy fibers. We collaborated with Dr. Greg Horwitz' lab to produce the two AAV vectors. For the first experiment, the AAV vector contains 1 kb fragment of the L7 promoter upstream of the ChR2(H143R) gene (AAV1–L7–ChR2–mCherry) [59], and was injected into the OMV. For the second, AAV1-hSyn-ArchT-EYFP was injected into the NRTP.

2.7.1 Experiment to Examine the Role of OMV Purkinje Cells Simple Spike Activity on Saccade Dynamics

We recently showed that SS population activity of OMV P-cells are correlated with saccade kinematics [40]. However, the moment-by-moment contribution of P-cell SS activity on the saccade generator while the movement is ongoing is unclear. We tested this contribution by examining of the effects of a brief perturbation of P-cell activity at different phases of a saccade. Optogenetics and real-time eye movement recording instrumentations are necessary to perform this experiment.

Identifying OMV Purkinje Cells

The OMV area (lobules VIc and VII, midline) of the cerebellum is identified by the present of saccade-related bursting activity in the background. In addition to resting activities, isolated neurons in OMV exhibit a variety of saccade-related phasic activity which includes bursting, pausing, and their combinations [38, 39, 60, 61]. Using iron coated microelectrode with impedance ~100 kΩ at 1 kHz [54], P-cell's action potentials usually have the largest voltage than that of other neurons. We confirm a neuron as a P-cell if it exhibits complex spikes which are followed by a 10 ms or longer pause of the simple spikes [62]. We map the left and right sides of the OMV by measuring the preferred error directional tuning of the complex spikes. The preferred error direction of a P-cell's complex spike is contraversive to its anatomical location [40, 63].

Injecting Vector Solution

After we fill the injectrode (Fig. 2) and its Teflon tubing with 40–50 µL viral vector solution (AAV1–L7–ChR2–mCherry) and place it in the guide cannula, we then run a recording track. Once we hear saccade-related activities in the background, we start the injection process. Because of the mechanical elasticity of the tissue, we let the tip of the injectrode stabilize for about 1–2 min every time we stop at a new depth. At each depth, we inject 1–2 µL (25–50 mm meniscus movement) at a rate of 1 µL per minute. During the injection the background activity usually becomes weaker. We wait for about 1–2 min before we move to the next depth that is busy with saccade-related activities. Usually the new depth with saccade-related activity is preceded by a silent pass. The viral solution is delivered at every new depth until we no longer hear saccade-related activities. We usually perform one or two more injections at sites ~1 mm away in the same day, so that the viral solution covers the midline and a more lateral side. We wait for 6–8 weeks before we start the experiment.

Technical Requirements

The duration of a 10° saccade of a rhesus monkey is <35 ms [64]. Therefore, an experiment that perturbs P-cell's SS activity at different phases of the movement requires fast responding instrumentations. Our CNC-engineering scleral search coil phase detector (Fig. 5a) is suitable for this experiment because it has a latency of ~100 µs and a bandwidth of 500 Hz. Next we must trigger the 450 nm laser source when the eye velocity reaches a certain threshold. To accomplish this, we used a homemade calibrated analog eye position signal differentiator [57] to generate the eye velocity signal. Our system noise allows for a reliable triggering threshold of 15°/s.

The Power 1401 recorded both the eye and target positions, as well as eye velocity signals. A sequencer program executed by the 1401's microprocessor performs the velocity threshold detection and generates the necessary pulse to turn on the laser source. In the data shown in Fig. 5b, c, the pulse is 10 ms long. A user adjustable delay can be added by the program before generating the pulse so that the light pulse may occur at a time near the peak saccade velocity (Fig. 5c).

The Effects of Perturbing SS Activity on Saccades

We used 24° saccades so that we had a longer duration (~50 ms) to work with. We randomized stimulated (10 ms long pulse, blue, cyan) and non-stimulated (black, gray) with equal probability. The stimulated trials consisted pulses that occurs at 7 (Fig. 5b, blue,

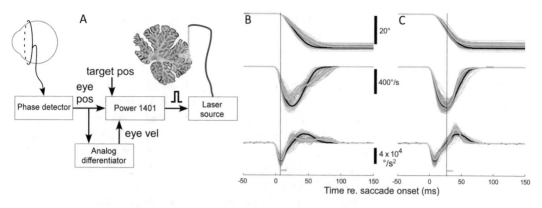

Fig. 5 (**a**) Block diagram of the system for triggering the laser source on an eye velocity threshold. The laser light pulse perturbs the SS activity of OMV P-cells. (**b** and **c**) effects of brief optical activation (blue) of P-cells on saccades. Laser turns ON (vertical lines) for 10 ms (blue, small dashes) starting at 7 ms (**b**) and 27 ms (**c**) after saccade onsets. Top-bottom, eye position, velocity, and acceleration traces. All traces aligned on saccade onset. Black, control trials, no activation. Thick traces, average traces

cyan), 17, 27 (Fig. 5c, blue, cyan), and 37 ms after a 15°/s threshold was detected. The timings were randomly presented. The 15°/s velocity threshold was also used to define the onset of the saccades.

Briefly turning the laser on (Fig. 5b, blue, cyan) for 10 ms at 7 ms after the onset of the saccades shortened the acceleration phase, which led to lower peak velocities. The acceleration changes occurred ~9 ms after the laser was on. The saccadic system compensated for the lower velocity by reducing deceleration and increasing its duration so that the average amplitude of the saccades was equal to that of the unstimulated ones (black, gray). Conversely, turning the blue laser pulse on near the peak velocity of the saccades increased the deceleration and shortened it, which resulted in smaller amplitude saccades (Fig. 5c, blue traces). In both 7 and 27 ms timings, increasing P-cell simple spike activity inhibited the movements, but when this brief inhibition occurred early during the movement, a compensation mechanism keeps the amplitude constant.

2.7.2 Experiment to Examine the Role of Mossy Fiber Inputs to OMV P-Cell Simple Spike Activity

The OMV receives major mossy fiber (MF) input from the NRTP which relays signal from the intermediate layer of the SC [24, 25, 65]. The NRTP also appears to be an important input to the cerebellum because inactivation of a part of the NRTP by muscimol produces severely hypometric saccades [66], whereas inactivation of any of the other weaker inputs to the OMV from saccade-related neurons in the omnipause region, nucleus prepositus hypoglossus, and paramedian pontine reticular formation has only a modest effect on saccade accuracy [5, 10, 19]. Here we describe an experiment to measure the partial contribution of NRTP mossy fibers on the SS activity of OMV P-cells. We express opsin ArchT (archaerhodopsin) in NRTP neurons and their mossy fibers. This technique allows us to deliver the light to the OMV, which is a much larger structure and easier to access than the dorsomedial NRTP.

Identifying the OMV and NRTP and Injecting AAV-ArchT

The OMV is located and mapped as described in the previous experiment. The dorsomedial NRTP is located by means of the characteristics of the activity of its saccade-related neurons. The neurons discharge similarly to the SC neurons that each neuron exhibits largest saccade-related bursts for movement with a certain amplitude and direction [45]. We then inject 3 μL of AAV1-hSyn-ArchT-EYFP solution into the NRTP using the injectrode described in Fig. 2. We confirm the location of the injectrode tip by recording nearby background activity. We wait 6–8 weeks for the ArchT expression before starting to record in the OMV.

Histological Verification of ArchT Expression in OMV

To verify that our injection of ArchT opsin is successfully expressed at the MF terminals in the OMV, but not P-cells or any other cells, we processed the cerebellar tissue histologically and check for EYFP fluorescence. Briefly, the animal is euthanized with an overdose of pentobarbital, then it is perfused through the heart with 4% paraformaldehyde. The perfusion continues with a gradient of sucrose in phosphate buffer (10, 20 and 30%). We extract the brain and place it in 30% sucrose. Sagittal sections (50 μm) of the cerebellum were cut on a sliding microtome. We then place the section under a fluorescence microscope. Fig. 6 shows fluorescence labeled MFs and their rosettes (inset) were found caudal to the primary fissure

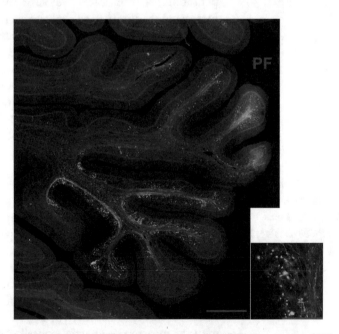

Fig. 6 Sagittal section of OMV histology showing EYFP fluorescence of MFs and MF terminals. Bar, 1 mm. PF, primary fissure. Inset, a magnified section of lobule VII. Bar, 100 μm

(PF) in lobules VI-VII, with the strongest fluorescence in lobule VII. Lobules VIc and VII constitute the OMV [18].

Identifying the Affected OMV Sites Electrophysiologically

We do not perform the electrophysiological verification of ArchT expression in the NRTP to avoid mechanical damage of the area. Instead, we perform this test in the OMV using the same double-barrel guide cannula holder system to drive down both a microelectrode and an optical fiber (Fig. 3). It is difficult to discern the reduction of background or resting SS activity caused by the inhibitory effects of ArchT on mossy fibers in response to 589 nm light illumination. We, therefore, exploit the rebound of activity that occurs after long inhibition by optogenetic opsins [67]. Figure 7a shows data where we aimed the laser at an OMV site whose mossy fibers possibly expressed ArchT. At this site, the extinction of laser illumination that had lasted at least 500 ms caused a rebound burst of SSs (rasters) in the recorded P-cell (Fig. 7a). The rebound could also be heard in the background activity. The rebound SS burst occurred 45 ± 6 ms after laser extinction. At this site and 3 others, such rebound bursts elicited saccades. We take these data as evidence that the mossy fibers within the illuminated area expressed ArchT. In contrast to the effects of laser extinction, constant illumination of the area during fixation had no direct effect on the baseline SS activity (Fig. 7a, left panel).

The Effects of Partial NRTP Mossy Fiber Inhibition on Purkinje Cell Activity

In the cerebellar cortex, mossy fiber signals are processed by a network that mostly composed of granule cells, whose axons give rise to parallel fibers, which synapse on P-cells, Golgi cells, and

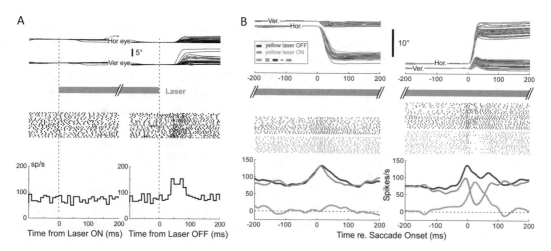

Fig. 7 (**a**) Rebound burst of a P-cell's SS activity after laser extinction. (**b**) Effects of optogenetic inhibition of MF on P-cell saccade-related bursts. Top-Bottom, Eye position traces, P-cell SS action potential raster, Spike density averages (10 ms bin histogram (**a**), 20 ms SD Gaussian (**b**)). Blue, control trials; Orange, 589 nm laser ON to inhibit a subset of MFs. Green, Differences between control and inhibited discharges. Laser was triggered 500 ms before a visual target step

other interneurons. Although the discharge patterns of NRTP cells have rather stereotypical Gaussian shaped bursts associated with saccades [45], the PC's simple spike (SS) discharge patterns associated with saccades are quite varied [38, 39]. One possible mechanism to produce the wide variety of SS saccade-related discharge patterns is that the cerebellar cortical circuitry segregates different subsets of MF inputs (and their granule cell targets) and processes them separately [68] to contribute to a specific SS discharge pattern [46].

In the experiment illustrated in Fig. 7b, the rasters show the SS activity of a P-cell before (blue) and after (orange) optogenetic inactivation of a subset of MF terminals from the NRTP. Prior to the inactivation, there was a single burst of SSs for 12° leftward saccades (blue data left column) and a double burst pattern for 12° rightward saccades (blue data right column). After inactivation of some of the MFs, and consequently of some of their target neurons in the OMV, the P-cell firing patterns changed. This was especially noticeable for rightward saccades where the pattern changed from burst–burst to smaller burst-pause (Fig. 7b right, spike density functions [SDFs] in lowest panel). A modest decrease in burst duration occurred for leftward saccades (Fig. 7b left, SDFs). The mossy fiber optogenetic inhibition produced a robust change of SS saccade-related activity and slightly reduced the variability of rightward saccades. However, the inhibition, on average, did not affect the amplitude of the saccades.

3 Conclusion

We described the techniques of two optogenetic experiments to better understand the neuronal signal processing that occurs in the oculomotor vermis during a saccade. First, we specifically expressed ChR2 in OMV P-cells to probe how changes of SS activity affect a saccade in real-time. For one saccade direction, this technique reveals a compensation for SS activity perturbation that occurs early during the movement. This compensation keeps the saccade accurate. The second technique involves expression of ArchT in mossy fibers that originate from the NRTP. Inhibition of a subset of these mossy fibers changed the pattern of P-cell's SS saccade-related discharges. This data suggest that the cerebellar cortical circuitry segregates different subsets of MF inputs and processes them separately to contribute to specific P-cell's SS discharge patterns.

Acknowledgments

Our studies described here were supported by National Institute of Health grants: EY028902 (RS), EY023277 (YK), OD010425, RR00166 (Washington National Primate Research Center), P30EY001730 (Vision Research Core of UW), and National Science Foundation BCS-1724176 (RS, YK).

References

1. Fuchs AF, Kaneko CR, Scudder CA (1985) Brainstem control of saccadic eye movements. Annu Rev Neurosci 8:307–337

2. Becker W (1989) The neurobiology of saccadic eye movements. Rev Oculomot Res 3:13–67

3. Gandhi NJ, Katnani HA (2011) Motor functions of the superior colliculus. Annu Rev Neurosci 34:205–231

4. Büttner-Ennever JA (2006) Neuroanatomy of the oculomotor system. Preface. Prog Brain Res 151:vii–viii

5. Kaneko CR (1997) Eye movement deficits after ibotenic acid lesions of the nucleus prepositus hypoglossi in monkeys. I. Saccades and fixation. J Neurophysiol 78(4):1753–1768

6. Goldman MS et al (2002) Linear regression of eye velocity on eye position and head velocity suggests a common oculomotor neural integrator. J Neurophysiol 88(2):659–665

7. Shadmehr R (2017) Distinct neural circuits for control of movement vs. holding still. J Neurophysiol 117(4):1431–1460

8. Scudder CA, Kaneko CS, Fuchs AF (2002) The brainstem burst generator for saccadic eye movements: a modern synthesis. Exp Brain Res 142(4):439–462

9. Sylvestre PA, Cullen KE (1999) Quantitative analysis of abducens neuron discharge dynamics during saccadic and slow eye movements. J Neurophysiol 82(5):2612–2632

10. Barton EJ et al (2003) Effects of partial lidocaine inactivation of the paramedian pontine reticular formation on saccades of macaques. J Neurophysiol 90(1):372–386

11. Sparks DL, Mays LE, Porter JD (1987) Eye movements induced by pontine stimulation: interaction with visually triggered saccades. J Neurophysiol 58(2):300–318

12. Keller EL, Gandhi NJ, Shieh JM (1996) Endpoint accuracy in saccades interrupted by stimulation in the omnipause region in monkey. Vis Neurosci 13(6):1059–1067

13. Scudder CA, Fuchs AF, Langer TP (1988) Characteristics and functional identification of saccadic inhibitory burst neurons in the alert monkey. J Neurophysiol 59(5):1430–1454

14. Fuchs AF, Scudder CA, Kaneko CR (1988) Discharge patterns and recruitment order of identified motoneurons and internuclear neurons in the monkey abducens nucleus. J Neurophysiol 60(6):1874–1895

15. Luschei ES, Fuchs AF (1972) Activity of brain stem neurons during eye movements of alert monkeys. J Neurophysiol 35(4):445–461

16. Scudder CA (1988) A new local feedback model of the saccadic burst generator. J Neurophysiol 59(5):1455–1475

17. Van Gisbergen JA, Robinson DA, Gielen S (1981) A quantitative analysis of generation of saccadic eye movements by burst neurons. J Neurophysiol 45(3):417–442

18. Mays LE, Sparks DL (1980) Dissociation of visual and saccade-related responses in superior colliculus neurons. J Neurophysiol 43(1):207–232

19. Soetedjo R, Kaneko CR, Fuchs AF (2002) Evidence that the superior colliculus participates in the feedback control of saccadic eye movements. J Neurophysiol 87(2):679–695

20. Lee C, Rohrer WH, Sparks DL (1988) Population coding of saccadic eye movements by neurons in the superior colliculus. Nature 332(6162):357–360

21. Vilis T, Hore J (1981) Characteristics of saccadic dysmetria in monkeys during reversible lesions of medial cerebellar nuclei. J Neurophysiol 46(4):828–838

22. Ritchie L (1976) Effects of cerebellar lesions on saccadic eye movements. J Neurophysiol 39(6):1246–1256

23. Noda H, Sugita S, Ikeda Y (1990) Afferent and efferent connections of the oculomotor region of the fastigial nucleus in the macaque monkey. J Comp Neurol 302(2):330–348

24. Yamada J, Noda H (1987) Afferent and efferent connections of the oculomotor cerebellar vermis in the macaque monkey. J Comp Neurol 265(2):224–241

25. Kralj-Hans I et al (2007) Independent roles for the dorsal paraflocculus and vermal lobule VII of the cerebellum in visuomotor coordination. Exp Brain Res 177(2):209–222

26. Thielert CD, Thier P (1993) Patterns of projections from the pontine nuclei and the nucleus reticularis tegmenti pontis to the posterior vermis in the rhesus monkey: a study using retrograde tracers. J Comp Neurol 337(1):113–126

27. Scudder CA, DM MG (2000) Connections of monkey saccade-related fastigial nucleus neurons revealed by anatomical and physiological methods. Soc Neurosci Abstr 26:97

28. Optican LM, Robinson DA (1980) Cerebellar-dependent adaptive control of primate saccadic system. J Neurophysiol 44(6):1058–1076

29. Barash S et al (1999) Saccadic dysmetria and adaptation after lesions of the cerebellar cortex. J Neurosci 19(24):10931–10939

30. Takagi M, Zee DS, Tamargo RJ (1998) Effects of lesions of the oculomotor vermis on eye movements in primate: saccades. J Neurophysiol 80(4):1911–1931

31. Robinson FR, Straube A, Fuchs AF (1993) Role of the caudal fastigial nucleus in saccade generation. II. Effects of muscimol inactivation. J Neurophysiol 70(5):1741–1758

32. Iwamoto Y, Yoshida K (2002) Saccadic dysmetria following inactivation of the primate fastigial oculomotor region. Neurosci Lett 325(3):211–215

33. Kojima Y, Soetedjo R, Fuchs AF (2010) Effects of GABA agonist and antagonist injections into the oculomotor vermis on horizontal saccades. Brain Res 1366:93–100

34. Noda H, Murakami S, Warabi T (1991) Effects of fastigial stimulation upon visually-directed saccades in macaque monkeys. Neurosci Res 10(3):188–199

35. Keller EL, Slakey DP, Crandall WF (1983) Microstimulation of the primate cerebellar vermis during saccadic eye movements. Brain Res 288(1–2):131–143

36. Ohtsuka K, Noda H (1991) Saccadic burst neurons in the oculomotor region of the fastigial nucleus of macaque monkeys. J Neurophysiol 65(6):1422–1434

37. Soetedjo R, Kojima Y, Fuchs AF (2019) How cerebellar motor learning keeps saccades accurate. J Neurophysiol 121(6):2153–2162

38. Kojima Y, Soetedjo R, Fuchs AF (2010) Changes in simple spike activity of some Purkinje cells in the oculomotor vermis during saccade adaptation are appropriate to participate in motor learning. J Neurosci 30(10):3715–3727

39. Ohtsuka K, Noda H (1995) Discharge properties of Purkinje cells in the oculomotor vermis during visually guided saccades in the macaque monkey. J Neurophysiol 74(5):1828–1840

40. Herzfeld DJ et al (2015) Encoding of action by the Purkinje cells of the cerebellum. Nature 526(7573):439–442

41. Sun Z et al (2016) Individual neurons in the caudal fastigial oculomotor region convey information on both macro- and microsaccades. Eur J Neurosci 44(8):2531–2542

42. Herzfeld DJ, Shadmehr R (2016) Cerebellar output encodes a corrective saccadic command (commentary on Sun et al.). Eur J Neurosci 44(8):2528–2530

43. Buzunov E et al (2013) When during horizontal saccades in monkey does cerebellar output affect movement? Brain Res 1503:33–42

44. Fuchs AF, Robinson FR, Straube A (1993) Role of the caudal fastigial nucleus in saccade generation. I. Neuronal discharge pattern. J Neurophysiol 70(5):1723–1740

45. Crandall WF, Keller EL (1985) Visual and oculomotor signals in nucleus reticularis tegmenti pontis in alert monkey. J Neurophysiol 54(5):1326–1345

46. Dean P, Porrill J (2011) Evaluating the adaptive-filter model of the cerebellum. J Physiol 589(Pt 14):3459–3470

47. Eccles JC, Ito M, Szentágothai J (1967) The cerebellum as a neuronal machine. Springer, Berlin Heidelberg

48. Isope P, Barbour B (2002) Properties of unitary granule cell-->Purkinje cell synapses in adult rat cerebellar slices. J Neurosci 22(22):9668–9678

49. Fuchs AF, Robinson DA (1966) A method for measuring horizontal and vertical eye movement chronically in the monkey. J Appl Physiol 21(3):1068–1070

50. Judge SJ, Richmond BJ, Chu FC (1980) Implantation of magnetic search coils for measurement of eye position: an improved method. Vis Res 20(6):535–538

51. Robinson DA (1963) A method of measuring eye movement using a scleral search coil in a magnetic field. IEEE Trans Biomed Eng 10:137–145

52. Meade ML (1983) Lock-in amplifiers : principles and applications. P. Peregrinus on behalf of the Institution of Electrical Engineers, London

53. McElligott JG, Loughnane MH, Mays LE (1979) The use of synchronous demodulation for the measurement of eye movements by means of an ocular magnetic search coil. IEEE Trans Biomed Eng BME-26(6):370–374

54. Soetedjo R, Kojima Y, Fuchs AF (2008) Complex spike activity in the oculomotor vermis of the cerebellum: a vectorial error signal for saccade motor learning? J Neurophysiol 100(4):1949–1966

55. National Research Council (2011) Guide for the Care and Use of Laboratory Animals, 8th edn. The National Academies Press, Washington, DC, p 246

56. Kozai TD, Vazquez AL (2015) Photoelectric artefact from optogenetics and imaging on microelectrodes and bioelectronics: new challenges and opportunities. J Mater Chem B 3(25):4965–4978

57. Horowitz P, Hill W (2015) The art of electronics. Cambridge University Press, Cambridge

58. Chkalov RD, Kochuev, Vasilchenkova D (2019) Precision medium-power laser diode drivers: design principles and functional features. in 2019 International Russian Automation Conference (RusAutoCon)

59. El-Shamayleh Y et al (2017) Selective optogenetic control of purkinje cells in monkey cerebellum. Neuron 95(1):51–62.e4

60. Prsa M et al (2009) Characteristics of responses of Golgi cells and mossy fibers to eye saccades and saccadic adaptation recorded from the posterior vermis of the cerebellum. J Neurosci 29(1):250–262

61. Kojima Y, Soetedjo R, Fuchs AF (2010) Behavior of the oculomotor vermis for five different types of saccade. J Neurophysiol 104(6):3667–3676

62. Sedaghat-Nejad E et al (2019) Behavioral training of marmosets and electrophysiological recording from the cerebellum. J Neurophysiol 122(4):1502–1517

63. Soetedjo R, Fuchs AF (2006) Complex spike activity of purkinje cells in the oculomotor vermis during behavioral adaptation of monkey saccades. J Neurosci 26(29):7741–7755

64. Fuchs AF (1967) Saccadic and smooth pursuit eye movements in the monkey. J Physiol 191(3):609–631

65. Harting JK (1977) Descending pathways from the superior collicullus: an autoradiographic analysis in the rhesus monkey (Macaca mulatta). J Comp Neurol 173(3):583–612

66. Kaneko CR, Fuchs AF (2006) Effect of pharmacological inactivation of nucleus reticularis tegmenti pontis on saccadic eye movements in the monkey. J Neurophysiol 95(6):3698–3711

67. Mattis J et al (2011) Principles for applying optogenetic tools derived from direct comparative analysis of microbial opsins. Nat Methods 9(2):159–172

68. Masoli S et al (2020) Parameter tuning differentiates granule cell subtypes enriching transmission properties at the cerebellum input stage. Commun Biol 3(1):222

Chapter 16

Clinical Assessment of the Cerebellum

Jason S. Gill, Jennifer Deger, and Roy V. Sillitoe

Abstract

There is increasing interest in the broad contribution of the cerebellum to all domains of human behavior. Neuroscientists and clinical neurologists can benefit from understanding how to recognize these different functional contributions of the cerebellum, knowledge that is essential for identifying the reflection of cerebellar dysfunction in the neurologic exam. Here, we describe components of the cerebellar exam, including classical methods to evaluate motor dysfunction of the cerebellum (ataxia, dysmetria, oculomotor dysfunction) as well as techniques to examine deficits in non-motor functions that are affected in cerebellar disease. Following the discussion on the cerebellar clinical exam, we will consider the relatively common cerebellar imaging methods that are currently used in the clinic. The goal of this chapter is thus to give the reader fundamental tools necessary to understand how human cerebellar dysfunction can be evaluated in the clinic.

Key words Cerebellum, Clinical assessment, Motor function, Non-motor behavior, Neurological diseases, Neuropsychiatric conditions

1 Introduction: Clinical Investigation of Cerebellar Dysfunction

The brain occupies a unique position among organs of the body, in the sense that for much of the history of western medicine, it remained obscured from direct scrutiny in the clinical exam. While the heart and lungs could be auscultated and the visceral organs could be palpated, the brain remained encased in the calvarium, with the optic disc being the only structure of the central nervous system amenable to direct evaluation. This unique anatomic positioning combined with the ubiquitous reflection of brain activity on nearly every aspect of human behavior, however, eventually led to the development of a more sophisticated neurological exam.

In the past half-century, the clinical exam has been supplemented initially with the computed tomography (CT) scan and then magnetic resonance imaging (MRI). These two imaging modalities provide the clinician with a virtual view inside the cranial vault to closely examine the vasculature and parenchyma of the brain itself

Roy V. Sillitoe (ed.), *Measuring Cerebellar Function*, Neuromethods, vol. 177, https://doi.org/10.1007/978-1-0716-2026-7_16,
© Springer Science+Business Media, LLC, part of Springer Nature 2022

in order to correlate clinical findings with brain lesions. The importance of neuroimaging to clinical neurology is hard to overstate and it has become an indispensable adjunct to the neurological exam.

Regarding the cerebellum, the historical reliance on the neurological exam has perhaps contributed to an oversimplified view that has obscured the remarkable diversity of functions in which the cerebellum is involved. Classically, the cerebellum has been taught to clinicians as a brain region devoted to the learning and coordination of motor function. This classical teaching reflects a major role of the cerebellum and cerebellar networks in motor function [1, 2]. While there may be a variety of explanations for this prioritization of the motor function of the cerebellum over its role in other functions, including cognitive, affective, and autonomic behavioral domains [3–6], the simplest explanation, and likely the most practical, is that the most obvious clinical manifestations of cerebellar injury are motor deficits, including ataxia, loss of coordination, and gait instability.

In this chapter, the clinical evaluation of human cerebellar deficits will be discussed, followed by a brief synopsis of clinical cerebellar neuroimaging and its use in the evaluation of classical cerebellar lesions.

2 Materials

The clinical evaluation of cerebellar deficits can largely be accomplished using only the experience and close observations of the examiner. In many cases, entailed examination diagrams can be improvised or recreated with little difficulty.

2.1 Assessment of Limb, Balance, and Gait Function

Motor function of the cerebellum can be evaluated without the use of any special instruments. Tremor is a very common movement disorder, with essential tremor being the most common clinical movement disorder [7, 8]. Tremor is described in terms of frequency of oscillation and timing of onset (during which type of movement or in which position, as will be described in more detail in Subheading 3). In regard to tremor originating from cerebellar circuit dysfunction, a low frequency (<4 Hz) intention tremor can be observed, although based on olivocerebellar function [1] tremor can also often occur in the 8-12 Hz range [9]. While for research purposes exact documentation of frequency of tremor may be noted, clinically tremor description often remains qualitative. Describing a tremor as high or low frequency and high or low amplitude is generally sufficient to differentiate diseases. The ability to make the qualitative judgment of tremor frequency does rely on the experience of the clinician and may be difficult for a naïve clinician to achieve without the help of a more experienced clinician or a reference video. Dysmetria, ataxia, titubation, gait/postural instability, and less commonly evaluated motor deficits such as dyssynergia are all evaluated with the aid only of a cooperative patient. In instances where a patient is non-cooperative, due either to coexisting neurologic deficits, diminished cognitive function, or developmental immaturity, clinical exam findings may be difficult to obtain. Often, however, observation may serve as a surrogate for a focused exam. For instance, observing a patient attend to certain activities of daily living (ADLs), such as eating or drinking, can be an excellent method to evaluate tremor, ataxia, and/or dysmetria. Similarly, watching an 18-month old child at free play may serve a similar function. While these findings may be clinically subtle, their impact on ADLs may be outsized. In particular, tremor has been associated with deficits in quality of life and mood alteration [10].

2.2 Evaluation of Oculomotor Dysfunction

As with the evaluation of limb, balance, and gait function, the evaluation of oculomotor abnormalities related to cerebellar dysfunction is often achieved without special instrumentation. In particular the floccular lobe, the ansiform lobule, and the cerebellar vermis receive mossy fiber input from various corticocerebellar and olivocerebellar loops to control the vestibuloocular reflex, saccades, and smooth pursuit [11]. Oculomotor dysfunction in cerebellar disease is evident in genetic (e.g. spinocerebellar ataxia 2), acquired (e.g. posterior fossa syndrome), and paraneoplastic (e.g. anti-Yo) disease. Especially in a cooperative patient, the examiner ought to be able to achieve nearly all aspects of the oculomotor evaluation by having the patient follow the examiner's finger or, in the case of young children, an alternative object of interest such as a stuffed animal or toy. One tool that can aid in the evaluation of deficits of smooth pursuit and generation of saccades is striped "optokinetic tape," which refers to a length of paper or fabric with alternating stripes of red and white.

2.3 Evaluation of Non-motor Cerebellar Dysfunction

Non-motor cerebellar dysfunction continues to be an underrecognized clinical entity. Canonically, the cerebellum is primarily thought of as a brain region devoted to sensorimotor integration. However, increasingly cerebellar dysfunction has been found to cause significant morbidity in affective and cognitive domains [12]. In particular, emotional lability and broad deficits in executive function have been associated with cerebellar injury [13, 14]. Evaluation of non-motor deficits may be difficult in the clinical setting due to the duration of time required for a thorough evaluation of all neurocognitive domains. Often, the history provided by a caretaker or cohabitant of the patient will provide invaluable information regarding changes in function over time and can serve as an important aid to the clinical examination. The examiner should inquire about changes in school or work performance, changes in interpersonal relationships, and personality changes, which may reflect alterations in affect. In order to screen for deficits in non-motor function, the mini-mental status exam (MMSE) and Montreal Cognitive Assessment (MOCA; http://www.mocatest.org) are two commonly used brief clinical exams that have been validated for identification of general cognitive impairment [15–17]. More recently, a tool (cerebellar cognitive affective/Schmahmann syndrome scale) has been devised to focus on the screening of non-motor deficits specifically associated with cerebellar dysfunction [18]. For a meta-analysis of the performance of patients with cerebellar lesions compared to controls on a wide variety of cognitive screening tools, please refer to Ahmadian et al. [19]. If there is a concern for non-motor dysfunction based on initial screening, referral for formal neuropsychologic testing from a trained psychologist would best characterize the precise domains of function that are compromised. In addition, in children or other individuals who cannot participate in screening tools, referral for formal testing may be of utility in the setting of cerebellar disease.

2.4 Computed Tomography Scanning

CT scanners are a nearly ubiquitous resource in health care. As of 2012, estimates were that over 6000 CT scanners had been purchased in the USA and over 30,000 worldwide [18]. In the context of neurological evaluation, CT scanning is generally an acute screening tool that is done to rule out emergent pathology that might warrant immediate intervention. As a result, for neuroimaging, the CT is generally performed in the emergency room, intensive care unit, or on the wards in case of acute mental status changes. Increasingly available, especially in tertiary care centers, are portable CT scanners that can be used for patients who may be too medically unstable to be transported to the radiology suite.

2.5 Magnetic Resonance Imaging

MRI machines are a very common diagnostic tool, though due to increased cost of the machine, associated image processing, and expertise required for interpretation of the images, they are less readily available in resource poor areas. Given the rapid acquisition of CT scans, and the ability of CT scans to adequately detect emergent pathology (hemorrhage, fracture, mass effect), MRI scans are more commonly used in medically stable patients. In addition, MRI scans are most often used for serial or surveillance scanning (for neoplasms, genetic conditions, and monitoring of disease progression in entities such as multiple sclerosis (MS)) due to their safety compared to CT scans. As a result, for neuroimaging, MRI scanners are typically available in both the inpatient and ambulatory settings.

3   Methods: Cerebellar Clinical Assessment

3.1   Clinical Examination

3.1.1   Motor Assessment

Loss of motor coordination is often the initial clinical finding that alerts the clinician to cerebellar deficits. This could include gait abnormalities, deficits in reaching and pointing, and problems with balance. As a result, the most basic evaluation of cerebellar function typically includes the testing of point-to-point movements (e.g. finger-to-nose or heel-to-shin), rapid alternating movements (RAMs), and evaluation of gait [20]. If these tests are abnormal, cerebellar abnormality should be added to the differential and a more thorough evaluation of cerebellar function may be undertaken on the focused exam. In this section, we will further discuss the motor complications of cerebellar damage.

Limb Findings

Dysmetria is defined as a loss of coordinated extremity movements [21] or disordered movement termination [22]. On a neurological exam, this can manifest as overshooting (hypermetria) or undershooting (hypometria) of limb movements to a target, such as in the finger-to-nose or heel-to-shin tests [23]. In the finger-to-nose test, the patient touches the clinician's index finger and then his or her nose, alternating several times as the clinician moves his or her finger so that the patient must change directions. The patient may initially overshoot the mark but then reach it in dysmetria. In the heel-to-shin test, the patient places one heel on the opposite knee and then slides it up and down the shin to the big toe, repeating the movement several times. The clinician must take care to instruct the patient to maintain contact between heel and shin. With dysmetria, the heel may overshoot the knee or foot and wobble from side to side along the shin [20].

    Tremor is defined as a rhythmic, oscillatory movement of a body part and is the most common movement disorder [24, 25]. Tremor can be identified by physical exam or by history taking. You can ask the patient about any shaking or general body movements that the patient is unable to control. If a tremor is present, observe or inquire as to when it occurs, worsens, and/or ceases: for instance, at rest, when standing, when reaching to grab an object, or with movement. Many types of tremor might indicate disease of the cerebellum or cerebellar networks. Tremor is commonly categorized by its timing, or when it is noticed. Tremors include postural tremor, intention/kinetic tremor, orthostatic tremor, and resting tremor [7]. Postural tremor is elicited by having the patient extend their arms against gravity. Intention tremors typically worsen with movement, especially deliberate movement toward a target. For example, in the finger-to-nose test, the patient's hand may have a more noticeable tremor as it approaches the examiner's finger or the patient's nose [20]. Orthostatic tremor will have onset with standing while resting tremor is noted when the patient is at rest and can be observed while interviewing the patient. Essential tremor, the most common cause of tremor, is usually a high-frequency, bilateral intention tremor that occurs with limb movement and is exacerbated by emotional distress. Essential tremor usually diminishes with relaxation or with alcohol. Head, voice, and leg tremor may also be present in essential tremor [25]. In addition, patients with essential tremor will often have a family history of tremor [20]. Although the pathogenesis of essential tremor is not well understood, some recent studies have suggested that some etiologies of essential tremor could result from neurodegenerative disease centered in the cerebellar cortex [25, 26]. Tremor can also be noted in the head (nodding), voice, and palate. In particular, myorhythmia is a rare tremor of 1–4 Hz that can affect cranial muscles. It is generally thought to be caused by diencephalon, brainstem, or cerebellar pathology [27, 28]. Specifically, myorhythmia has been attributed to cerebellar degeneration secondary to multiple sclerosis and chronic alcoholism [29]. However, based on the normal operation of olivocerebellar functional connectivity, tremor predicted to involve this circuit can also occur in the 8–12 Hz range.

Dysdiadochokinesia is the diminished ability to perform rapid alternating movements (RAMs). This can be tested by asking the patient to strike his or her thigh with the palm of the hand, then the back of the hand, and then repeating this sequence as rapidly as possible. In another variation of this test, the patient may tap the tip of the index finger with the distal joint of the thumb rapidly. In cerebellar disease, when these movements are slow and irregular rather than rapid and rhythmic, the finding is called dysdiadochokinesia [20].

Dyssynergia is defined as poor coordination or synthesis of movement. Ramsay Hunt first described dyssynergia in relation to cerebellar abnormalities in 1915. He observed a "general disturbance of motility" and an inability to "measure, regulate, and harmonize voluntary movements" in patients with this disorder. He defined dyssynergia as difficulty controlling the *association* of isolated, single movements [30]. Distinguishing dyssynergia from other motor signs such as tremor requires practice and close observation. The clinician asks the patient to perform a rapid, repetitive movement of the extremities. For example, they could ask the patient to stretch his or her arms and tap on the clinician's own palms quickly. Dyssynergia may be present if they see incoordination of these movements, especially if they see that multi-joint movements are decomposing into a sequence of single-joint movements [23].

Balance and Gait Findings

Titubation is a rhythmic, slow-frequency (3–7 Hz) tremor affecting the head or trunk associated with impaired postural innervation. The head and hips often move in antiphase [23, 31]. Various kinds of titubation have been observed in different cerebellar abnormalities. Truncal titubation has been observed in children with congenital cerebellar hypoplasia [32, 33]. A "yes/yes" head tremor without any tremor of the extremities was observed weeks after a cerebellar infarction in one patient [34]. In another case, isolated head titubation was observed in the acute phase of hemorrhagic cerebellar infarction [31]. And one patient with anti-mGluR1 autoantibody-associated cerebellitis presented with head titubation that consisted of continuous, large-amplitude oscillatory head movements [35]. Observation of the patient at rest will in most cases reveal if titubation is present.

Postural sway is increased sway when sitting or standing and is increased in the classic cerebellar syndrome [36]. Researchers have used more exhaustive means of measuring postural sway in the past, such as asking patients to stand on a force-measuring platform and recording displacements in pressure. This "posturography" was sensitive enough to differentiate cerebellar diseases such as Friedrich's ataxia, anterior cerebellar lobe atrophy, localized lesions to the cerebellar hemispheres, etc., based on unique swaying patterns [37]. Other studies have corroborated that the location of the cerebellar lesion may affect the predominant direction of the sway [23]. In a normal clinical setting, asking the patient to stand with feet together and observing should detect postural sway.

Instability during volitional movements is poor coordination between a volitional, or decisive, movement and the postural adjustments accompanying the movement [23]. Normally, during a volitional movement, the nervous system will anticipate any postural destabilization and compensate before the volitional movement is actually carried out [38]. The cerebral cortex, basal ganglia, and cerebellum contribute to this postural equilibrium [39]. The cerebellum is thought to predict the consequences of our actions and compare them to reality, implementing an error-based learning/error correction mechanism. Without a fully functioning cerebellum, reactive mechanisms may not adequately compensate for postural disturbances during volitional movements [40]. Thus, patients with cerebellar lesions may experience instability during everyday activities. A thorough history about all possible motor symptoms may help point to instability during volitional movements. The clinician may also observe this finding during portions of the neurological exam that require decisive movements, such as during the finger-to-nose or the heel-to-shin tests.

Ataxia generally refers to problems with gait or balance. In cerebellar ataxia, the gait may be widely based, staggering, and unsteady, with difficulty on turns [20]. Observing the patient's normal gait can identify these signs. You may also ask the patient to walk on heels, toes, or in tandem (heel to toe). Patients with cerebellar lesions may be unable to do any of these [41]. Cerebellar ataxia also often comes with a loss of positional sense, and the patient cannot stand with feet together, even with the eyes open [20]. Ataxia can arise from genetic or acquired cerebellar lesions. A common genetic etiology of cerebellar ataxia are the spinocerebellar ataxias. Spinocerebellar ataxias (SCAs) are a group of generally autosomal dominant disorders that present with a combination of ataxia and a variety of accompanying neurological symptoms including dystonia, rigidity, oculomotor dysfunction, and epilepsy depending on the particular SCA subtype. In SCAs, a thorough history may reveal a positive family history, though as with the genetic anticipation involved in repeat expansion disorders, this may not be present. Ataxia has many other genetic and non-genetic causes, including hereditary spastic paraplegias, multiple sclerosis, vitamin B deficiency, and alcoholism [42]. Again, a comprehensive history can discover these causes of ataxia and assist in diagnosis.

Oculomotor Findings

Dysmetric saccades are when saccades overshoot or undershoot a target [23]. Normally, saccades are rapid eye movements meant to bring objects of visual interest in line with the fovea [43]. Dysmetric saccades resemble dysmetria of the limbs, wherein a patient may overshoot or undershoot a target at first, but then correct and meet the target. The cerebellum plays an important role in controlling these precise eye movements. In particular, the posterior vermis and its major output, the fastigial nucleus, control saccadic accuracy and adaptation [44, 45]. In addition, the cerebellar cortex constantly recalibrates the saccadic system and compensates for muscle fatigue [46]. Thus, when the cerebellum is damaged, you may observe dysmetric saccades. Saccades may be clinically tested in various ways. The clinician may ask the patient to make repeated saccades between two visual targets, e.g. the left and right index fingers placed to the right and left of the patient's central fixation. This kind of test can be done in a self-paced manner or upon verbal or visual command, such as saying "left, right" or wiggling your fingers, respectively. The clinician must carefully observe the initiation of the patient's saccades, range of motion, conjugacy, speed, and accuracy [43]. Although oculography is the gold standard for oculomotor assessment, it is rarely available in the clinical setting. Thus, one group of researchers proposed a simple reading test composed of 120 regularly spaced single-digits, which can be completed in under 2 min, as a potential biomarker for oculomotor abnormalities [47].

"**Saccadic" smooth pursuit** is when the eyes pursue a moving target but lag behind, catching up with saccades [23]. Physical examination can reveal impaired smooth pursuit to the observant diagnostician. The clinician may ask the patient to track their finger across his or her visual field and watch carefully for catch-up saccades. They may also examine eye-head tracking by asking the patient to fixate on a target and rotate his or her head simultaneously with the target. Visual fixation will suppress the vestibuloocular reflex if the patient has impaired smooth pursuit, and thus the clinician will see that the patient's eyes are continuously taken off target and must catch up with corrective saccades [48]. As with many of the signs discussed in this chapter, impaired or "saccadic" smooth pursuit is not observed exclusively in cerebellar diseases, but it can certainly help diagnose a subtle cerebellar syndrome [49].

Nystagmus is a rhythmic oscillation of the eyes, not unlike a tremor. It occurs normally when a person watches a rapidly moving object, such as passing cars, and it can also signal the presence of several underlying pathologies. Early vision impairment, drug toxicity, internuclear ophthalmoplegia, disorders of the labyrinth system, and cerebellar disease can cause nystagmus. In cerebellar disease, the nystagmus increases with retinal fixation, whereas in vestibular disorders, it decreases with retinal fixation [20]. Nystagmus can vary in its characteristics, and clinicians must be astute to notice and distinguish it [50, 51].

Downbeat nystagmus is involuntary oscillation of the eyes wherein the eyes drift upwards and are brought back down by a corrective saccade [49]. The upwards motion is slow phase and the downward is fast phase [23]. One study observed that after flocculectomy, monkeys developed downward beating nystagmus during fixation in the primary position [50]. Downbeat nystagmus is associated with cerebellar lesions, especially of the floccular and parafloccular lobules, but it can also be experimentally produced by lesioning the medulla [50, 52]. Idiopathic downbeat nystagmus is also common [53]. As far as detecting downbeat nystagmus in the clinic, this sign may be present in all gaze directions, but it may become more noticeable upon extreme deviation of the eyes [20]. For downbeat nystagmus, asking the patient to look to the side and down may make the tremor of the eyes more noticeable [53]. Interestingly, hyperventilation is believed to increase the velocity of the slow phase of downbeat nystagmus in cerebellar patients [54]. However, it is not recommended to ask patients to hyperventilate in a routine clinical scenario.

Gaze-Evoked Nystagmus is another involuntary oscillation of the eyes. The eyes drift centripetally in the slow phase and are brought back by a corrective saccade in the fast phase [49]. This is usually brought on when sustaining a gaze to the side [23]. Gaze-evoked nystagmus is one of the most common forms of central nystagmus and is not specific to cerebellar disease [49]. It can also be produced by lesioning the medulla. Most individual cerebellar signs can be produced by brainstem lesions, but the combination of signs can suggest a cerebellar localization [50]. To look for its presence, the clinician asks the patient to sustain a lateral gaze and then they watch closely for any involuntary jerking of the eyes [20].

Diplopia, or double vision, is seen with weakness of the extraocular muscles, lesions of the brainstem, and lesions of the cerebellum [20]. Damage to the cerebellum may cause asymmetric input to the ocular motor nuclei, producing "skew deviation" where the eyes look in opposite directions [55]. This "skew deviation" causes diplopia. Monocular diplopia, or diplopia that is present when the patient covers one eye, is likely a problem with the cornea or lens. Binocular diplopia, present with both eyes open, points to a central problem in the cranial nerves, brainstem, or cerebellum. To differentiate monocular versus binocular diplopia, the clinician has to make sure to ask the patient to cover one eye at a time when assessing eyesight. They may hold up a number on their fingers and ask the patient to tell them the number as well as if they see double. A thorough history may also assist in the clinical assessment of diplopia. The patient is asked if he or she has experienced double vision recently. Diplopia is a subjective visual complaint and must therefore be elicited from the patient. In patients who cannot communicate effectively, falling may be an indication that diplopia may be present [20].

3.1.2 Non-motor Complications and Cerebellar Cognitive Affective Syndrome

Damage to the cerebellum has also been linked to the disruption of widespread non-motor function. The constellation of non-motor symptoms resulting from cerebellar lesions has been referred to as the cerebellar cognitive affective syndrome (CCAS), sometimes referred to by the eponym, Schmahmann syndrome [56, 57]. Patients with CCAS may have deficits in executive function, visuospatial cognition, language, affect regulation, and other areas.

Executive findings following cerebellar lesions may include impairments in several behavioral domains, such as attention, emotion, and social cognition [58, 59]. Deficits in executive functions may be the sum of the earliest and most pervasive symptoms of CCAS [57].

Visuospatial abilities may be impaired by cerebellar lesions [57, 60, 61]. There are many ways to test a patient's visuospatial ability, but we will only discuss a handful that has been used in studies specifically regarding the cerebellum. In the Block Design Test, used to assess spatial visualization and motor skills, the patient is asked to rearrange differently colored blocks to match a pattern [19]. In the Visual Search Test, the patient is usually shown a computer screen with a display such as the one shown in Fig. 1a. With the computer mouse, the patient must click all of the figures matching the "target figure" (a red "X" in Fig. 1a) as quickly as possible [61]. In a standard administration of another test, the Rey Figure Copy Test, the patient is instructed to copy a complex figure (example of such a drawing is shown in Fig. 1b), first with the figure

A

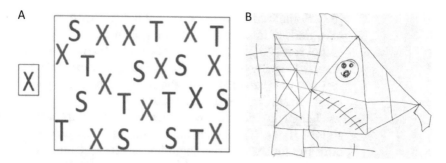

B

Fig. 1 Tests of visuospatial function. (**a**) Example of a Visual Search Test, from [93] permission under creative commons license. (**b**) The Rey±Osterrieth Complex Figure (Example image from https:/en.wikipedia.org/wiki/Rey-Osterrieth_complex_figure obtained under creative commons license)

in view and then, without warning, without the figure, from memory [62]. Poor performance on any of these tests may indicate visuospatial impairment, which in turn may indicate a cerebellar lesion. Simpler ways to assess visuospatial abnormalities in the clinic may be to ask about any perceived difficulties with vision, including car accidents that may have been caused by visual deficits, and to perform a physical exam of vision, though these exams may confound visuospatial processing defects with field defects.

Working memory is a type of short-term memory that requires recall for several minutes to complete a task [63]. Deficits in working memory have been widely reported in patients with cerebellar lesions, often suspected to arise due to deficient "silent rehearsal" of verbal information and/or reduced attentional resources [18, 64, 65]. As with visuospatial abilities, clinicians have many methods of testing working memory. One example is the Wechsler Memory Scale. This method of assessment has gone through many revisions, but all versions provide insight into the subject's working memory abilities [66]. In the fourth edition of the Wechsler Memory Scale, test takers are asked to select and place cards on a 4 × 4 grid to match a previously viewed image [67]. Poor performance on this task in the presence of other cerebellar findings may indicate a cerebellar abnormality [57]. Specifically, lesions of the tonsil, the inferior semilunar lobule, and portions of the vermal pyramid specifically impair performance when to-be-remembered targets are presented with task-irrelevant items [68]. Asking the patient about trouble completing tasks may also reveal impairment in working memory.

Planning, related to working memory, may also be impaired following cerebellar lesions. Patients with cerebellar lesions have problems planning speech and coordination, leading to some of the speech and motor difficulties discussed in this chapter [69]. But, cerebellar damage may also impair higher cognitive planning, leading to disorganization, in schooled aged children to problems at school, and to adults difficulties in the workplace, or at home

[57]. Some studies have shown that patients with cerebellar atrophy have deficits in cognitive planning as measured by the Tower of Hanoi test [70], but taking a thorough social history could suffice for assessing any difficulty with planning. For example, the clinician may ask the patient if he or she has experienced a particular difficulty with keeping activities organized.

Language deficits may present in patients with cerebellar abnormalities. The cerebellum is thought to play a key role in both the motor and the cognitive aspects of language [2, 71]. In dysarthria, the meaning of language remains intact, but due to impaired muscular control of the lips, tongue, palate, etc., words may be nasal, slurred, or indistinct. Dysarthria can be caused by cerebellar disease, motor-related lesions of the central or peripheral nervous system, and parkinsonism [20]. Children who undergo surgical excision of cerebellar tumors often experience post-operative dysarthria and mutism, with acute onset (12–48 h post-operatively) and spontaneous resolution (4–6 months) in most cases [72]. Impaired metalinguistic ability, related to the more cognitive aspects of language, is also observed in cerebellar syndrome. Patients with impaired metalinguistic ability may have trouble understanding metaphor, ambiguity, and inferential thinking. They may also have trouble verbally expressing their thoughts [73]. Other language difficulties that have been observed in cerebellar syndrome include dysprosodia (impaired intonation patterns; robotic speech), agrammatism (telegraphic speech), and mild anomia (inability to name familiar objects) [49]. Listening to the patient's speech when the history is taken and during the exam will likely reveal obvious language deficits. Does the patient have trouble forming words? Is it difficult for the patient to find the words they want to speak? If the clinician suspects that the patient may have difficulty with the cognitive aspects of speech, assessing the patient's understanding of proverbs or idioms may also be useful in further assessing cognitive function [74].

Affective findings such as emotional lability and depression can present in patients with cerebellar disease. Some patients may experience personality change with blunted affect and inappropriate behaviors [49]. Interestingly, lesions in the evolutionarily most ancient cerebellar structures seem to cause more severe disorder. For example, damage to the cerebellar vermis causes greater affective impairments [75]. In addition, the concept of a "limbic cerebellum" has been proposed, which the functional abnormalities likely stemming from circuits centered in the vermis and fastigial nucleus. These midline structures are activated during states of panic, sadness, pain, and other extreme emotional states. Thus, the cerebellum is thought to be integral in the neural circuits controlling emotional behavior [58].

Emotional lability may be observed in patients with cerebellar disease due to impaired emotional control. Patients may experience anxiety, impulsiveness, pathological laughing or crying, or dysphoria [59]. Both emotional fragility and rigidity are frequently observed in patients with CCAS, indicating the complicated effect the cerebellum has on emotional state [75]. Asking the patient about mood swings, relationship or work problems, and other problems related to emotional state can help reveal this finding during a clinical assessment.

Depression is frequently observed in cases of cerebellar damage, particularly when there is dysfunction of the cerebellar vermis [76]. Understanding the etiology of depression in patients with any kind of chronic illness poses a challenge to clinicians and researchers due to the difficulty in distinguishing organic brain dysfunction from environmental factors associated with the concomitant loss of ADLs, disruption of socialization, and inability to maintain the previous lifestyle. Even still, patients with degenerative cerebellar disorders exhibit an unusually high prevalence of depression [58]. Recognizing depression in the setting of broader neurologic dysfunction may be a challenge and warrants focused examination. A patient with depression may lose his or her typical affect, give short answers to questions, and seem generally subdued. Direct questions about symptoms of depression, such as mood changes, fatigue, unusual tearfulness, appetite or weight changes, insomnia, and vague somatic complaints may be warranted if any concern for depression arises. Validated screening questions for depression include: "Over the past two weeks, have you felt little interest or pleasure in doing things?" and "Over the past two weeks, have you felt down, depressed, or hopeless?" If the patient answers "yes" to either of these questions, further inquiry about the severity of the depression and risk of suicide is warranted [20].

Autonomic responses, closely tied to emotional responses, may also be altered in cerebellar disease. The cerebellum is connected to various centers involved in autonomic control, such as the hypothalamus, cerebral cortex, and the medullary reticular formation [58, 77]. An example of this is a patient who had a stroke involving the fastigial nucleus and paravermis areas of the cerebellar cortex. This patient suffered from spells of hiccupping and coughing, followed by bradycardia and syncope [57]. In cats, cerebellar stimulation leads to autonomic hypothalamic outbursts [78]. Besides questions about emotional state, the clinician may also ask the patient about any palpitations, light-headedness, sweating, etc. to determine if the patient is experiencing changes in his or her autonomic control.

3.2 Neuroimaging

Neuroimaging is a crucial part of the neurological exam and directly informs about the development of a differential diagnosis in the clinical evaluation of neurological disease. This section will elaborate on the methodology and application of the current imaging modalities that are commonly applied to the investigation of cerebellar disease.

3.2.1 Conventional Angiography

The first truly modern neuroimaging technique is cerebral angiography (CA). Often called "conventional" or "4-vessel" angiography, this technique involves direct injection of an iodine containing contrast agent into the arterial circulation via a catheter. Upon injection of the contrast agent, plain film X-rays are taken to capture circulation arising from the 4 great cerebral vessels, namely the left and right carotid and vertebral arteries. In general, circulation of each vessel is captured in two planes, lateral (sagittal) and antero-posterior (coronal). CA is common in clinical neurology today, though it is increasingly used as a gold standard test in special circumstances as less invasive Computed tomography angiography (CTA) or magnetic resonance angiography (MRA) can be used in its place for screening purposes. As implied by its name, conventional angiography is generally used to evaluate the extra- and intracranial vasculature and associated diseases. While this category of disease more often affects the anterior circulation, which does not supply blood to the cerebellum, CA can be used to interrogate concerns of compromised posterior circulation, including vertebral artery dissection, vertebrobasilar insufficiency, and occlusion of the cerebellar arteries, namely the superior cerebellar artery (SCA), the anterior inferior cerebellar artery (AICA), and the posterior inferior cerebellar artery (PICA). Conventional angiography is also useful in characterization of brain neoplasms, which are especially common in the posterior fossa/cerebellum in the pediatric population. While CA is often considered the gold standard for imaging the vasculature of the central nervous system, its use is gradually decreasing due to a combination of the technological advances that have been made for CTA and MRA and the invasive nature of conventional angiography itself. Conventional angiography is complicated by low rates of stroke and contrast associated non-neurologic morbidity [79].

3.2.2 Computed Tomography (CT) Scan

The CT scan, which began its clinical use in the 1970s, is a technological innovation based upon conventional tomography, an X-ray based technique developed by Italian Radiologist Alessandro Vallebona, which was in turn a technological innovation upon plain film X-rays [80]. Today CT scans are used primarily in the detection of hemorrhage, bony abnormality (fracture), or mass effect in the brain due to the rapid acquisition of the CT, the acuity of these pathologies, and the superiority of MRI in evaluating soft tissue anatomy. CT scans are read based on signal density, by which regions are described as hyperdense or hypodense. On CT, scans of bone and blood are hyperdense, appearing bright on the image, while cerebrospinal fluid (CSF) is hypodense, appearing dark (*see*

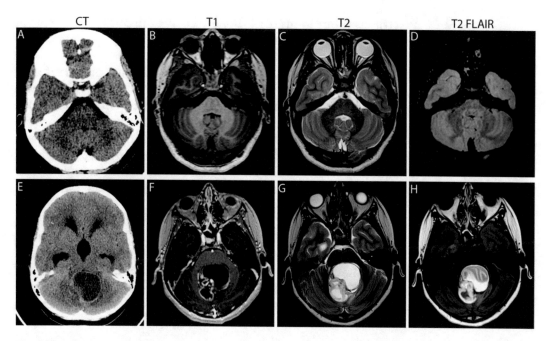

Fig. 2 Neuroimaging modalities. Upper panel (**a-d**) from an 11-year-old boy with normal brain imaging. Lower panels (**e-h**) are from a 13-year-old girl with pilocytic astrocytoma prior to tumor resection. All images are axial cuts at level of the pons, though concordance between images varies slightly based on plane obtained and patient position. MRI images (**b-d**; **f-h**) obtained using 3T magnet (**a**) CT scan, note streak artifacts from beam hardening. Bilateral hypodense areas in cerebellum reflect white matter tracts. Note poor resolution of CSF spaces of fourth ventricle. (**b**) T1 weighted post-contrast image. Note excellent resolution of cerebellar gray (dark) and white (light) matter. Additionally, note the bilaterally hyperintense internal carotid arteries, which reflects contrast. (**c**) T2 weighted image. Note bright CSF surrounding the basilar artery and inversion of gray and white matter intensities as compared to T1 weighted image. (**d**) T2 FLAIR image. Note the attenuation of bright CSF signal compared to T2 weighted image in **c**. (**e**) CT scan in patient with pilocytic astrocytoma on initial presentation. Effacement of fourth ventricle is noted with concomitant dilation of lateral and third ventricles. Note poor resolution of tumor compared to MRI images. (**f**) T1 post-contrast image. Solid and cystic components with rim enhancing mural nodule classic for Pilocytic Astrocytoma are evident. Rim around tumor reflects breakdown of the blood–brain barrier. (**g**) T2 image allows for visualization of effacement of basal cistern, compare thin layer of hyperintense CSF anterior to the pons in **g** to the much fuller CSF space anterior to the pons in **c**. (**h**) T2 FLAIR image. The fluid attenuation of this sequence allows excellent visualization of the heterogeneous solid and cystic components of the tumor, compare the variegated appearance of the largest cystic components in **h** to the same structure in **f** and **g**. (Deidentified images were obtained from Texas Children's Hospital Electronic Medical Records)

Fig. 2a, e for normal and pathologic CT scans). Brain parenchyma has intermediate density, with gray matter appearing slightly brighter than white matter. Hemorrhagic stroke will appear bright white on CT scan and can be parenchymal (contained within the brain parenchyma) or subarachnoid (overlaying the brain proper). Epidural and subdural hematoma are less common in the posterior fossa. Today, the role of the CT scan in evaluation of acute stroke is primarily in ruling OUT hemorrhage or neoplasm. Regarding cerebellar pathology, CT is less adept at visualizing posterior fossa structures due to the dense bone overlying this portion of the brain, which increases beam hardening artifacts (Fig. 2a) and limits evaluation of the underlying soft tissue [81].

Figure 2e shows the utility of CT in the acute presentation of intractable headache with emesis. In this patient, the CT scan was performed based on the patient's clinical presentation and revealed a mass effect from a hypodense space occupying lesion in the posterior fossa. Associated dilation of the ventricles and effacement of the fourth ventricle and basal cisterns revealed that obstructive hydrocephalus was the etiology of the presenting complaint and allowed for emergent placement of an external ventricular drain (EVD).

Complications to CT scanning include significant exposure to ionizing radiation, which limits its use in the pediatric population, in pregnant women, and for serial use in monitoring of intracranial pathology. Given the high frequency of posterior fossa tumors in the pediatric population, this is a concerning limitation. However, optimized dose limited scans for acute evaluation may be used to limit radiation exposure even in these populations.

3.2.3 Magnetic Resonance Imaging (MRI)

Less than a decade after introduction of the CT scan magnetic resonance imaging was developed. MRI, alongside the neurologic exam, has become perhaps the most indispensable tool in the diagnostic evaluation of neurologic disease. While the physics of MRI are complex and outside the scope of this chapter, the basics of the technique will be briefly described (for a more in-depth discussion *see* [82]). MRI uses a powerful uniform magnetic field to force the parallel or antiparallel alignment of the magnetic fields of individual atoms. After alignment of the atoms, radiofrequency (RF) energy pulses are used to force some low energy (parallel) nuclei into a high energy (antiparallel) state. The net effect of the excitation is to shift the net magnetization field into the transverse plane. Once the RF pulses cease, there is a "relaxation" of the atoms back into the low energy (parallel) configuration, which can be measured as the magnetization field returns to equilibrium from the transverse plane. The time to return to equilibrium is termed the relaxation time. There are two independent measurable relaxation times, longitudinal and transverse, that form the basis of the two primary MRI sequences, termed T1 and T2, respectively. T1 measures the relaxation of the transverse field back to equilibrium, while T2 refers to the dephasing of phase matched spin of individual atoms after RF excitation. The measurement of the loss of homogeneity of the spin forms the basis of the T2 signal. For brain (and most body) imaging, the target atom is the hydrogen atom, which is abundant in both water and fat, two key components of brain tissue.

To create an image from the relaxation times, a gradient is applied to the magnetic field such that any 2–8 mm slice of tissue can be targeted by a particular RF excitation frequency (the plane of the slice can be altered by positioning of the machine). Within that plane, phase and frequency encoding (which will not be discussed here) measurements are taken in an orthogonal fashion to create a grid where each point reflects a voxel on the final image. After Fournier transformation of the phase and frequency encoding measurements the intensity of each pixel on the resulting image then reflects the properties of the targeted tissue. Different tissues in the human body have unique signatures on both T1 and T2 sequences, which then allow for a detailed image of the tissue of the human body to be produced.

In general MRI images are described in terms of signal intensity (pixel saturation on grayscale) that is hyperintense, hypointense, and isointense (in contrast to "density" in CT scanning). On T1 sequences, CSF is hypointense (dark), white matter is hyperintense (light), and gray matter is isointense (Fig. 2b, f). In this sequence, the intensity of the brain parenchyma corresponds to the colloquial description used for brain tissue such that neuronal cell body rich areas (cerebellar cortex and cerebellar/brainstem nuclei) appear darker, while areas of white matter appear lighter. On T2 sequences the intensities are largely reversed, such that CSF appears hyperintense, white matter is hypointense, and gray matter is hyperintense (Fig. 2c, g). A third very common sequencing modality that is a variation of the T2 sequence is T2 FLAIR imaging. FLAIR stands for FLuid Attenuated Inversion Recovery. In this modality the pulse sequence is adjusted to alter the net magnetization of the fluid, which results in an image where fluid (CSF) remains hypointense, while the white and gray matter signals appear as in T2. This allows for a sharp contrast between CSF and neighboring gray matter [83] and subsequently improves the ability for the clinician to resolve parenchymal and dural lesions (Fig. 2d, h). Another basic MRI sequence is diffusion weighted imaging (DWI), which is often used in conjunction with apparent diffusion coefficient imaging (ADC), in the evaluation of acute ischemia [84]. DWI and ADC are low contrast rapidly acquired imaging sequences that measure water diffusivity. Their main application is in the detection of acute ischemia, which is especially important in the current era of effective rapid interventions in the treatment, prevention, and reversal of tissue damage in the setting of ischemic stroke. Finally, contrast enhancement with gadolinium is an indispensable tool in clinical neuroimaging. Contrasted MRI imaging uses IV gadolinium to (1) visualize vasculature in magnetic resonance angiography (MRA) and (2) detect break down/permeability of the blood–brain barrier, which is useful in inflammatory, infectious, and neoplastic brain lesions (Fig. 2f).

MRI has several advantages over the previously described imaging techniques that make it ideal for imaging of the cerebellum. These include excellent visualization of soft tissue (brain parenchyma), absence of beam hardening artifact from the thick bone overlying the posterior fossa, absence of ionizing radiation and associated side effects (which makes MRI a preferred modality in the pediatric population). MRI has no directly associated known health risks. However, the technique can be difficult to tolerate for those with claustrophobia, difficult to tolerate for children due to the length of the procedure and the requirement for the patient to stay relatively still, and the machine can be very loud, which may also impede the tolerability of the procedure for patients. However, MRI does have some associated indirect risks including increasing concern about the deposition of gadolinium in the brain with contrasted studies (though as of yet despite evidence for tissue deposition there is no conclusive evidence for associated neurotoxicity [85]), concerns about the long term effects of sedation on young children [86], who require anesthesia in order to obtain studies of adequate quality for clinical interpretation, and individuals with implantable medical devices often require documentation to prove compatibility with MRI, as metallic objects can be displaced, damaged, or malfunction in the presence of the magnetic field generated by MRI.

3.2.4 Application of Neuroimaging Modalities in Cerebellar Disease

Together, CT scan, T1, T2 Flair, and angiography (generally CTA or MRA as first line studies) comprise the initial suite of neuroimaging studies performed for the evaluation of neurologic disease. CT scanning is the primary tool for evaluation of acute onset of cerebellar dysfunction (which can be evaluated as described in the preceding section) and is used to rule out hemorrhage or mass effect, both of which may require urgent medical or surgical intervention to prevent brain herniation (*see* discussion above and compare Fig. 2a, e). Such concerns are especially important in regard to the posterior fossa given its limited volume and proximity to critical brainstem structures. If the CT scan is unremarkable, follow-up MRI can be used to evaluate for smaller ischemic changes. The most frequently used clinical scale for the evaluation of ischemic stroke, the NIH stroke scale, is more effective for the evaluation of ischemic changes in the anterior circulation [87, 88]. Therefore, rapid acquisition of MRI with DWI and ADC mapping can be a useful adjunct to the clinical exam and may provide important information guiding the use of thrombolytic therapy in posterior circulation stroke. In the case of insidious symptom onset, MRI with or without contrast (depending on whether there is subacute or chronic symptom onset and accompanying exam and laboratory findings) is likely to be a higher yield study than CT scanning. In particular, T1 weighted sequencing is the sequence most well-suited to the evaluation of brain anatomy and is an excellent modality for the evaluation of both fixed anatomic abnormalities (such as Chiari I/II malformation and Dandy-Walker malformation) and atrophy associated with progressive cerebellar disease (including the spinocerebellar ataxias and hereditary spastic paraplegias). T2 flair can be useful for the detection of inflammatory disease, which frequently involves the cerebellum (including entities such as multiple sclerosis (MS), acute demyelinating encephalomyelitis (ADEM), chronic lymphocytic inflammation with pontine perivascular enhancement responsive to steroids (CLIPPERS), and paraneoplastic disease (anti-GAD and opsoclonus-myoclonus ataxia)). In the case of inflammatory disease, the addition of contrast to the study can be useful to differentiate chronic from active lesions, while in the case of infectious etiologies, contrast is useful for the detection of leptomeningeal disease and differentiation of abscess from alternative lesions.

A final MRI based imaging technique that will be discussed is diffusion tensor imaging (DTI), which is used for tractography. This sequence is increasingly acquired in clinical practice, though its use by the neurologist in clinical evaluations is still extremely limited. Briefly, DTI uses the diffusivity properties of water to generate images. In particular it takes advantage of the restricted diffusivity of water in white matter tracts (where water is more likely to diffuse parallel to the tract rather than orthogonal to the tract) to generate images that reflect alterations in microstructure of white matter [89]. As mentioned, in the clinical practice of the neurologist, the application of this technique remains somewhat limited. However, in clinical research DTI has been applied in the surgical planning and prognostication of outcomes in posterior fossa tumor resection [90].

4 Conclusions

The cerebellum occupies an interesting position in clinical neurology. While it has been a crucial structure of investigation in the neurosciences for centuries and was central to Ramon y Cajal's early studies on neuroanatomy, it would not be an overstatement to say that it has taken a backseat in both clinical neurology training and evaluation. This is despite the identification of cerebellar pathogenesis in a wide variety of "cortical" diseases including epilepsy and stroke [91, 92]. Our hope is that current and future generations of neurologists place greater consideration on the effects of cerebellar disease and insult on function across behavioral domains. Furthermore, we predict that the increasing application of sophisticated imaging techniques will further illuminate the role of the cerebellum in widespread cortical function. It continues to be an exciting time for clinical and basic neuroscientists interested in cerebellar motor and non-motor function.

Acknowledgements

This work was supported by funds from Baylor College of Medicine (BCM) and Texas Children's Hospital. RVS received support from The Hamill Foundation, BCM IDDRC U54HD083092 (Intellectual and Developmental Disabilities Research Center), and the National Institutes of Neurological Disorders and Stroke (NINDS) R01NS089664 and R01NS100874. JSG received support from T32NS043124.

References

1. Lang EJ et al (2017) The roles of the Olivocerebellar pathway in motor learning and motor control. A consensus paper. Cerebellum 16(1): 230–252

2. Manto M et al (2012) Consensus paper: roles of the cerebellum in motor control-the diversity of ideas on cerebellar involvement in movement. Cerebellum 11(2):457–487

3. Adamaszek M et al (2017) Consensus paper: cerebellum and emotion. Cerebellum 16(2): 552–576

4. Schmahmann JD (1991) An emerging concept: the cerebellar contribution to higher function. Arch Neurol 48(11):1178–1187

5. Baumann O et al (2015) Consensus paper: the role of the cerebellum in perceptual processes. Cerebellum 14(2):197–220

6. Koziol LF et al (2014) Consensus paper: the cerebellum's role in movement and cognition. Cerebellum 13(1):151–177

7. Louis ED (2019) Tremor. Continuum (Minneap Minn). 25(4):959–975. https://doi.org/10.1212/CON.0000000000000748. PMID: 31356289

8. Louis ED (2005) Essential tremor. Lancet Neurol 4(2):100–110

9. Louis ED, Faust PL (2020) Essential tremor within the broader context of other forms of cerebellar degeneration. Cerebellum 19: 879–896

10. Lorenz D, Schwieger D, Moises H, Deuschl G (2006) Quality of life and personality in essential tremor patients. Mov Disord 21(8): 1114–1118

11. Voogd J, Schraa-Tam CKL, Van Der Geest JN, De Zeeuw CI (2012) Visuomotor cerebellum in human and nonhuman primates. Cerebellum 11(2):392–410

12. Gill JS, Sillitoe RV (2019) Functional outcomes of cerebellar malformations. Front Cell Neurosci 13:1–20

13. Palmer SL et al (2010) Neurocognitive outcome 12 months following cerebellar mutism syndrome in pediatric patients with medulloblastoma. Neuro-Oncology 12(12): 1311–1317

14. Levisohn L, Cronin-golomb A, Schmahmann JD (2000) Neuropsychological consequences of cerebellar tumour resection in children cerebellar cognitive affective syndrome in a paediatric population. Brain 123:1041–1050

15. Roalf DR, Moberg PJ, Xie SX, Wolk DA, Moelter S, Arnold SE (2014) Comparative accuracies of two common screening instruments for the classification of Alzheimer's disease, mild cognitive impairment and healthy aging. Alzheimers Dement 9(5):529–537

16. Folstein M, Folstein S, McHugh P (1975) 'Mini-mental state': a practical method for grading the cognitive state of patients for the clinician. J Psychiatry Res 12:189–198

17. Nasreddine Z et al (2005) The montreal cognitive assessment, MoCA: a brief screening. J Am Geriatr Soc 53:695–699

18. Hoche F, Guell X, Vangel MG, Sherman JC, Schmahmann JD (2018) The cerebellar cognitive affective/Schmahmann syndrome scale. Brain 141(1):248–270

19. Ahmadian N, van Baarsen K, van Zandvoort M, Robe P (2019) The cerebellar cognitive affective syndrome—a meta-analysis. Cerebellum 18:941–950

20. Bickley LS, Szilagyi PG (1988) Guide to physical examination and history taking. Ann Intern Med 108(3):500

21. Koeppen AH (2018) The neuropathology of the adult cerebellum. Handb Clin Neurol 154: 129–149

22. Timmann D, Kolb FP, Diener HC (1999) Pathophysiology of cerebellar ataxia. Klin Neurophysiol 30(2):128–144

23. Marsden JF (2018) Cerebellar ataxia. In: Handbook of Clinical Neurology, vol 159, 1st edn. Elsevier B.V., Amsterdam, pp 261–281

24. Elias WJ, Shah BB (2014) Tremor. J Am Med Assoc 311(9):948–954

25. Benito-León J (2014) Essential tremor: a neurodegenerative disease? Tremor Other Hyperkinet Mov (N Y) 4:252

26. Louis ED (2016) Linking essential tremor to the cerebellum: neuropathological evidence. Cerebellum 15(3):235–242

27. Bhatia KP et al (2018) Consensus statement on the classification of tremors. From the task force on tremor of the International Parkinson and Movement Disorder Society. Mov Disord 33(1):75–87

28. Baizabal-Carvallo JF, Cardoso F, Jankovic J (2015) Myorhythmia: phenomenology, etiology, and treatment. Mov Disord 30(2): 171–179

29. Masucci EF, Kurtzke JF, Saini N (1984) Myorhythmia: a widespread movement disorder: Clinicopathological correlations. Brain 107(1):53–79

30. Hunt R (1914) Dyssynergia cerebellaris progressiva. Brain 37(2):247–268

31. Ueno T, Nishijima H, Arai A, Tomiyama M (2017) Acute hemorrhagic cerebellar infarction presenting with isolated head titubation. J Neurol Sci 372:456–458

32. Sarnat HB, Alcala H (1980) Hypoplasia syndrome of diverse causes. Arch Neurol 37(5): 300–305

33. Wassmer E, Davies P, Whitehouse WP, Green SH (2003) Clinical spectrum associated with cerebellar hypoplasia. Pediatr Neurol 28(5): 347–351

34. Finsterer J, Muellbacher W, Mamoli B (1996) Yes/yes head tremor without appendicular tremor after bilateral cerebellar infarction. J Neurol Sci 139(2):242–245

35. Pedroso JL, Dutra LA, Espay AJ, Höftberger R, Barsottini OGP (2018) Video NeuroImages: head titubation in anti-mGluR1 autoantibody-associated cerebellitis. Neurology 90(16):746–747

36. Manto M, Habas C (2016) Cerebellar disorders: clinical/radiologic findings and modern imaging tools. In: Handbook of Clinical Neurology, vol 135, 1st edn. Elsevier B.V., Amsterdam, pp 479–491

37. Diener HC, Dichgans J, Bacher M, Gompf B (1984) Quantification of postural sway in normals and patients with cerebellar diseases. Electroencephalogr Clin Neurophysiol 57(2): 134–142

38. Massion J (1992) Movement, posture and equilibrium: interaction and coordination. Prog Neurobiol 38(1):35–56

39. Bouisset S, Zattara M (1981) A sequence of postural movements precedes voluntary movement. Neurosci Lett 22(3):263–270

40. Dakin CJ, Bolton DAE (2018) Forecast or fall: Prediction's importance to postural control. Front Neurol 9:1–10

41. Hollman H (2007) Neuroanatomy in Clinical Context, vol 22, 9th edn. Wolters Kluwer, Alphen aan den Rijn

42. Hack N, Kass J (2021) Ataxia. In: Ferri's Clinical Advisor, Web, pp 181.e2–181.e4

43. Termsarasab P, Thammongkolchai T, Rucker JC, Frucht SJ (2015) The diagnostic value of saccades in movement disorder patients: a practical guide and review. J Clin Mov Disord 2(1): 1–10

44. Bötzel K, Rottach K, Büttner U (1993) Normal and pathological saccadic dysmetria. Brain 116(2):337–353

45. Sato H, Noda H (1992) Saccadic dysmetria induced by transient functional decortication of the cerebellar vermis. Exp Brain Res 89(3): 690

46. Barash S, Melikyan A, Sivakov A, Zhang M, Glickstein M, Thier P (1999) Seccadic dysmetria and adaptation after lesions of the cerebellar cortex. J Neurosci 19(24):10931–10939

47. Oh AJ, Chen T, Shariati MA, Jehangir N, Hwang TN, Liao YJ (2018) A simple saccadic reading test to assess ocular motor function in cerebellar ataxia. PLoS One 13(11):1–18

48. Sharpe JA (2008) Neurophysiology and neuroanatomy of smooth pursuit: lesion studies. Brain Cogn 68(3):241–254

49. Bodranghien F et al (2016) Consensus paper: revisiting the symptoms and signs of cerebellar syndrome. Cerebellum 15(3):369–391

50. Zee DS, Yamazaki A, Butler PH, Gucer G (1981) Effects of ablation of flocculus and paraflocculus on eye movements in primate. J Neurophysiol 46(4):878–899

51. Nam J, Kim S, Huh Y, Kim JS (2009) Ageotropic central positional nystagmus in nodular infarction. Neurology 73(14):1163

52. de Jong JMBV, Cohen B, Matsuo V, Uemura T (1980) Midsagittal pontomedullary brain stem section: effects on ocular adduction and nystagmus. Exp Neurol 68(3):420–442

53. Wagner JN, Glaser M, Brandt T, Strupp M (2008) Downbeat nystagmus: Aetiology and comorbidity in 117 patients. J Neurol Neurosurg Psychiatry 79(6):672–677

54. Walker MF, Zee DS (1999) The effect of hyperventilation on downbeat nystagmus in cerebellar disorders. Neurology 53(7):1576–1579

55. McGee S (2018) Chapter 59—nerves of the eye muscles (III, IV, and VI): approach to diplopia. In: Evidence based physical diagnosis, 4th edn. Elsevier, Amsterdam

56. Manto M, Mariën P (2015) Schmahmann's syndrome - identification of the third cornerstone of clinical ataxiology. Cerebellum Ataxias 2(1):1–5

57. Hoche F, Daly MP, Chutake YK, Valera E, Sherman JC, Schmahmann JD (2019) The cerebellar cognitive affective syndrome in ataxia-telangiectasia. Cerebellum 18(2):225–244

58. Schmahmann JD, Weilburg JB, Sherman JC (2007) The neuropsychiatry of the cerebellum - insights from the clinic. Cerebellum 6(3):254–267

59. Hoche F, Guell X, Sherman JC, Vangel MG, Schmahmann JD (2016) Cerebellar contribution to social cognition. Cerebellum 15(6):732–743

60. Wallesch CW, Horn A (1990) Long-term effects of cerebellar pathology on cognitive functions. Brain Cogn 14(1):19–25

61. Fabbro F, Tavano A, Corti S, Bresolin N, De Fabritiis P, Borgatti R (2004) Long-term neuropsychological deficits after cerebellar infarctions in two young adult twins. Neuropsychologia 42(4):536–545

62. Kirkwood MW, Weiler MD, Bernstein JH, Forbes PW, Waber DP (2001) Sources of poor performance on the Rey-Osterrieth complex figure test among children with learning difficulties: a dynamic assessment approach. Clin Neuropsychol 15(3):345–356

63. Flemming KD (2006) Essential neuroscience. Mayo Clin Proc 81(10):1409

64. Ravizza SM, McCormick CA, Schlerf JE, Justus T, Ivry RB, Fiez JA (2006) Cerebellar damage produces selective deficits in verbal working memory. Brain 129(2):306–320

65. Justus T, Ravizza SM, Fiez JA, Ivry RB (2005) Reduced phonological similarity effects in patients with damage to the cerebellum. Brain Lang 95(2):304–318

66. Kent P (2013) The evolution of the Wechsler memory scale: a selective review. Appl Neuropsychol 20(4):277–291

67. Martin PK, Schroeder RW (2014) Chance performance and floor effects: threats to the validity of the wechsler memory scale - fourth edition designs subtest. Arch Clin Neuropsychol 29(4):385–390

68. Baier B, Müller NG, Dieterich M (2014) What part of the cerebellum contributes to a visuospatial working memory task? Ann Neurol 76(5):754–757

69. Mariën P, Verhoeven J, Engelborghs S, Rooker S, Pickut BA, De Deyn PP (2006) A role for the cerebellum in motor speech planning: evidence from foreign accent syndrome. Clin Neurol Neurosurg 108(5):518–522

70. Grafman J, Litvan I, Massaquoi S, Stewart M, Sirigu A, Hallett M (1992) Cognitive planning deficit in patients with cerebellar atrophy. Neurology 42(8):1493–1496

71. Leiner HC, Leiner AL, Dow RS (1993) Cognitive and language functions of the human cerebellum. Trends Neurosci 16(11):444–447

72. Mastronardi L (1996) Mutism and pseudobulbar symptoms after resection of posterior fossa tumors in children: Incidence and pathophysiology and transient cerebellar mutism after posterior fossa surgery in children. Neurosurgery 38(5):1066

73. Guell X, Hoche F, Schmahmann JD (2015) Metalinguistic deficits in patients with cerebellar dysfunction: empirical support for the Dysmetria of thought theory. Cerebellum 14(1):50–58

74. Van Lancker D (1990) The neurology of proverbs. Behav Neurol 3(3):169–187

75. Tavano A et al (2007) Disorders of cognitive and affective development in cerebellar malformations. Brain 130(10):2646–2660

76. Depping MS, Schmitgen MM, Kubera KM, Wolf RC (2018) Cerebellar contributions to major depression. Front Psychiatry 9:1–5

77. Dietrichs E, Haines DE (2002) Possible pathways for cerebellar modulation of autonomic responses: Micturition. Scand J Urol Nephrol Suppl 36(210):16–20

78. Zanchetti A, Zoccolini A (1954) Autonomic hypothalamic outbursts elicited by cerebellar stimulation. J Neurophysiol 17(5):475–483

79. Ahn SH, Prince EA, Dubel GJ (2013) Basic neuroangiography: review of technique and perioperative patient care. Semin Intervent Radiol 1(212):225–233

80. Bull J (1980) The history of computed tomography. In: Caille J, Salamon G (eds) Computerized tomography. Berlin Heidelberg, Springer, pp 1–3

81. Hwang DY, Silva GS, Furie KL, Greer DM (2012) Comparative sensitivity of computed tomography vs. magnetic resonance imaging for detecting acute posterior fossa infarct. J Emerg Med 42(5):559–565

82. Van Geuns RJM et al (1999) Basic principles of magnetic resonance imaging. Prog Cardiovasc Dis 42(2):149–156

83. Saranathan M, Worters PW, Rettmann DW, Winegar B, Becker J (2017) Physics for clinicians : fluid-attenuated inversion recovery (FLAIR) and double inversion recovery (DIR) imaging. J Magn Reson Imaging 46(6):1590–1600

84. Norris CD, Quick SE, Parker JG (2020) Diffusion MR imaging in the head and neck principles and applications. Neuroimaging Clin N Am 30(3):261–282

85. Olchowy C, Cebulski K, Łasecki M, Chaber R (2017) The presence of the gadolinium-based contrast agent depositions in the brain and symptoms of gadolinium neurotoxicity - A systematic review. PLoS One 12:1–14

86. Barton K, Nickerson JP (2018) Pediatric anesthesia and neurotoxicity: what the radiologist needs to know. Pediatr Radiol 48:31–36

87. Sommer P et al (2018) Is functional outcome different in posterior and anterior circulation stroke? Stroke 49:2728–2732

88. Schneck MJ (2018) Current stroke scales may be partly responsible for worse outcomes in posterior circulation stroke. Stroke 49: 2565–2566

89. Alexander A, Lee JE, Lazar M, Field A (2007) Diffusion tensor imaging of the brain. Neurotherapeutics 4(3):316–329

90. Morris EB et al (2009) Proximal dentatothalamocortical tract involvement in posterior fossa syndrome. Brain 132(11):3087–3095

91. Hermann BP, Bayless K, Hansen R, Parrish J, Seidenberg M (2005) Cerebellar atrophy in temporal lobe epilepsy. Epilepsy Behav 7(2): 279–287

92. Wessel MJ, Hummel FC (2018) Non-invasive cerebellar stimulation: a promising approach for stroke recovery? Cerebellum 17(3): 359–371

93. Courchesne V, Meilleur AAS, Poulin-Lord MP, Dawson M, Soulières I (2015) Autistic children at risk of being underestimated: School-based pilot study of a strength-informed assessment. Mol Autism 6(1):12

INDEX

Roy V. Sillitoe (ed.), *Measuring Cerebellar Function*, Neuromethods, vol. 177, https://doi.org/10.1007/978-1-0716-2026-7,
© Springer Science+Business Media, LLC, part of Springer Nature 2022

Printed in the United States
by Baker & Taylor Publisher Services